Understanding Climatology

Understanding Climatology

Salvador Poole

Callisto Reference

www.callistoreference.com

Callisto Reference,
118-35 Queens Blvd., Suite 400,
Forest Hills, NY 11375, USA

Visit us on the World Wide Web at:
www.callistoreference.com

ISBN: 978-1-64116-238-8 (Hardback)

Printed in China.

Cataloging-in-Publication Data

Understanding climatology / Salvador Poole.
p. cm.
Includes bibliographical references and index.
ISBN 978-1-64116-238-8
1. Climatology. 2. Meteorology. I. Poole, Salvador.
QC981 .U53 2019
551.6--dc23

Table of Contents

Preface

Climate refers to the statistical measurement of weather conditions over extended periods of time. The environmental conditions for a given region, such as temperature, pressure, precipitation and humidity are measured. The study of climate and all its diverse aspects is approached from a branch of atmospheric science called climatology. The climate of a region is formed by the five chief components namely, the atmosphere, hydrosphere, biosphere, lithosphere and the cryosphere. Other factors determining the climatic condition of a region are its latitude, altitude, terrain and its physical geography. Various climate models, which are based on mathematical formulations of past, present and future climates are proposed for the study of climates. This textbook is a valuable compilation of topics, ranging from the basic to the most complex theories and principles in the field of climatology. It aims to shed light on some of the unexplored aspects of climatology. It will serve as a resource guide for students and experts alike and contribute to the growth of the discipline.

A detailed account of the significant topics covered in this book is provided below:

Chapter 1- As a branch of atmospheric science, climatology refers to the study of weather conditions averaged over time. Climate models proposed by climatological studies are used for studying the dynamics of climate and weather as well as project future climates. This is an introductory chapter, which will introduce briefly all the significant aspects of climate and climatology.

Chapter 2- Global climatology involves the study of the layers of gases surrounding the Earth. This chapter examines the chief atmospheric gases of the Earth such as nitrogen, oxygen, carbon dioxide, etc. and their effects on the global climate. It also discusses in detail the structure and composition of the Earth's atmosphere.

Chapter 3- Köppen climate classification is a famous climate classification system first introduced by Wladimir Köppen. According to this system, climate is divided into five major categories or groups, which are tropical, dry, continental, temperate and polar. Each group is represented by a letter of the English alphabet. This chapter introduces the different climates under the Köppen climate classification.

Chapter 4- Global warming refers to the rise in earth's average temperature and the effects it has on the environment. Climate change is the change in weather patterns over an extended time period. This chapter presents in detail the phenomenon of climate change, global warming and greenhouse gases. It also explores the physical impacts of climate change, its effect on ecosystems, plant biodiversity, etc.

Chapter 5- The ozone layer is the region in the stratosphere of the Earth, which absorbs most of the ultraviolet radiation coming from the Sun. Ozone depletion refers to the steady decrease in stratospheric ozone. This chapter has been carefully written to provide an easy understanding of the ozone layer and the phenomenon of ozone layer depletion.

I would like to make a special mention of my publisher who considered me worthy of this opportunity and also supported me throughout the process. I would also like to thank the editing team at the back-end who extended their help whenever required.

Salvador Poole

Chapter 1

Climatology: An Introduction

As a branch of atmospheric science, climatology refers to the study of weather conditions averaged over time. Climate models proposed by climatological studies are used for studying the dynamics of climate and weather as well as project future climates. This is an introductory chapter, which will introduce briefly all the significant aspects of climate and climatology.

Climate

The climate of a place may be defined as a "composite" of the long-term prevailing weather that occurs at that location. In a sense, climate is "average weather". Climate can be measured quantitatively by calculating the long term averages of different climate elements such as temperature and rainfall. Extremes in the weather however, also help us define the climate of a particular area.

We can study climate on a range of geographical scales. At the smallest scale, local climates influence areas maybe only a few miles or tens of miles across. Examples of local climatic phenomena include sea breezes and urban heating. At larger scales, regional climates provide a picture of particular patterns of weather within individual countries, or within climate zones that exist at different latitudes on the Earth. Climate zones include tropical, subtropical, desert, Savannah, temperate and polar climates. Different climate zones reveal variable patterns of temperature and rainfall. At the largest scale of all, climatologists study the global climate, for example when they are interested in how it has changed in the past, and how it may change in the future.

There are many influences on the climate of a particular place, the most important of which is latitude. Places nearer the equator receive more sunlight and are much warmer than places nearer poles. This temperature difference between low and high latitudes drives the general circulation of the atmosphere, transferring heat away from the equator towards the poles. The general circulation is split into a number of circulation cells which define separate pressure zones and belts of wind. These wind belts are sometimes called the prevailing winds. They blow more east-west than north-south however, because they are deflected by the Coriolis force, due to the Earth's rotation. The wind belts shift with the seasons, and so too, therefore, do the climate zones.

Other influences on climate include the oceans, land-sea distribution and mountain ranges. Ocean currents have a significant effect on regional and even global climate. The Gulf Stream in the Atlantic influences the weather of Northwest Europe. The periodic El Niño current in the equatorial Pacific can have drastic consequences for the weather in parts of South America and Australasia. Coastal regions generally experience mild and humid maritime climates, whilst the interiors of large land masses have more continental climates, with warmer summers and colder winters. Mountains not only affect local climates but can influence regional climate as well, by diverting the course of the prevailing winds.

The climate of the British Isles is temperate and maritime, with few weather extremes and a small annual temperature range. The British climate is influenced by the convergence of warm subtropical and cold polar air masses, which generates a zone of cyclonic low pressure and an abundance of rain-bearing depressions. The western side is generally milder in winter and receives more precipitation than the east, on account of its proximity to the Atlantic Gulf Stream and the greater presence of mountains.

The first scientific attempt in climatology was made by Edmund Halley in 1686. Halley attempted to map the trade winds. He asserted that there is a relationship between solar heating and atmospheric change. His work introduced the idea of weather and climate systems as interacting systems with the physical features of the Earth. Putting a step forward George Hadley in 1735 postulated that easterly trade winds are caused due to earth's rotation. It was further added that greater heating from the sun over the equator causes air to rise. Similarly, to balance the ascent of air, there must be an equivalent region of air descent away from tropics. The region is the cause of westerly winds. Hadley's circulation model, which is now considered the basic part of the climate system, was developed to explain the air motion in and around tropics.

Edmond Halley's mapping of Trade Winds

Gaspard-Gustave de Coriolis developed a theory of the effects of rotation in 1835. He explained how objects move within a rotating frame of reference and the forces operating upon them.Because of the rotational force, storms swirl counter clockwise in northern hemisphere and clock wise in southern hemisphere. It is now popularly known as 'Coriolis Effect'. William Ferrel offered explanation about the global circulation and the westerly wind system that characterize mid-latitude climates in 1856. Ferrel is also credited to have devised a law that winds deflect towards right from normal in northern hemisphere and towards left from normal in southern hemisphere due to earth's rotation.

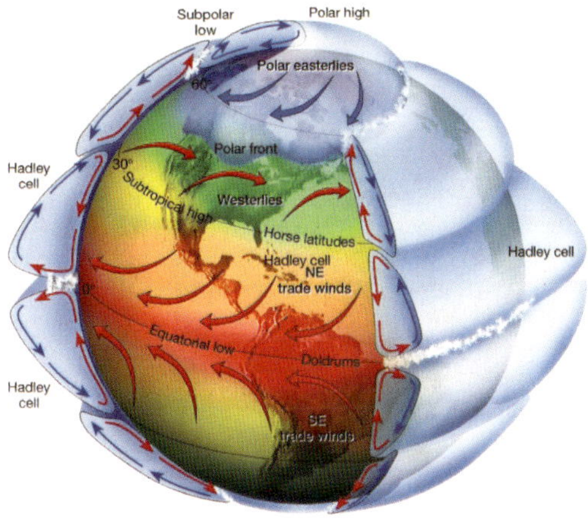

Hadley and Ferrel's Contribution to Cell Formation in Atmospheric Circulation

The role of earth's rotation in the movement of waves and winds was further improved by Carl Gustaf Rossby who introduced the concept of absolute vorticity and its conservation in adiabatic conditions. The theory of planetary waves was proposed by him. It forms the basis for dynamic climatology and oceanography. In 1837 Louis Agassiz proposed the hypothesis that Europe had experienced past "Ice Ages". It was based on the discovery of similar glacier features in Scotland. He found scratches on a rocky hill side near Edinburgh like those made by glaciers. He argued that Scotland had recently been covered in a thick ice sheet. Later on some scholars at University of Bristol refuted Louse Agasiz claim of ice age, by saying that it could also be a Wave cut platform and a coastal land feature. There has been a growing awareness that long term climate observations can lead to the understanding the mechanisms that affect climate change.

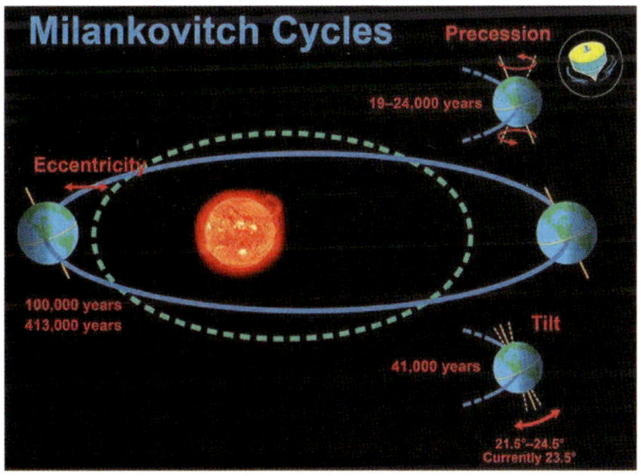

Milankovitch Cycles of Climate Change: Eccentricity, Precession and Tilt

The scientific study of patterns of solar energy reaching Earth's surface due to changes in the Earth's orbital parameters has a long history. The Earth's orbital calculations of Milankovitch (1941) led to the astronomical theory (or Milankovitch theory) of climate change and they are well known today. The Serbian astronomer and mathematician who is generally credited with calculating their magnitude particularly variations in the Earth's eccentricity, axial tilt, and precession comprise the three dominant cycles, collectively known as the Milankovitch Cycles.

John Tyndall demonstrated in the late 1850s, that earth's atmosphere has greenhouse effect. He explained that GHG were both emitters as well as absorbers of infrared radiation. The fact is vital for understanding the surface energy budget. Svante Arrhenius calculated the influence of carbon dioxide (CO_2) on earth's surface temperature in 1896. He asserted that any doubling of the percentage of carbon dioxide in the air would raise the temperature of earth surface by 4times. Henry Blanford noted in 1875 that order and regularity are the two prominent characteristics of India's atmospheric phenomena. Sir Gilbert Walker is credited to have identified the Southern Oscillation and its association with failures of the Indian Monsoon in 1904.

It was the early 20th century that the understanding of climatic elements with natural objects became well established. The researches moved beyond observation to the analysis and synthesis. Major impetus for change in climatology in 20th century was World War II. It has demonstrated the advantage of predictive climatology to modern warfare. The system was used to forecast the weather for military operations. The need to understand weather patterns and climates as they related to the possibility of nuclear war and expanding industrial, agricultural, communication and transport technologies. Considering the significance of climatology funding for training, research and education increased significantly.

Based on the principles of hydrodynamics and thermodynamics, Vilhelm Bjerknes's worked on mid latitude cyclones. Along with Lewis Fry Rechardson; V. Bjerknes developed a rudimentary three dimensional atmospheric model. It was an equation-based weather prediction model. The complexity of equations and magnitude of calculations involved made it limited to use by a small section of scientists.

Major contributions towards classification of regional and world climates came from H.E. Blanford (1889), W. Koppen (1918, 1931 1936), C.W. Thornthwaite (1931, 1933, 1948), L.D. Stamp and W.G. Kendrew (1953) and G.T. Trewartha (1954). Scholars have placed stress on working out temperature effectiveness, precipitation effectiveness, seasonal deficit and surplus of moisture, water balance and ground water recharge. World Meteorological Organization (WMO) based at Geneva, was established in 1950. WMO is engaged in collecting data globally and sharing its results to meteorological offices operating at regional levels.

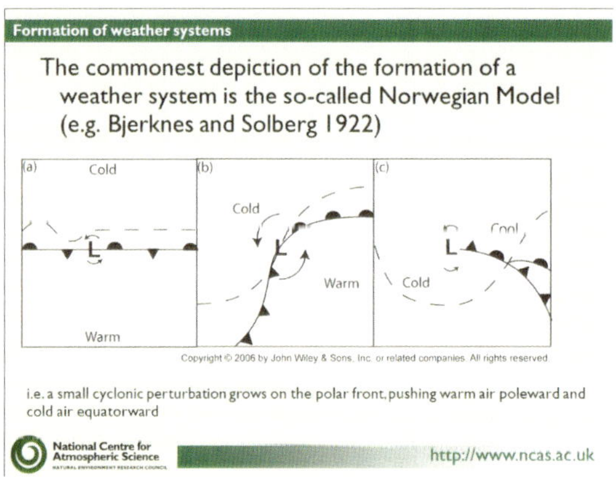

Norwegian Model of Cyclones by Bjerknes and Solberg

The invention and availability of high-speed computers in 1950s made numerical modeling of climate system possible. Later on, launching of resource satellites in 1960s further improved and diversified the data base for monitoring and forecasting of weather elements. New sub disciplines like dynamic meteorology and dynamic climatology have further improved the modeling and monitoring of climate systems.

Definition

Climate is commonly defined as the weather averaged over a long period. The standard averaging period is 30 years, but other periods may be used depending on the purpose. Climate also includes statistics other than the average, such as the magnitudes of day-to-day or year-to-year variations. The Intergovernmental Panel on Climate Change (IPCC) 2001 glossary definition is as follows:

Climate in a narrow sense is usually defined as the "average weather," or more rigorously, as the statistical description in terms of the mean and variability of relevant quantities over a period ranging from months to thousands or millions of years. The classical period is 30 years, as defined by the World Meteorological Organization (WMO). These quantities are most often surface variables such as temperature, precipitation, and wind. Climate in a wider sense is the state, including a statistical description, of the climate system.

The World Meteorological Organization (WMO) describes climate "normals" as "reference points used by climatologists to compare current climatological trends to that of the past or what is considered 'normal'. A Normal is defined as the arithmetic average of a climate element (e.g. temperature) over a 30-year period. A 30 year period is used, as it is long enough to filter out any interannual variation or anomalies, but also short enough to be able to show longer climatic trends." The WMO originated from the International Meteorological Organization which set up a technical commission for climatology in 1929. At its 1934 Wiesbaden meeting the technical commission designated the thirty-year period from 1901 to 1930 as the reference time frame for climatological standard normals. In 1982 the WMO agreed to update climate normals, and these were subsequently completed on the basis of climate data from 1 January 1961 to 31 December 1990.

The difference between climate and weather is usefully summarized by the popular phrase "Climate is what you expect, weather is what you get." Over historical time spans there are a number of nearly constant variables that determine climate, including latitude, altitude, proportion of land to water, and proximity to oceans and mountains. These change only over periods of millions of years due to processes such as plate tectonics. Other climate determinants are more dynamic: the thermohaline circulation of the ocean leads to a 5 °C (9 °F) warming of the northern Atlantic Ocean compared to other ocean basins. Other ocean currents redistribute heat between land and water on a more regional scale. The density and type of vegetation coverage affects solar heat absorption, water retention, and rainfall on a regional level. Alterations in the quantity of atmospheric greenhouse gases determines the amount of solar energy retained by the planet, leading to global warming or global cooling. The variables which determine climate are numerous and the interactions complex, but there is general agreement that the broad outlines are understood, at least insofar as the determinants of historical climate change are concerned.

Climate Classification

There are several ways to classify climates into similar regimes. Originally, climes were defined in Ancient Greece to describe the weather depending upon a location's latitude. Modern climate classification methods can be broadly divided into *genetic* methods, which focus on the causes of climate, and *empiric* methods, which focus on the effects of climate. Examples of genetic classification include methods based on the relative frequency of different air mass types or locations within synoptic weather disturbances. Examples of empiric classifications include climate zones defined by plant hardiness, evapotranspiration, or more generally the Köppen climate classification which was originally designed to identify the climates associated with certain biomes. A common shortcoming of these classification schemes is that they produce distinct boundaries between the zones they define, rather than the gradual transition of climate properties more common in nature.

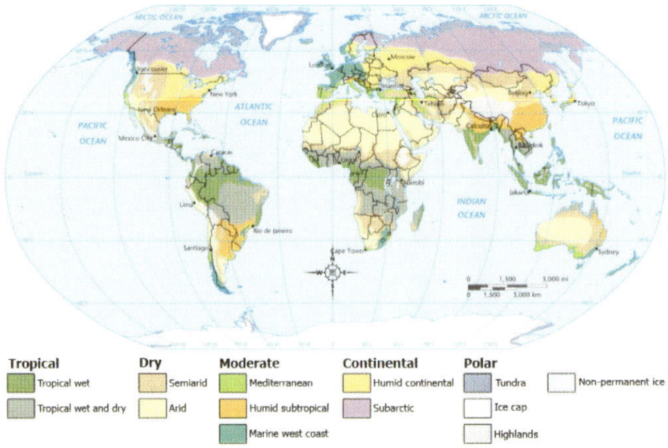

Worldwide climate classifications

Bergeron and Spatial Synoptic

The simplest classification is that involving air masses. The Bergeron classification is the most widely accepted form of air mass classification. Air mass classification involves three letters. The first letter describes its moisture properties, with c used for continental air masses (dry) and m for maritime air masses (moist). The second letter describes the thermal characteristic of its source region: T for tropical, P for polar, A for Arctic or Antarctic, M for monsoon, E for equatorial, and S for superior air (dry air formed by significant downward motion in the atmosphere). The third letter is used to designate the stability of the atmosphere. If the air mass is colder than the ground below it, it is labeled k. If the air mass is warmer than the ground below it, it is labeled w. While air mass identification was originally used in weather forecasting during the 1950s, climatologists began to establish synoptic climatologies based on this idea in 1973.

Based upon the Bergeron classification scheme is the Spatial Synoptic Classification system (SSC). There are six categories within the SSC scheme: Dry Polar (similar to continental polar), Dry Moderate (similar to maritime superior), Dry Tropical (similar to continental tropical), Moist Polar (similar to maritime polar), Moist Moderate (a hybrid between maritime polar and maritime tropical), and Moist Tropical (similar to maritime tropical, maritime monsoon, or maritime equatorial).

Köppen

The Köppen classification depends on average monthly values of temperature and precipitation. The most commonly used form of the Köppen classification has five primary types labeled A through E. These primary types are A) tropical, B) dry, C) mild mid-latitude, D) cold mid-latitude, and E) polar. The five primary classifications can be further divided into secondary classifications such as rainforest, monsoon, tropical savanna, humid subtropical, humid continental, oceanic climate, Mediterranean climate, desert, steppe, subarctic climate, tundra, and polar ice cap.

Rainforests are characterized by high rainfall, with definitions setting minimum normal annual rainfall between 1,750 millimetres (69 in) and 2,000 millimetres (79 in). Mean monthly temperatures exceed 18 °C (64 °F) during all months of the year.

A monsoon is a seasonal prevailing wind which lasts for several months, ushering in a region's rainy season. Regions within North America, South America, Sub-Saharan Africa, Australia and East Asia are monsoon regimes.

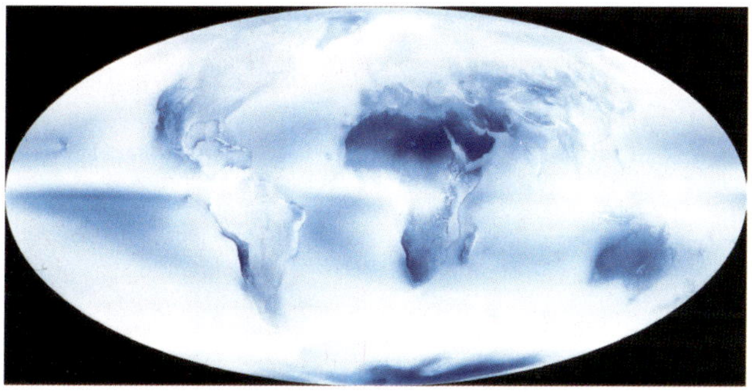

The world's cloudy and sunny spots. NASA Earth Observatory map

A tropical savanna is a grassland biome located in semiarid to semi-humid climate regions of subtropical and tropical latitudes, with average temperatures remain at or above 18 °C (64 °F) year round and rainfall between 750 millimetres (30 in) and 1,270 millimetres (50 in) a year. They are widespread on Africa, and are found in India, the northern parts of South America, Malaysia, and Australia.

The humid subtropical climate zone where winter rainfall (and sometimes snowfall) is associated with large storms that the westerlies steer from west to east. Most summer rainfall occurs during thunderstorms and from occasional tropical cyclones. Humid subtropical climates lie on the east side of continents, roughly between latitudes 20° and 40° degrees away from the equator.

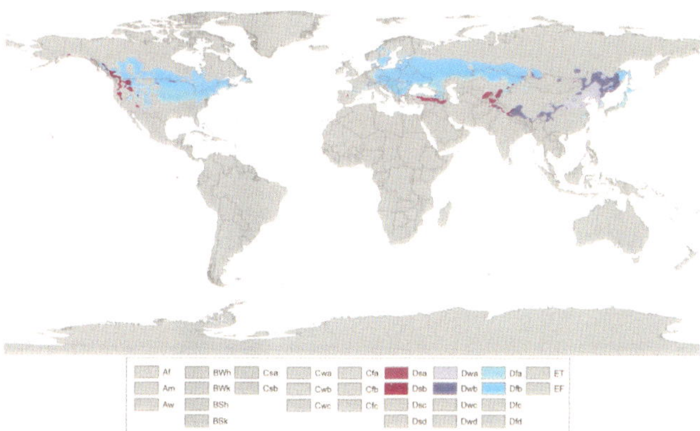

Humid continental climate, worldwide

A humid continental climate is marked by variable weather patterns and a large seasonal temperature variance. Places with more than three months of average daily temperatures above 10 °C

(50 °F) and a coldest month temperature below −3 °C (27 °F) and which do not meet the criteria for an arid or semiarid climate, are classified as continental.

An oceanic climate is typically found along the west coasts at the middle latitudes of all the world's continents, and in southeastern Australia, and is accompanied by plentiful precipitation year-round.

The Mediterranean climate regime resembles the climate of the lands in the Mediterranean Basin, parts of western North America, parts of Western and South Australia, in southwestern South Africa and in parts of central Chile. The climate is characterized by hot, dry summers and cool, wet winters.

A steppe is a dry grassland with an annual temperature range in the summer of up to 40 °C (104 °F) and during the winter down to −40 °C (−40 °F).

A subarctic climate has little precipitation, and monthly temperatures which are above 10 °C (50 °F) for one to three months of the year, with permafrost in large parts of the area due to the cold winters. Winters within subarctic climates usually include up to six months of temperatures averaging below 0 °C (32 °F).

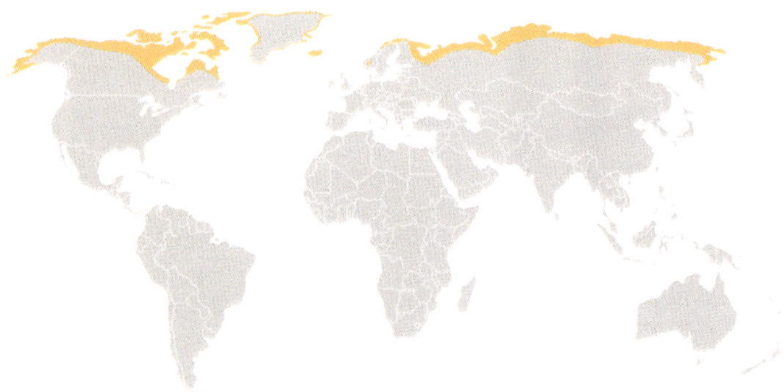

Map of arctic tundra

Tundra occurs in the far Northern Hemisphere, north of the taiga belt, including vast areas of northern Russia and Canada.

A polar ice cap, or polar ice sheet, is a high-latitude region of a planet or moon that is covered in ice. Ice caps form because high-latitude regions receive less energy as solar radiation from the sun than equatorial regions, resulting in lower surface temperatures.

A desert is a landscape form or region that receives very little precipitation. Deserts usually have a large diurnal and seasonal temperature range, with high or low, depending on location daytime temperatures (in summer up to 45 °C or 113 °F), and low nighttime temperatures (in winter down to 0 °C or 32 °F) due to extremely low humidity. Many deserts are formed by rain shadows, as mountains block the path of moisture and precipitation to the desert.

Thornthwaite

Devised by the American climatologist and geographer C. W. Thornthwaite, this climate classification method monitors the soil water budget using evapotranspiration. It monitors the portion of total precipitation used to nourish vegetation over a certain area. It uses indices such as a humidity index and an aridity index to determine an area's moisture regime based upon its average temperature, average rainfall, and average vegetation type. The lower the value of the index in any given area, the drier the area is.

The moisture classification includes climatic classes with descriptors such as hyperhumid, humid, subhumid, subarid, semi-arid (values of −20 to −40), and arid (values below −40). Humid regions experience more precipitation than evaporation each year, while arid regions experience greater evaporation than precipitation on an annual basis. A total of 33 percent of the Earth's landmass is considered either arid or semi-arid, including southwest North America, southwest South America, most of northern and a small part of southern Africa, southwest and portions of eastern Asia, as well as much of Australia. Studies suggest that precipitation effectiveness (PE) within the Thornthwaite moisture index is overestimated in the summer and underestimated in the winter. This index can be effectively used to determine the number of herbivore and mammal species numbers within a given area. The index is also used in studies of climate change.

Thermal classifications within the Thornthwaite scheme include microthermal, mesothermal, and megathermal regimes. A microthermal climate is one of low annual mean temperatures, generally between 0 °C (32 °F) and 14 °C (57 °F) which experiences short summers and has a potential evaporation between 14 centimetres (5.5 in) and 43 centimetres (17 in). A mesothermal climate lacks persistent heat or persistent cold, with potential evaporation between 57 centimetres (22 in) and 114 centimetres (45 in). A megathermal climate is one with persistent high temperatures and abundant rainfall, with potential annual evaporation in excess of 114 centimetres (45 in).

Record

Modern

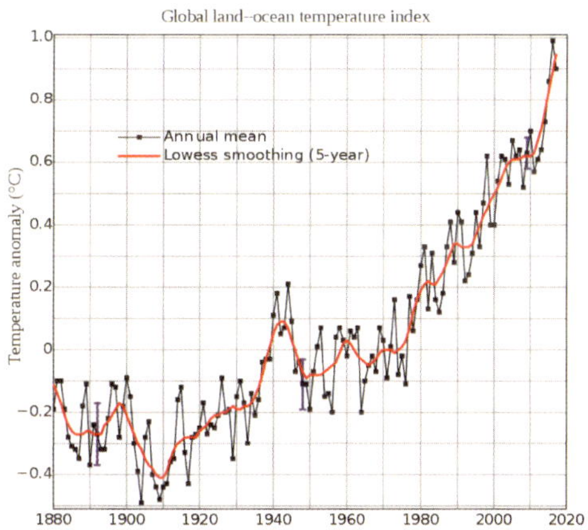

Global mean surface temperature change

Details of the modern climate record are known through the taking of measurements from such weather instruments as thermometers, barometers, and anemometers during the past few centuries. The instruments used to study weather over the modern time scale, their known error, their immediate environment, and their exposure have changed over the years, which must be considered when studying the climate of centuries past.

Paleoclimatology

Paleoclimatology is the study of past climate over a great period of the Earth's history. It uses evidence from ice sheets, tree rings, sediments, coral, and rocks to determine the past state of the climate. It demonstrates periods of stability and periods of change and can indicate whether changes follow patterns such as regular cycles.

Climate Change

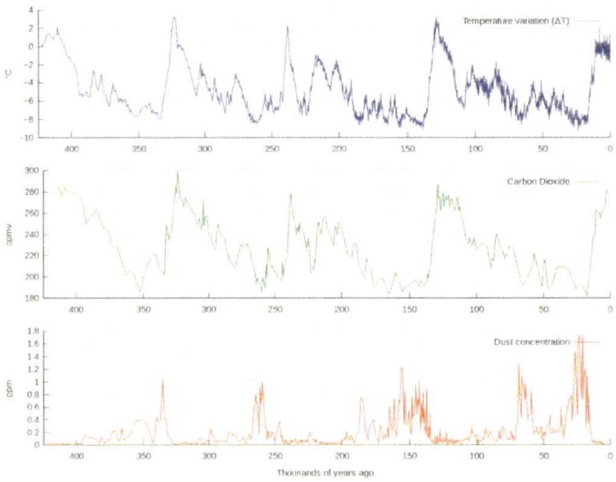

Variations in CO_2, temperature and dust from the Vostok ice core over the past 450,000 years

Climate change is the variation in global or regional climates over time. It reflects changes in the variability or average state of the atmosphere over time scales ranging from decades to millions of years. These changes can be caused by processes internal to the Earth, external forces (e.g. variations in sunlight intensity) or, more recently, human activities.

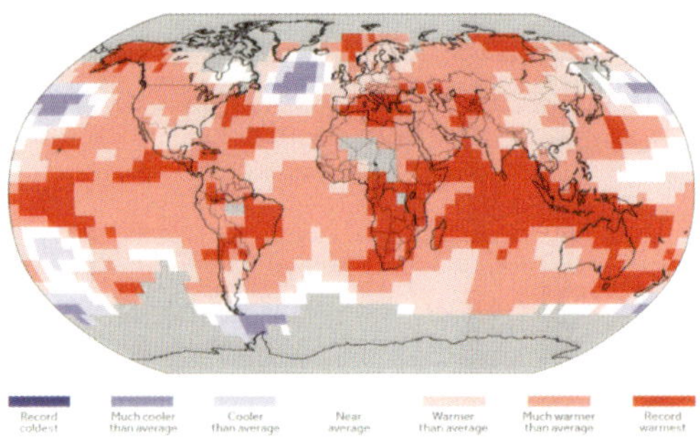

Colors indicate temperature anomalies

In recent usage, especially in the context of environmental policy, the term "climate change" often refers only to changes in modern climate, including the rise in average surface temperature known as global warming. In some cases, the term is also used with a presumption of human causation, as in the United Nations Framework Convention on Climate Change (UNFCCC). The UNFCCC uses "climate variability" for non-human caused variations.

Earth has undergone periodic climate shifts in the past, including four major ice ages. These consisting of glacial periods where conditions are colder than normal, separated by interglacial periods. The accumulation of snow and ice during a glacial period increases the surface albedo, reflecting more of the Sun's energy into space and maintaining a lower atmospheric temperature. Increases in greenhouse gases, such as by volcanic activity, can increase the global temperature and produce an interglacial period. Suggested causes of ice age periods include the positions of the continents, variations in the Earth's orbit, changes in the solar output, and volcanism.

Climate Models

Climate models use quantitative methods to simulate the interactions of the atmosphere, oceans, land surface and ice. They are used for a variety of purposes; from the study of the dynamics of the weather and climate system, to projections of future climate. All climate models balance, or very nearly balance, incoming energy as short wave (including visible) electromagnetic radiation to the earth with outgoing energy as long wave (infrared) electromagnetic radiation from the earth. Any imbalance results in a change in the average temperature of the earth.

The most talked-about applications of these models in recent years have been their use to infer the consequences of increasing greenhouse gases in the atmosphere, primarily carbon dioxide. These models predict an upward trend in the global mean surface temperature, with the most rapid increase in temperature being projected for the higher latitudes of the Northern Hemisphere.

Models can range from relatively simple to quite complex:

- Simple radiant heat transfer model that treats the earth as a single point and averages outgoing energy.
- This can be expanded vertically (radiative-convective models), or horizontally.
- Finally, (coupled) atmosphere–ocean–sea ice global climate models discretise and solve the full equations for mass and energy transfer and radiant exchange.

Climate forecasting is a way by some scientists are using to predict climate change. In 1997 the prediction division of the International Research Institute for Climate and Society at Columbia University began generating seasonal climate forecasts on a real-time basis. To produce these forecasts an extensive suite of forecasting tools was developed, including a multimodel ensemble approach that required thorough validation of each model's accuracy level in simulating interannual climate variability.

Climatology

Climatology is a branch of atmospheric sciences. It is devoted to the description and analysis of climates on the surface of the earth and its interaction with plants and animal life forms. Climatology is defined as the average weather condition of a place over a long period of time. Climate has to do with the atmospheric conditions over an extended to indefinite period of time. Atmospheric phenomenon always exercised a decisive role in the origin, development and activities of all life forms. Humans were always interested in decoding the mystery of atmospheric elements. Initially in oral tradition and later in written form, atmospheric elements occupied a central place in describing and creating awareness about the natural phenomenon.

The Ancient Greeks believed that climates were nothing more than temperature gradients varying along latitude belts. Aristotle's seminal work 'Meteorologica' presents behavior of weather across altitudes and latitudes. Aristotle is most often credited with the beginning of scientific thinking in the western world and he based his conclusions on empirical evidence. His influence is still felt today in many of the sciences, especially in medicine and the "natural" sciences such as earth science, biology, geology, and sub-disciplines. A brief treatise of habitable world is also explained in this monumental work. During medieval period, Arab's had a comprehensive knowledge of alternating wind system (Mausim) in Arabian Sea and Indian Ocean. The wisdom was used to take advantage of winds for sailing the ships and boats. The wind system, later on, came to be known as 'Monsoon Wind System'. Climatology has about 600 year's history. But it has come into wide spread scientific use for the past about 150 years.

Climatology was primarily an observational speculation prior to the scientific age. Gradually, devices for measuring and studying weather were invented and the keeping of systematic weather records began. Climatology, thus, began as the observation and description of weather on sub-continental and continental levels. The early nineteenth century marks the beginning of the scientific discovery of climate when ice ages in paleo-climate were first suspected. The natural greenhouse effect was also identified as an element of climate change. During first and second world war, effects of ground and upper air circulations were recognized. This led to the need for the statistical treatment of weather elements and its predictions. Earth observation through satellites and availability of huge quantum of data led to model the weather conditions and monitor the climate elements both at regional and global levels. Tracking of cyclones and early warnings for the mitigation of effects became possible through regular observations. It is now possible to make very precise prediction of extreme weather phenomenon. People's trust on climate forecasts for various activities has increased over the years. It is considered to be an important input for initiating any program.

Under the umbrella of Earth and Atmospheric Sciences, the global climatic processes and their impacts are studied in the subjects like Meteorology and Climatology. The word climate, is derived from Greek word *klimat*, denaoting inclination or latitude. It originally stood for a simple band of latitude on earth. When the scientists began to ponder around the possibility of climate change, their thoughts naturally turned towards the Sun, its radiation and the role of earth's atmosphere, hydrosphere and biosphere, in terms of climate and weather with reference to latitudes. The birth of the subject started from these issues. It was stated very clearly by the philosophers that the "Changing Sun is Changing the Climate". It holds good while studying climatology and meteorology, the twin-subjects of earth and atmospheric sciences.

Climatology is the science of studying the average atmospheric conditions of a region, on long-term perspective. It includes a systematic study of the atmospheric conditions pertaining to both weather and climate. Climatology also aims to study the nature , causes and interpretations of its spatial variation and its association with the other spheres. It is closely related to both meteorology and geography. All the global atmospheric processes are spatially distributed and are varying with reference to time. It deals with all climatic components that are responsible in controlling these processes.

Climatic Components

The term Climate can be defined as the average weather conditions for a particular location and for a specific period of time. The average weather conditions are reflected in the form of climatologically parameters or variables. The very essential climatic parameters are temperature, precipitation, atmospheric pressure, humidity, and velocity and direction of winds. The elements of weather are recorded and analysed by the scientists called climatologists. This is one of the oldest sciences of nature. And it is an ancient science, too.

The Ancient Science

These ancient literature contain enormous information about the processes of cloud formation, occurrence of rain and the seasonal cycles that are caused by the movement of earth round the sun. Initially, the subject of meteorology was given much importance by the ancient Indians. This has happened as early as 3000 BC. There is a chronology behind the growth of climatological studies, in the world.

The Chronology

If we look back to the ancient observations, we may be able to find the core concepts and thoughts of this subject. It was around 600 BC, Thales described the water cycle. He was the first Greek meteorologist to work on this line. In 400 BC, Democritus predicted the changes in the weather with proper evidences. The first written records about the atmosphere were given by Hippocrates in the year 400 B.C. Aristotle writes about Meteorology, in the year 350 BC. Aristotle was the first to bring out the ever first compendium of Meteorology, as a weather science. It was mainly based on the works of the Greeks, the world was divided into three temperate zones. The tropical zone, temperate zone and the frigid zone are the basic divisions of the globe. Indian scholars and philosophers contributed significantly to this subject.

The Work of Indians

In India, the reference to meteorological aspects can be traced from the works of the Upanishads. The ancient people were the first to record the climatic observations in the form of charts, maps, drawings, rock paintings and canvas paintings. In the year 500 AD, Varāhamihira's classical work, called as Brihatsamhita, provided a clear evidence that a deep knowledge of atmospheric processes existed, in India, even in those times. During the 7th century, Kalidasa, in his epic Meghaduta, mentions about the date of onset of Monsoon over central India and traces the path of monsoon clouds. These paved the way for all modern works in climatology.

The Modern Works

The oldest known Irish diary containing the weather data was of the year 1682/83. The middle ages (from 7 th century to 17th century) have shown a lot of outstanding analysis about global weather and climate. The first classification of clouds was done as early as 1803. The oldest preserved diary of meteorological observations was found in an observatory's archives belonging to the period between 1826 to 1828. The newspaper "Inland" published the weather information regularly every year since 1839. Over a period of time, the twin-subjects, climatology and meteorology have grown as very important scientific disciplines.

Climatology & Meteorology

The root for climatology lies in the subject of meteorology. It was the Greeks who tried to understand the atmosphere and its influence on life. The 20th century has witnessed a lot of advancements in the development of tools and techniques in order to obtain the regular and accurate data of weather and climate. Technological advancements have helped to operationalise many of the applications. Satellite based, real-time and continuous monitoring of all the segments of the whole globe, helps to understand the trends of weather and climate, well in advance.

Differences with Meteorology

In contrast to meteorology, which focuses on short term weather systems lasting up to a few weeks, climatology studies the frequency and trends of those systems. It studies the periodicity of weather events over years to millennia, as well as changes in long-term average weather patterns, in relation to atmospheric conditions. Climatologists study both the nature of climates – local, regional or global – and the natural or human-induced factors that cause climates to change. Climatology considers the past and can help predict future climate change.

Phenomena of climatological interest include the atmospheric boundary layer, circulation patterns, heat transfer (radiative, convective and latent), interactions between the atmosphere and the oceans and land surface (particularly vegetation, land use and topography), and the chemical and physical composition of the atmosphere.

Aim of Studying Climatology

The primary goal of Climatology is to study the unique characteristics of atmosphere, its structure and composition in controlling the global climate, origin and types of climates, causes and processes influencing the climatic variations, elements of weather and the impact of climate on human and other life. Agricultural climate, aviation climate, bio climate, medical climate, macro climate, micro climate, meso climate, Palaeo climate and Arctic climate are some areas of special studies coming under the broad science of climatology.

An Inter-disciplinary Science

Climatology is an interdisciplinary subject. It involves the basic principles and application areas of physics, chemistry, geology, mathematics, electronics, botany, zoology and environmental science. It has direct links with the subjects like oceanography, forestry, hydrology, glaciology, agriculture, animal husbandry, fisheries, horticulture, floriculture, marine biology, microbiology. Today, climatology is a well-grown field of science. It has several major and minor branches.

Major and Minor Branches

The major branches are:

1) Historical Climatology

2) Physical and Dynamic Climatology

3) Regional Climatology

4) Satellite climatology

5) Tropical Climatology

6) Palaeoclimatology

7) Applied Climatology

Diversified Fields

Historical climatology deals with the study of historical changes in climate and their effect on human history and development. The Physical Climatology is mainly concerned with the various atmospheric phenomena and their resultant climates. Regional Climatology deals with the regional variations of climatic conditions which include Micro climate, Local climate, Meso Climate and Macro Climate. Paleoclimatology is yet another branch which encompasses the effects of climatic change over the entire history of the earth. Applied Climatology deals with climatic controls of human activities and the application of climatic knowledge in solving unique problems faced by human population. In this respect, climatology is fully related to ecology.

Climatology and Ecology

Ecology is the study of the mutual relationship between organisms and their environment. The study of earth's ecosystems involve both biotic and abiotic components. The environment of living organisms is directly affected by weather and climate, including those changes in climate that are gradually being made by human beings. Climatology has a direct bearing on ecology, ecosystems and the well-being of all life. They are all based on the global climate and cyclical processes.

Global Climate and Cyclical Processes

The global climate and many cyclical processes on the earth are all controlled by the atmosphere and its processes. These processes have both direct and indirect influence on several impacts on the world. Among those, the Weather and Climate, are the two distinctive features, that are having very direct influence on the whole globe and the life. The science of climatology draws full attention, due to these impacts on the earth and life. The four dimensions of the subject are, Structure and dynamics of the atmosphere, Suns' radiation, global cyclic processes and their impacts on all segments of the earth. The first one is related to the atmosphere.

Climatology and Atmosphere

Atmosphere is an important segment of the globe, containing air, water vapour and particulate matter. Atmosphere is an envelope is attached with the globe due to the gravitational pull of the earth. The life on the Earth exists because of the presence of atmosphere. It protects all life by blocking out the some of the harmful radiation coming from the sun. Atmosphere is fully inter-linked with all other spheres like hydrosphere, lithosphere and biosphere. The weather and climate are a part of the atmospheric effects, which control the life and life support systems. The primary aspect in the subject of climatology is in analyzing the atmospheric variables and data acquisition. In order to do that, it is necessary to understand the Atmosphere, atmospheric chemistry, the Gas Laws, the Vertical Structure of the Earth's Atmosphere.

Importance of Atmosphere

Atmosphere is the multilayered gaseous envelope attached to the planet earth. It is a unique sphere of air, gases and water vapour. The height of this huge mass is estimated to range from 16000 to 29000 KM from the sea level. This is an essential environmental segment for the life support systems of biosphere and hydrosphere. The vertical variation in temperature and concentration of gases have a significant impact on global scales. The layer concerning with weather and climate, is trophosphere. Atmosphere is composed of three major constituents as gases, water vapor and aerosols. Some are permanent (or) constant gases and some are variable gases. They obey Boyle's gas Law, Lussac's Law, Charle's Law and combined gas law.

The Study of Atmosphere

Under the umbrella of Climatology, the structure and composition of the atmosphere are studied first. An understanding of the thickness and disposition of different layers like troposphere, stratosphere and mesosphere are needed. Under the conditions of the atmosphere the following aspects are to be understood.

- Elements of Weather and Climate
- Structure and Composition of the Atmosphere
- Insolation and Energy balance
- Atmospheric Temperature
- Atmospheric Pressure and Motion

Climatology and Global Energy

Solar energy and gravitational energy are the fundamental sources of energy for the Earth's climate system. The Sun is the source of the energy that causes the motion of the atmosphere and thereby controls weather and climate. Any change in the energy from the Sun received at the Earth's surface will therefore affect climate. During stable conditions there has to be a balance between the energy received from the Sun and the energy that the Earth radiates back into Space. This energy is mainly radiated in the form of long wave radiation corresponding to the mean temperature of the Earth.

Climatology Involves the Study of

- Insolation
- Flow of Energy
- Absorption of radiation
- Emission of Radiation
- Distribution of Radiation
- Heat Budget (or) Energy balance
- Human influence on radiation balance

Solar Radiation

The Sun is the ultimate source of energy on earth. Radiant energy is the energy propagated in the form of electromagnetic waves. The proportion of Solar Radiation falling on a non-luminous body which the latter reflects is known as Albedo. It is the reflective quality of the surface. Albedo of

Earth is 0.3 (about 30% of the Solar radiation is reflected back into the space). These waves do not need molecules to propagate them, and in a vacuum they travel at nearly 3,00,000 Km per second. Insolation is maximum at tropics and decreases towards poles (75% at equator, 50% along Arctic/ Antarctic circles and 40% at poles).

The Factors Influencing Insolation are

- The Angle of Incidence
- The Length of the day
- The transparency of the atmosphere
- Heating and Cooling of the atmosphere

The Nature of Radiation

Climatology analyses the nature of sun's radiation. The Solar Source, the details about Photosphere, Chromospheres and Sunspots are studied in detail. The Earth's Surface and Solar Energy are in continuous contacts as a never ending cycle. The Atmosphere and Solar Radiation in terms of Reflection, Scattering and Transmission are the basic elements of study in climatology.

The Energy Balance

The Energy system of the Earth involves, Solar Radiation, gravity and geothermal sources. The Solar energy plays a very significant role in heating the earth's surface, and the atmosphere. Because of these, it is responsible for the movement of air and oceanic currents. The radiant energy is the driving force for the global water cycle, rock and the biogeochemical cycles. Without Sunlight, there is no photosynthesis in plants, and there may not be any food for the other life.

Study of Weather and Climate

The term weather refers to the sum total of short term variations in atmospheric conditions in terms of temperature, pressure, wind, water vapour, clouds, precipitation and visibility. The term climate refers to the long-term aggregate of weather conditions of a region (or) the globe, in terms of several atmospheric parameters.

Elements of Weather and Climate

The following are the elements of weather and climate:

- Temperature
- Air Pressure
- Wind
- Humidity
- Precipitation
- Cloudiness

These elements are controlled by various factors like latitude, altitudes, unequal distribution of land and water, air-sea interface, ocean water circulation and geomorphological conditions.

Temperature is the First Element

Atmospheric temperatures, seasons, daily temperature changes, daytime heating, nighttime cooling, daily temperature range, seasonal lag and extreme temperatures, factors influencing the vertical distribution of temperature, factors influencing the horizontal distribution of temperature, the global heat budget, impact of latitude, surface properties of the earth, slope and topography, the dynamic factors of climate and monitoring of temperatures over the earth's surface, are the very essential parts of study in climatology.

Atmospheric Temperature

Climatology involves the study of

- Transfer of heat energy
- Heating and Cooling of the atmosphere
- Mean Temperatures
- Distribution of Temperature
- Inversion of Temperature
- Role of temperature in other environmental segments like Hydrosphere and biosphere

Uneven Distribution of Temperature

The distribution of temperature varies horizontally and vertically. The uneven distribution of temperature comes due to

- Latitude
- Distribution of land and sea
- Relief and altitude
- Ocean Currents
- Planctary Winds
- Vegetal Cover
- Nature of the Soil
- Slope and aspect

The Three Processes

Sun is the major source of temperature in the atmosphere and also in the oceans. Heating and Cooling of the atmosphere is controlled by direct Sunlight. There is also a transfer of energy from the earth through convection, conduction and re-radiation.

The processes of heating of the atmosphere are:

Convection - is the upward movement of air.

Conduction - is from high temperature zone to low temperature zones.

Radiation - is radiant energy emitted by a body in the form of heat. Re-radiation also heats the air.

Climatology & Global Pressure

Climatology studies the Pressure Gradient, Pressure Types, Variations in atmospheric pressure, Horizontal distribution of air pressure, Pressure belts and shifting of Pressure belts, Atmospheric motion and air circulation, Geostrophic wind and Gradient Wind, Wind direction and speed and Classification of Wind.

Distribution of Air Pressure

The horizontal distribution of air pressure depends on

- Air temperature
- Earth's rotation and
- Zones of low (or) high pressure

Climatology and Atmospheric Circulation

The large-scale structure of the atmospheric circulation varies from year to year, but the basic climatological structure remains fairly constant. The movement of air in the atmosphere is known as atmosphere circulation. Atmospheric circulation denotes the motion of air due to pressure gradient from local to global scales. It has spatial and temporal patterns of circulation components. It involves

- Primary Circulation (or) long term circulation
- Secondary Circulation
- Seasonal Circulation
- Diurnal Circulation

Types of Circulation

There are four types of air circulation in the atmosphere, as primary, secondary, seasonal and diurnal. The Primary Circulation of air is related to the trade winds, westerly winds and polar winds. The Secondary Circulation is related to Cyclones and anti-cyclones. The Seasonal Circulation is related to Monsoons and air masses. The diurnal Circulation is related to Land/Sea breezes, mountain (or) valley breezes.

Forces Controlling Circulation

Atmospheric circulation, speed and direction of winds are controlled by

- Gravitational force

- Pressure gravitational force
- Coriolis force
- Friction force

Atmospheric Pressure and Motion

Air is mixture of gases and hence has its own weight. Air exerts pressure. Atmospheric pressure is maximum at sea level. It decreases with increasing altitudes. Air pressure varies vertically and horizontally. The distribution of atmospheric pressure is controlled by altitude, temperature, air circulation, earth's rotation, water vapour and other factors.

Dynamics of the Atmosphere

Under the Concept of Dynamics of the Atmosphere the following aspects are to be understood.

- The Atmospheric Circulation
- The Local and Seasonal Winds
- Humidity and Condensation
- Fogs, their origin and types
- Clouds, their origin and types
- Precipitation and its distribution

Local and Seasonal Winds

Wind is the circulation of air relative to the surface of the earth. Climatology analyses the local and seasonal winds in detail. The periodic local winds include the breezes between land and sea and the breezes from mountains and valleys. The non-periodic local winds include the hot local winds and Cold Local winds. There are some more types of local winds as beneficial local winds and non-beneficial local winds.

Climatology involves the evaluation of

- Zonal Circulation of atmosphere as global wind belts
- Tropical Circulation (Doldrums, Westerlies and trade winds)
- Mid-latitude Circulation
- Polar Circulation
- Meridonial Circulation
- Jet Streams
- El Nina and LA NINA

Circulation Features

Earth's rotation causes many circulation features. The resulting Coriolis effect causes the meridional flow to be disrupted as winds are deflected to the right of their course in the Northern

Hemisphere and to the left of their course in the Southern Hemisphere. Atmospheric circulation can be divided into three convection cells in each hemisphere. From equator to the poles these cells are the Hadley cell, Ferrel cell, and Polar cell. These are latitudinal circulation features. The Longitudinal circulation features include Walker circulation and El Niño – Southern Oscillation circulation.

Moisture in the Atmosphere

Atmosphere contains water vapour as moisture in air. The changes of state of matter in the hydrologic cycle, relative humidity, evaporation and transpiration, condensation near the ground, distribution of dew, mist, and fog, foggy places, condensation above the surface, origin of clouds, their vertical motion in the atmosphere, adiabatic heating and cooling, lapse rates, the regional variation in precipitation, the variation in precipitation through time and climatic water balance are the very essential aspects involved in climatology.

Atmospheric Humidity and Condensation

The term Humidity refers to the amount of water vapour present in a parcel of air having a unit volume and temperature. Water Vapour is released during the process of evaporation from water bodies of land and oceans. Since water vapour is very important constituent of the atmosphere, it plays a significant role in controlling the climate. It is present upto a height of 5 Km from the earth's surface. Condensation is the process of converting Water Vapour into liquid and solid forms. Atmospheric water under goes phase changes through the exchange (or) transfer of heat energy. The energy in the form of heat required to carry out all these conversions is called as the Latent Heat.

Climatology Analyses Humidity

Climatology deals with the elements of the hydrologic cycle, including Humidity Capacity of air, Types of Humidity and their measurement, Evaporation & its rate, Evapotranspiration, Condensation & forms of Condensation, (Cooling of air and adiabatic change of temperature), Stability and instability of the atmosphere.

Fogs, their Origin and Types

Fogs are special types of clouds consisting minute water droplets existing in suspension in the air near the ground surface. Fogs generally reduce the horizontal visibility. Fog is formed when the moist air becomes saturated and reaches its due point. Further cooling of this mass make the water vapour to condense around the dust particles and create the dense smoky Fog. There are a dozen bases of classification available in climatology to study the types of Fogs.

Types of Fogs

The major types of Fogs studied are

- Radiation Fog - Fog produced over land when radiational cooling reduces the air temperature to or below its dew point. It is also known as ground Fog and Valley fog.

- Steam Fog – Evaporation (mixing) Fog.
- Upslope Fog – Fog formed as moist stable air flows upward over a topographic barrier.
- Tropical Airmass Fog.
- Frontal Fogs.

Fogs are distributed in different parts of the world. Climatology also analyses the effects of Fogs, and the natural (or) Man induced dispersal of Fogs.

Clouds, their Origin and Types

Clouds are the aggregates of innumerable tiny water droplets in the atmosphere above the ground surface. Clouds are formed due to condensation of water vapour. Global patterns of precipitation depend on the clouds. Clouds play a significant role in controlling the heat budget of the earth's surface. The study of clouds involved the classification of clouds and the characteristic features of clouds. The structure, altitude, occurrence and appearance of the following types of clouds are studied in climatology.

Monsoons and Climatology

Climatology involves the study of Monsoons, their origin, types and distribution. Monsoon is a complex phenomena of the atmosphere. The Origin of Monsoons is based on Thermal concept and Dynamic concept. Monsoon refers to the annual cycle of dryness and wetness, with seasonally shifting winds produced by changing atmosphere, pressure systems. The study of Seasonal Winds provides detailed large-scale information on Monsoons. Monsoon has a flow pattern of general atmospheric circulation over a wide geographical area. Monsoons have a clear and dominant wind in one direction during one season and in reverse direction during the other season.

Seasons

The seasons are the result of the tilt of the Earth's axis. The Earth's axis is tilted from perpendicular to the plane of the ecliptic globe, by 23.45°. This tilting is what gives us the four seasons of the year - spring, summer, autumn (fall) and winter. Since the axis is tilted, different parts of the globe are oriented towards the Sun at different times of the year. Summer is warmer than winter (in each hemisphere) because the Sun's rays hit the Earth at a more direct angle during summer than during winter. It also reflects it, in the duration of the days that are much longer than the nights during the summer. During the winter, the Sun's rays hit on the Earth at an extreme angle, and due to which the days are very short. These effects are due to the tilt of the Earth's axis. Seasons are very essential aspects in this subject. These provide us the rainfall, snowfall and other forms of precipitation.

Precipitation and its Distribution

The term precipitation refers to the fall of atmospheric moisture either in the form of liquids or solids. Climatology includes the study of precipitation, its measurement over space and time. The forms of precipitation includes rain, drizzle, snow, ice pellets, hail, and sleet. Among these, rain is the most wide spread form of precipitation. Climatology analyses the types of precipitation, controls of precipitation and various rainfall regimes.

The Rainfall Regimes Studied are

a) Equatorial Rainfall Regime

b) Tropical Rainfall Regime

c) Monsoonal Rainfall Regime

d) Continental Rainfall Regime

e) Mediterranean Rainfall Regime

f) Maritime Rainfall Regime

These are considered as contributions of the atmosphere. In addition to these, there are a lot of disturbances happen on the earth due to global atmospheric processes.

Atmospheric Disturbances

Atmospheric disturbances includes the basic understanding of air masses, and extreme events in the atmosphere. The term Air Mass is used to denote a large body of air whose dominant Physical properties are more or less uniform horizontally and vertically for hundreds of Km. Temperature and moisture are the two major properties. Air mass controls the weather conditions of an area. It plays a major role in the transfer of energy in the atmosphere through its movement. Air Masses are subjected to Thermodynamic and Mechanical Modification. Due to these, air masses are divided into cold air and warm air masses. Climatology attempts to classify the air masses, based on thermodynamic and geographical factors. An Air is defined as the transition zone between two air masses of different density. Hence, climatology deals with the distribution of Air fronts.

Air is a Poor Conductor

One of the important properties of air is that it is a poor conductor of energy. This means that when two different bodies of air come together, they do not readily mix each other. Rather, each body of air will retain its individual properties, and a boundary forms between them. When two large air masses meet, the boundary that separates them is called as an air front. Fronts represent fairly abrupt transitions between two large air masses.

Air Fronts

The term Air Front is used to denote the sloping boundary between two opposing air masses having contrasting physical properties. These properties are temperature, humidity, pressure and wind direction. Air fronts differ from each other in terms of their location and spread. Air fronts are classified based on their characteristics as Warm Front, Cold Front, Occluded Front and Stationary Front. Earth's surface has three major frontal zones as Polar frontal Zone, Arctic frontal Zone and Inter tropical frontal Zone. The Weather associated with warm and cold fronts, are also studied in Climatology. Air Fronts extend not only in the horizontal direction, but in the vertical as well. Air fronts appear and disappear in the atmosphere. The processes of formation of new fronts, and decaying of old fronts is known as Frontgenesis. It is an important aspect in climatology.

Frontogenesis

Frontogenesis is the generation or intensification of a front. It occurs when warm air converges onto colder air, and the horizontal temperature gradient amplifies by at least an order of magnitude. The conditions responsible for frontogenesis are:

a) Temperature difference and

b) Opposite directions of air masses.

Extreme Events in Atmosphere

Extreme events are those which are originated by natural or human factors causing severe impact on the environment like a hazard. These events exceed the tolerable magnitude in the atmosphere causing irreparable loss to life and properties. Atmospheric extreme events include Severe Storms, lightning, hailstorm and thunderstorms. Cyclones are also very dangerous natural hazards, generated by the atmosphere.

Cyclones

Climatology includes the study of cyclones. Cyclones are the centres of low atmospheric pressure in which air blows inwards in anticlockwise direction in the northern hemisphere and clockwise direction in the Southern hemisphere. Cyclones are considered to be the major atmospheric disturbances. They are classified into

a) Extra tropical cyclones (Temperate) and

b) Tropical Cyclones – One of the most powerful and destructive types of cyclonic storms is the tropical cyclone. It is also known as Hurricane or Typhoon.

Anticyclones are opposite in character to those of cyclones. It shows divergent air circulation in all directions from a high pressure centre. The other issues are the occurrence of hurricanes.

Hurricanes

Tropical cyclones are low pressure systems having a diameter of 650 Km moving with an energy equivalent to 10000 atomic bombs. They are called as hurricanes in the North Atlantic Ocean and as Typhoons in the North Pacific Ocean. The wind speed may be as high as 180 to 400 Km per hour, giving a rainfall of 2000 mm per day, as seen in Philippines. The other extreme event is the thunderstorm.

Thunderstorms

A thunderstorm is a form of turbulentweather. It is also known as an electrical storm, a lightning storm, thundershower or simply a storm. A thunderstorm develops in three stages as Cumulus Stage, Mature Stage and Dissipating Stage. Stronger thunderstorm cells are capable of producing tornadoes and waterspouts.

Tornadoes

Tornadoes are funnel shaped violent dark storms where in the upper portion is like an umbrella and the lower portion is like a pipe touching the ground surface. It is a small, intense cyclone in which the air is spiraling at tremendous velocity. It appears as a dark funnel cloud. Climatology also includes the impacts of hazards and their side effects like floods.

World Climatic Types

Climatic parameters have a profound influence on vegetation, soil and fauna of different regions of the world. World climates depend on the nature of parameters like temperature and precipitation and their influence over fauna and flora of ecosystems. Under the concept of World Climatic Types, the distribution of Climatic Types and Biomes and classification of Climatic Types are studied in detail. The world is rich in holding many endless variety of climatic type. The factors that produce the climate in any given place are called as climatic controls.

Classification of Climates

The classification of Climatic Types are studied in detail. The aspects considered in this are:

- Intensity of sunshine and its variation with latitude
- Distribution of land and water
- Ocean currents
- Prevailing winds
- Positions of high and low pressure areas
- Mountain barriers and
- Altitude

Notable Climatic Classifications

Climatic classification of regions are very essential for several human activities. Climatic variations are seen with reference to latitudes and altitudes. Both Vertical and horizontal variation of climate are analysed in Climatology. Several climatic zones have been identified so far. World climatic zones are classified based on empirical, numerical and genetic approaches. The important classification schemes studied are:

- Koppen's multi-tiered classification system
- Thornthwaite's Classification
- Genetic Classification and
- Composite Classification

Climatic Types and Biomes

Climatology helps in classifying the global biomes. A biome is a large natural identical ecosystem having unique assemblage of plant and animal communities having a definite distributional patterns in terms of climatic and soil conditions. The world biomes are also classified into 3 major groups as

a) Tropical biomes (Rain forest, Monsoon Forest, Savanna, Sahara)

b) Temperate biomes (Forests, Grassland, Mediterranean, Humid, Steppe, Boreal)

c) Tundra Biomes (Arctic, Alpine) and these include more number of sub orders

Regional Climatology

Climatology attempts to evaluate the regional climatic conditions. The types of climates at regional level are microclimate, mesoclimate, macroclimate and global climate.

Microclimate, is related to the climatic conditions of a small climatic region.

Mesoclimate, is related to the climate of a small area of the Earth's Surface (an acre to several Square Kilometers).

Macroclimate, is related to the climate of a much larger area (state/country).

Global climate, refers the climate related to the entire Globe.

Climatic Change

The global climate is always changing. Evidence shows that climate has changed in the past and there are also evidences that show that it will change in future, too. Variation in Earth's climatic conditions is seen in both space and time. The reason behind this variation is the dynamic nature of atmosphere. Global climate may change due to natural causes and man-made activities. The change may occur drastically or gradually. The notable modifications may come in moisture, cloudiness precipitation, and in the heat balance of the earth-atmosphere system. The changes may be cyclic and rhythmic. Climate change can adversely affect the lives of millions. Due to these, climatology analyses the causes and effects of climatic change and find remedial measures.

The Indicators of Climatic Change

There are several theories and causes of climatic changes. In Climatology all of them are studied to understand the past and predict the future. There are several kinds of indictors reflecting the climatic changes. The notable ones are:

1. Biological Indicators (Floral, Faunal remain in fossils)
2. Geological Indicators (Sedimentary deposits)
3. Cryogenic Indicators (ice ages, glaciers, ice sheets)
4. Tectonic Indicators (Sea floor spreading and sea level changes)
5. Geomorphological Indicators
6. Historical records of droughts and floods and migration of man and animals

Dendrochronology is one aspect coming under these analyses. The Global Warming refers to increase in global surface and atmospheric temperature affecting the global radiation balance. It needs a sound weather forecast methodology. Weather forecasting is a part of climatic studies.

Weather Forecast

Weather Forecasting is a major field of predicting the weather conditions comprising the temperature of atmospheric air, movement of wind, humidity, sunshine hours, cloudiness, precipitation and atmospheric disturbances over a place or region, well in advance. Different sources of weather and climatological data are collected on daily, weekly, monthly and annual bases from global network of observatories and used to forecast.

Methods in Weather Forecast

The weather forecasting methods include

a) Synoptic forecasting methods

b) Statistical forecasting method and

c) Numerical forecasting method.

Weather Forecasts are of various types as

a) Short-range weather forecasts (upto 48 hours)

b) Medium-range weather forecasts (3 days to 2 weeks)

c) Long-range weather forecasts (2 weeks to seasons)

Weather warning, is a quick forecast indicating that hazardous weather is either imminent (or) actually occurring within the specified forecast area. Weather watch is a forecast indicating the instant atmospheric conditions and to tell whether it is favourable or hazardous.

Satellite Climatology

Remote sensing refers to the acquisition of information about a target or phenomenon from a distance. The climate is reflected in the form of low-frequency signal of weather that involves interactions among Earth's environmental systems. Both remote sensing and climate are related to radiation transfer through Earth's atmosphere. Combining these two disciplines forms Satellite Climatology. It emphasizes understanding the different techniques used to determine, from space-borne platforms, the atmospheric, oceanic and land surface conditions important to climate and its variations, and the interpretation of these remotely sensed data in the context of "climate dynamics" and "synoptic climatology."

Satellite Climatology Applications

The most common type of satellite climatology is a cloud composite, which is a measure of the frequency and persistence of clouds over a particular region. Cloud composites can be derived from geostationary or polar orbiting satellites. The primary focus is on the use of satellite imageries.

Because of the great spatial coverage, diurnal composites of imagery over monthly, seasonal, and annual time scales can reveal the amplitude of diurnal cloudiness in relation to geographic features, such as land/water boundaries and changes in their elevation. It is a very interesting study in climatology.

Tropical Climatology

This subject, tropical climatology, is of great relevance to the tropics as climate in many ways controls the lives and economic activities of the approximately 2400 million people living in tropical regions. The tropical climates have profound effects that reach far beyond the limits of the regions where they actually prevail. The global general circulation is largely driven by the export of considerable amounts of heat energy from tropical to extratropical latitudes. A large part of all atmospheric water content originates from the tropics, and intermittent tropical phenomena, like El Nino Southern Oscillation (ENSO). These not only influence the climates over extensive tropical areas but many parts of the extratropics. The subject of tropical climatology is an important aspect in the subject of climatology.

Paleoclimatology

Paleoclimatology refers to the study of changes in climate that has taken place on a large scale in the entire history of Earth. It uses a variety of indirect methods from the Earth and life sciences to obtain data previously preserved within the rocks, sediments, ice sheets, tree rings, corals, shells and microfossils. Palaeoclimatologists use these records to determine the past states of the earth's various climate regions. Studies of the past changes, in the environment and biodiversity, can reflect on the current situation, and the impact of climate on mass extinctions and biotic recovery.

Leading Role of Climatologists

Climatologists, all over the world, play a very significant role in studying the atmospheric phenomena by using meteorological data. The meteorological data are gathered by surface observation stations, air stations, using sensors in satellites, and also by using radar. Climatologists interpret the meteorological data and prepare the daily, weekly, monthly and annual weather reports. Forecasting the climatic changes, for the knowledge of the public and their attention, is a regular work of the climatologists.

Climatic Maps and Charts

Weather and climatic maps and charts are prepared by the climatologists, for routine use. In addition, climatologists conduct numerical simulations to understand and predict the global, regional and local weather patterns. They apply the basic principles of meteorology to solve practical problems in the field of agriculture, pollution control, and water management.

Problem-solving Professionals

Climatologists are problem-solving professionals. Tackling the issues like global warming , climate change, ozone depletion, desertification, cyclones, hurricanes and other natural hazards are part of their professional work. Climatologists evaluate the long-term and short-term weather conditions.

Different Approach

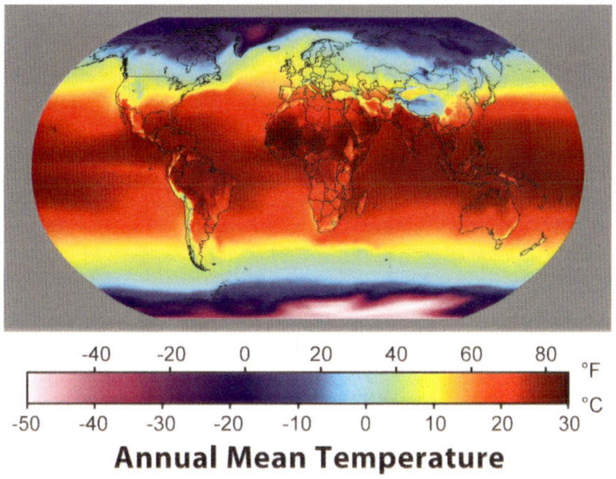

Map of the average temperature over 30 years. Data sets formed from the long-term average of historical weather parameters are sometimes called a "climatology".

Climatology is approached in various ways such as Paleoclimatology, which seeks to reconstruct past climates by examining records such as ice cores and tree rings (dendroclimatology). Paleotempestology uses these same records to help determine hurricane frequency over millennia. The study of contemporary climates incorporates meteorological data accumulated over many years, such as records of rainfall, temperature and atmospheric composition. Knowledge of the atmosphere and its dynamics is also embodied in models, either statistical or mathematical, which help by integrating different observations and testing how they fit together. Modeling is used for understanding past, present and potential future climates. Historical climatology is the study of climate as related to human history and thus focuses only on the last few thousand years.

Climate research is made difficult by the large scale, long time periods, and complex processes which govern climate. Climate is governed by physical laws which can be expressed as differential equations. These equations are coupled and nonlinear, so that approximate solutions are obtained by using numerical methods to create global climate models. Climate is sometimes modeled as a stochastic process but this is generally accepted as an approximation to processes that are otherwise too complicated to analyze.

Indices

Scientists use climate indices based on several climate patterns (known as modes of variability) in their attempt to characterize and understand the various climate mechanisms that culminate in our daily weather. Much in the way the Dow Jones Industrial Average, which is based on the stock prices of 30 companies, is used to represent the fluctuations in the stock market as a whole, climate indices are used to represent the essential elements of climate. Climate indices are generally devised with the twin objectives of simplicity and completeness, and each index typically represents the status and timing of the climate factor it represents. By their very nature, indices are simple, and combine many details into a generalized, overall description of the atmosphere or ocean which can be used to characterize the factors which impact the global climate system.

El Niño–Southern Oscillation

El Niño–Southern Oscillation (ENSO) is a global coupled ocean-atmosphere phenomenon. The Pacific Ocean signatures, El Niño and La Niña are important temperature fluctuations in surface waters of the tropical Eastern Pacific Ocean. The name El Niño, from the Spanish for "the little boy", refers to the Christ child, because the phenomenon is usually noticed around Christmas time in the Pacific Ocean off the west coast of South America. La Niña means "the little girl". Their effect on climate in the subtropics and the tropics are profound. The atmospheric signature, the Southern Oscillation (SO) reflects the monthly or seasonal fluctuations in the air pressure difference between Tahiti and Darwin. The most recent occurrence of El Niño started in September 2006 and lasted until early 2007.

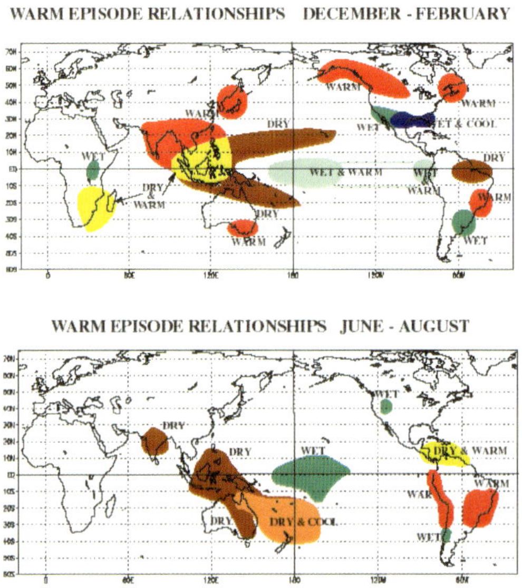

El Niño impacts

ENSO is a set of interacting parts of a single global system of coupled ocean-atmosphere climate fluctuations that come about as a consequence of oceanic and atmospheric circulation. ENSO is the most prominent known source of inter-annual variability in weather and climate around the world. The cycle occurs every two to seven years, with El Niño lasting nine months to two years within the longer term cycle, though not all areas globally are affected. ENSO has signatures in the Pacific, Atlantic and Indian Oceans.

In the Pacific, during major warm events, El Niño warming extends over much of the tropical Pacific and becomes clearly linked to the SO intensity. While ENSO events are basically in phase between the Pacific and Indian Oceans, ENSO events in the Atlantic Ocean lag behind those in the Pacific by 12–18 months. Many of the countries most affected by ENSO events are developing countries within tropical sections of continents with economies that are largely dependent upon their agricultural and fishery sectors as a major source of food supply, employment, and foreign exchange. New capabilities to predict the onset of ENSO events in the three oceans can have global socio-economic impacts. While ENSO is a global and natural part of the Earth's climate, whether its intensity or frequency may change as a result of global warming is an important concern. Low-frequency variability has been evidenced: the quasi-decadal oscillation (QDO). Inter-decadal (ID) modulation of ENSO (from PDO or IPO) might exist. This could explain the so-called protracted ENSO of the early 1990s.

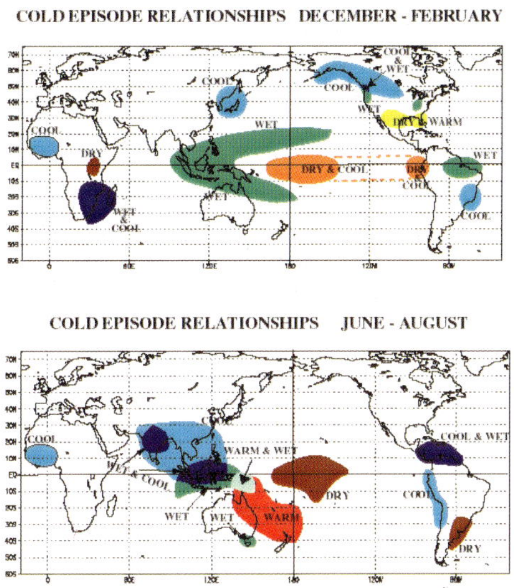

La Niña impacts

Madden–Julian Oscillation

The Madden–Julian oscillation (MJO) is an equatorial traveling pattern of anomalous rainfall that is planetary in scale. It is characterized by an eastward progression of large regions of both enhanced and suppressed tropical rainfall, observed mainly over the Indian and Pacific Oceans. The anomalous rainfall is usually first evident over the western Indian Ocean, and remains evident as it propagates over the very warm ocean waters of the western and central tropical Pacific. This pattern of tropical rainfall then generally becomes very nondescript as it moves over the cooler ocean waters of the eastern Pacific but reappears over the tropical Atlantic and Indian Oceans. The wet phase of enhanced convection and precipitation is followed by a dry phase where convection is suppressed. Each cycle lasts approximately 30–60 days. The MJO is also known as the 30- to 60-day oscillation, 30- to 60-day wave, or the intraseasonal oscillation.

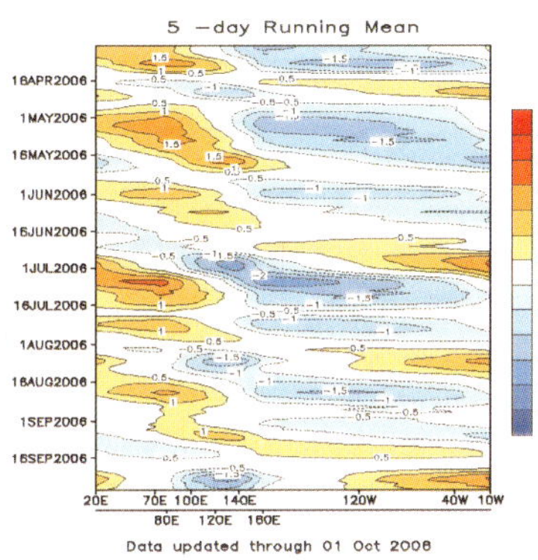

The MJO moves eastward with time

North Atlantic Oscillation (NAO)

Indices of the NAO are based on the difference of normalized sea level pressure (SLP) between Ponta Delgada, Azores and Stykkisholmur/Reykjavik, Iceland. The SLP anomalies at each station were normalized by division of each seasonal mean pressure by the long-term mean (1865–1984) standard deviation. Normalization is done to avoid the series of being dominated by the greater variability of the northern of the two stations. Positive values of the index indicate stronger-than-average westerlies over the middle latitudes.

Northern Annular Mode (NAM) or Arctic oscillation (AO)

The NAM, or AO, is defined as the first EOF of northern hemisphere winter SLP data from the tropics and subtropics. It explains 23% of the average winter (December–March) variance, and it is dominated by the NAO structure in the Atlantic. Although there are some subtle differences from the regional pattern over the Atlantic and Arctic, the main difference is larger amplitude anomalies over the North Pacific of the same sign as those over the Atlantic. This feature gives the NAM a more annular (or zonally symmetric) structure.

Northern Pacific (NP) Index

The NP Index is the area-weighted sea level pressure over the region 30N–65N, 160E–140W.

Pacific Decadal Oscillation (PDO)

The PDO is a pattern of Pacific climate variability that shifts phases on at least inter-decadal time scale, usually about 20 to 30 years. The PDO is detected as warm or cool surface waters in the Pacific Ocean, north of 20° N. During a "warm", or "positive", phase, the west Pacific becomes cool and part of the eastern ocean warms; during a "cool" or "negative" phase, the opposite pattern occurs. The mechanism by which the pattern lasts over several years has not been identified; one suggestion is that a thin layer of warm water during summer may shield deeper cold waters. A PDO signal has been reconstructed to 1661 through tree-ring chronologies in the Baja California area.

Interdecadal Pacific Oscillation (IPO)

The Interdecadal Pacific oscillation (IPO or ID) display similar sea surface temperature (SST) and sea level pressure patterns to the PDO, with a cycle of 15–30 years, but affects both the north and south Pacific. In the tropical Pacific, maximum SST anomalies are found away from the equator. This is quite different from the quasi-decadal oscillation (QDO) with a period of 8–12 years and maximum SST anomalies straddling the equator, thus resembling ENSO.

Climatology Pioneers

Several key figures emerged in the academic world whose contributions enhanced the development of climatology as a science. Carl – Gustav Rossby is credited to have established the Department of Meteorology at the University of Chicago in 1942. Researchers at University of Chicago under the leadership of Carl- Gustav Rossby developed the first physical climate model in which entire planet was conceived to be an integrated physical system.

Reid Bryson, a World War II military meteorologist left the Department of Geography at University of Wisconsin, Madison to develop a separate Department of Meteorology there. A Center of Climate Research was established in 1962 at Madison. Edward N. Lorenz, another military meteorologist of

World War II, worked with Massachusetts Institute of Technology, developed a Chaos Theory to the atmosphere in 1960s, and theories were integrated into the increasingly complex atmospheric modeling relegated to computers. One skeptic claim is that if weather is chaotic, then surely the climate must also be chaotic. A major concern was that for small changes in starting conditions, the system takes a cyclic turn. Beside cyclic turn, it keeps on increasing the intensity and complexity of elements overtime. As such, any chaotic development has to be studied as complex function with increasing intensity in cyclic manner. It has a multiplier and cumulative causation effect ranging over a long period of time from a small chaos to a complex chaos. Weather Bureau of United States viewed climatology as an effective tool for weather prediction and forecasting.

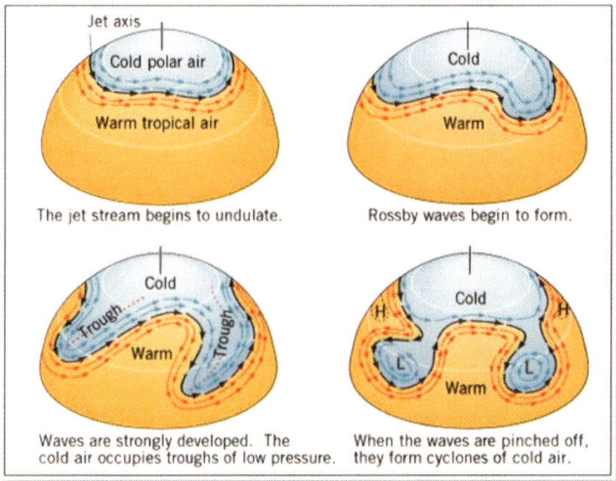

Planetary Flow Patterns in the Atmosphere by Rossby Waves

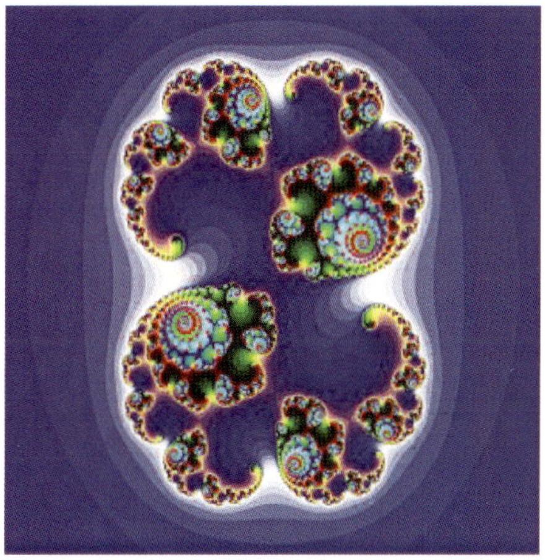

Chaos Theory of Climate Change by Edward N. Lorenz

References

- Young, Davis A. (1995). The biblical Flood: a case study of the Church's response to extrabiblical evidence. Grand Rapids, Mich: Eerdmans. ISBN 0-8028-0719-4
- Planton, Serge (France; editor) (2013). "Annex III. Glossary: IPCC – Intergovernmental Panel on Climate Change" (PDF). IPCC Fifth Assessment Report. p. 1450. Retrieved 25 July 2016
- Fredlund, D.G.; Rahardjo, H. (1993). Soil Mechanics for Unsaturated Soils (PDF). Wiley-Interscience. ISBN 978-0-471-85008-3. OCLC 26543184. Retrieved 2008-05-21
- Intergovernmental Panel on Climate Change. Appendix I: Glossary. Archived 2017-01-26 at the Wayback Machine. Retrieved on 2007-06-01
- J. Bowden; Cynthia V. Burek; C. V. Burek; Richard Wilding (2005). History of palaeobotany: selected essays. Geological Society. p. 293. ISBN 978-1-86239-174-1. Retrieved 3 April 2013
- Robert E. Davis, L. Sitka, D. M. Hondula, S. Gawtry, D. Knight, T. Lee, and J. Stenger. J1.10 A preliminary back-trajectory and air mass climatology for the Shenandoah Valley (Formerly J3.16 for Applied Climatology). Retrieved on 2008-05-21
- J. Bowden; Cynthia V. Burek; C. V. Burek; Richard Wilding (2005). History of palaeobotany: selected essays. Geological Society. p. 293. ISBN 978-1-86239-174-1

Chapter 2

Global Climatology

Global climatology involves the study of the layers of gases surrounding the Earth. This chapter examines the chief atmospheric gases of the Earth such as nitrogen, oxygen, carbon dioxide, etc. and their effects on the global climate. It also discusses in detail the composition and structure of the Earth's atmosphere.

Atmosphere

The atmosphere of the earth is a vast expanse of gases enveloping our entire earth. Within this envelope, we are surviving and all our activities are confined. It is made up of several gases, water vapour and minute particles suspended in the gaseous substance of air. The atmosphere extends several hundred km above the earth surface. It is not uniform at every height we climb above, but it has drastic changes with height. The atmosphere is composed of several layers. At the transition zones of the layers, the change is very sharp but within a layer, the changes are slow. In this chap-ter, we will discuss about the composition and structure of the atmosphere. Apart from these, our concern will also be there to through light on its importance and utility of the atmosphere for us as well as for the entire living organisms.

Composition of Atmosphere

The literal meaning of composition is 'ingredients' or 'constituents' of something. In another words, it is a manner by which something is made up of. When we apply the same meaning with atmosphere, it signify the items or the elements with which our atmosphere is composed. Our atmosphere is composed of numerous gases and other substances, hence, it is a mechanical mixture of the gases, water vapour and dust particles. Let us discuss about the composition of atmosphere.

The envelope of atmosphere around the earth, a mechanical mixtures of numerous gases and other substances are very important to all living organisms of the planet. The four major gases – nitrogen, oxygen, argon and carbon dioxide together constitute 99.99% of the total volume of dry air. The maximum concentration is of nitrogen with more than 78 percent while the oxygen is little less than 21 percent.

Table: Atmospheric Gases

Groups	Gases	Volume % of dry air
Major Gases	1. Nitrogen	78.084
	2. Oxygen	20.9476
	3. Argon	0.934
	4. *CO2	0.04
Minor Gases (* are also variable gases)	5. *Methane	0.002
	6. Neon	0.001818
	7. Helium	0.000524
	8. Krypton	0.000114
	9. Hydrogen	0.00005
	10. Xenon	0.0000087
	11. *Ozone	0.00006
Variable Gasses (CO2, methane and ozone gases are also variable)	12. Water vapour	Variable amount
	13. Dust particles	Variable amount
	14. Aerosols	Variableamount

Nitrogen

Nitrogen is the most abundant found in atmosphere constituting 78.084 percent to the total volume of dry gases. This is almost chemically inactive and have nothing to do with any sort of chemical actions in the atmosphere. It does not combine freely with other elements, hence, it is termed as neutral substance. This gas is found beyond a height of 100 km, but its concentration is below 50 km height from the sea level. This gas is significant for the growth and reproduction in plants and animals. Certain bacteria in the soil are capable of converting a very small amount of atmospheric nitrogen into nitrates and fix it to the soils and water bodies to be consumed by animals and plants. This process is called as Nitrogen fixation. The nitrogen fixed in the earth's surface is again converted and sent back to the atmosphere by bacterial action through a chemical reaction called denitrification.

Nitrogen Cycle

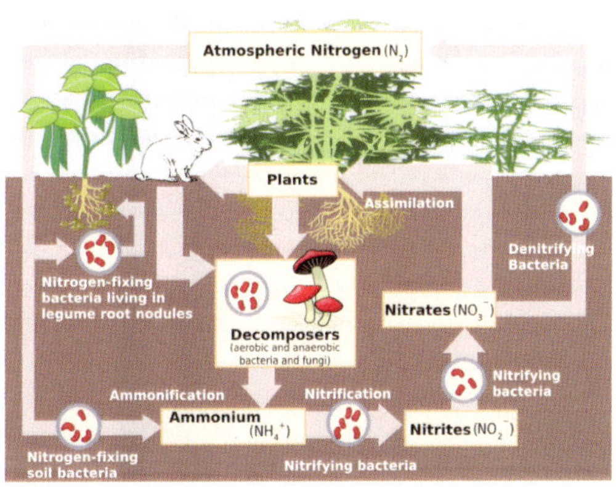

Schematic representation of the flow of nitrogen through the land environment. The importance of bacteria in the cycle is immediately recognized as being a key element in the cycle, providing different forms of nitrogen compounds assimilable by higher organisms

The nitrogen cycle is the biogeochemical cycle by which nitrogen is converted into multiple chemical forms as it circulates among the atmosphere, terrestrial, and marine ecosystems. The conversion of nitrogen can be carried out through both biological and physical processes. Important processes in the nitrogen cycle include fixation, ammonification, nitrification, and denitrification. The majority of Earth's atmosphere (78%) is nitrogen, making it the largest source of nitrogen. However, atmospheric nitrogen has limited availability for biological use, leading to a scarcity of usable nitrogen in many types of ecosystems.

The nitrogen cycle is of particular interest to ecologists because nitrogen availability can affect the rate of key ecosystem processes, including primary production and decomposition. Human activities such as fossil fuel combustion, use of artificial nitrogen fertilizers, and release of nitrogen in wastewater have dramatically altered the global nitrogen cycle.

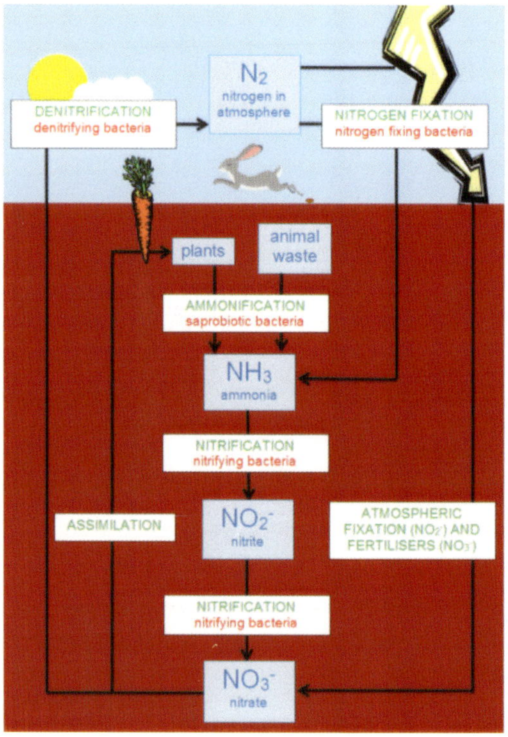

A simple diagram of the nitrogen cycle. The blue boxes represent stores of nitrogen, the green writing is for processes that occur to move the nitrogen from one place to another and the red writing are all the bacteria involved

Processes

Nitrogen is present in the environment in a wide variety of chemical forms including organic nitrogen, ammonium (NH_4^+), nitrite (NO_2^-), nitrate (NO_3^-), nitrous oxide (N_2O), nitric oxide (NO) or inorganic nitrogen gas (N_2). Organic nitrogen may be in the form of a living organism, humus or in the intermediate products of organic matter decomposition. The processes of the nitrogen cycle transform nitrogen from one form to another. Many of those processes are carried out by microbes, either in their effort to harvest energy or to accumulate nitrogen in a form needed for their growth. For example, the nitrogenous wastes in animal urine are broken down by nitrifying bacteria in the soil to be used as new. The diagram alongside shows how these processes fit together to form the nitrogen cycle.

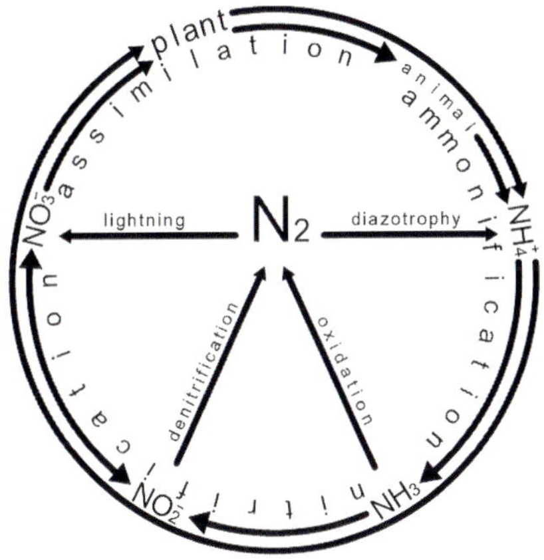

Classical representation of nitrogen cycle

Nitrogen Fixation

The conversion of nitrogen gas (N_2) into nitrates and nitrites through atmospheric, industrial and biological processes is called nitrogen fixation. Atmospheric nitrogen must be processed, or "fixed", into a usable form to be taken up by plants. Between $5x10^{12}$ and $10x10^{12}$ g per year are fixed by lightning strikes, but most fixation is done by free-living or symbiotic bacteria known as diazotrophs. These bacteria have the nitrogenase enzyme that combines gaseous nitrogen with hydrogen to produce ammonia, which is converted by the bacteria into other organic compounds. Most biological nitrogen fixation occurs by the activity of Mo-nitrogenase, found in a wide variety of bacteria and some Archaea. Mo-nitrogenase is a complex two-component enzyme that has multiple metal-containing prosthetic groups. An example of free-living bacteria is *Azotobacter*. Symbiotic nitrogen-fixing bacteria such as *Rhizobium* usually live in the root nodules of legumes (such as peas, alfalfa, and locust trees). Here they form a mutualistic relationship with the plant, producing ammonia in exchange for carbohydrates. Because of this relationship, legumes will often increase the nitrogen content of nitrogen-poor soils. A few non-legumes can also form such symbioses. Today, about 30% of the total fixed nitrogen is produced industrially using the Haber-Bosch process, which uses high temperatures and pressures to convert nitrogen gas and a hydrogen source (natural gas or petroleum) into ammonia.

Assimilation

Plants can absorb nitrate or ammonium from the soil via their root hairs. If nitrate is absorbed, it is first reduced to nitrite ions and then ammonium ions for incorporation into amino acids, nucleic acids, and chlorophyll. In plants that have a symbiotic relationship with rhizobia, some nitrogen is assimilated in the form of ammonium ions directly from the nodules. It is now known that there is a more complex cycling of amino acids between *Rhizobia* bacteroids and plants. The plant provides amino acids to the bacteroids so ammonia assimilation is not required and the bacteroids pass amino acids (with the newly fixed nitrogen) back to the plant, thus forming an interdependent relationship. While many animals, fungi, and other heterotrophic organisms obtain nitrogen by ingestion of amino acids, nucleotides, and other small organic molecules, other heterotrophs (including many bacteria) are able to utilize inorganic compounds, such as ammonium as sole N sources. Utilization of various N sources is carefully regulated in all organisms.

Ammonification

When a plant or animal dies or an animal expels waste, the initial form of nitrogen is organic. Bacteria or fungi convert the organic nitrogen within the remains back into ammonium (NH_4^+), a process called ammonification or mineralization. Enzymes involved are:

- GS: Gln Synthetase (Cytosolic & Plastic)
- GOGAT: Glu 2-oxoglutarate aminotransferase (Ferredoxin & NADH-dependent)
- GDH: Glu Dehydrogenase:
    - Minor Role in ammonium assimilation.
    - Important in amino acid catabolism.

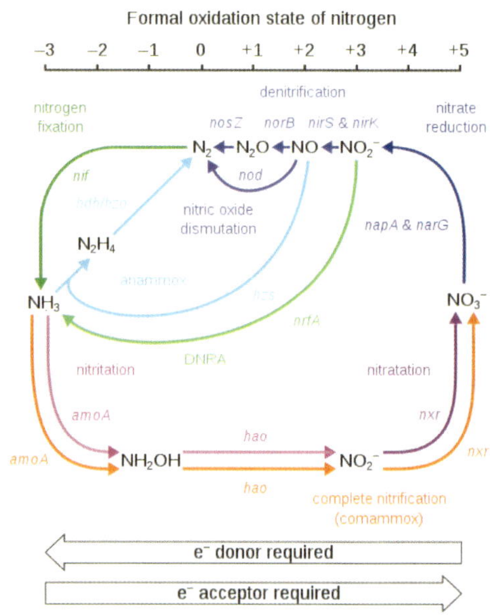

A schematic representation of the microbial nitrogen cycle. ANAMMOX is anaerobic ammonium oxidation, DNRA is dissimilatory nitrate reduction to ammonium, and COMMAMOX is complete ammonium oxidation

Nitrification

The conversion of ammonium to nitrate is performed primarily by soil-living bacteria and other nitrifying bacteria. In the primary stage of nitrification, the oxidation of ammonium (NH_4^+) is performed by bacteria such as the *Nitrosomonas* species, which converts ammonia to nitrites (NO_2^-). Other bacterial species such as *Nitrobacter*, are responsible for the oxidation of the nitrites (NO_2^-) into nitrates (NO_3^-). It is important for the ammonia (NH_3) to be converted to nitrates or nitrites because ammonia gas is toxic to plants.

Due to their very high solubility and because soils are highly unable to retain anions, nitrates can enter groundwater. Elevated nitrate in groundwater is a concern for drinking water use because nitrate can interfere with blood-oxygen levels in infants and cause methemoglobinemia or blue-baby syndrome. Where groundwater recharges stream flow, nitrate-enriched groundwater can contribute to eutrophication, a process that leads to high algal population and growth, especially blue-green algal populations. While not directly toxic to fish life, like ammonia, nitrate can have indirect effects on fish if it contributes to this eutrophication. Nitrogen has contributed to severe eutrophication problems in some water bodies. Since 2006, the application of nitrogen fertilizer has been increasingly controlled in Britain and the United States. This is occurring along the same lines as control of phosphorus fertilizer, restriction of which is normally considered essential to the recovery of eutrophied waterbodies.

Denitrification

Denitrification is the reduction of nitrates back into nitrogen gas (N_2), completing the nitrogen cycle. This process is performed by bacterial species such as Pseudomonas and Clostridium in anaerobic conditions. They use the nitrate as an electron acceptor in the place of oxygen during respiration. These facultatively anaerobic bacteria can also live in aerobic conditions. Denitrification happens in anaerobic conditions e.g. waterlogged soils. The denitrifying bacteria use nitrates in the soil to carry out respiration and consequently produce nitrogen gas, which is inert and unavailable to plants.

Dissimilatory Nitrate Reduction to Ammonium

Dissimilatory nitrate reduction to ammonium (DNRA), or nitrate/nitrite ammonification, is an anaerobic respiration process. Microbes which undertake DNRA oxidise organic matter and use nitrate as an electron acceptor, reducing it to nitrite, then ammonium ($NO_3^- \rightarrow NO_2^- \rightarrow NH_4^+$). Both denitrifying and nitrate ammonification bacteria will be competing for nitrate in the environment, although DNRA acts to conserve bioavailable nitrogen as soluble ammonium rather than producing dinitrogen gas.

Anaerobic Ammonia Oxidation

In this biological process, nitrite and ammonia are converted directly into molecular nitrogen (N_2) gas. This process makes up a major proportion of nitrogen conversion in the oceans. The balanced formula for this "anammox" chemical reaction is: $NH_4^+ + NO_2^- \rightarrow N_2 + 2H_2O$ ($\Delta G° = -357$ kJ mol^{-1}).

Other Processes

Though nitrogen fixation is the primary source of plant-available nitrogen in most ecosystems, in areas with nitrogen-rich bedrock, the breakdown of this rock also serves as a nitrogen source.

Marine Nitrogen Cycle

The nitrogen cycle is an important process in the ocean as well. While the overall cycle is similar, there are different players and modes of transfer for nitrogen in the ocean. Nitrogen enters the water through the precipitation, runoff, or as N_2 from the atmosphere. Nitrogen cannot be utilized by phytoplankton as N_2 so it must undergo nitrogen fixation which is performed predominately by cyanobacteria. Without supplies of fixed nitrogen entering the marine cycle, the fixed nitrogen would be used up in about 2000 years. Phytoplankton need nitrogen in biologically available forms for the initial synthesis of organic matter. Ammonia and urea are released into the water by excretion from plankton. Nitrogen sources are removed from the euphotic zone by the downward movement of the organic matter. This can occur from sinking of phytoplankton, vertical mixing, or sinking of waste of vertical migrators. The sinking results in ammonia being introduced at lower depths below the euphotic zone. Bacteria are able to convert ammonia to nitrite and nitrate but they are inhibited by light so this must occur below the euphotic zone. Ammonification or Mineralization is performed by bacteria to convert organic nitrogen to ammonia. Nitrification can then occur to convert the ammonium to nitrite and nitrate. Nitrate can be returned to the euphotic zone by vertical mixing and upwelling where it can be taken up by phytoplankton to continue the cycle. N_2 can be returned to the atmosphere through denitrification.

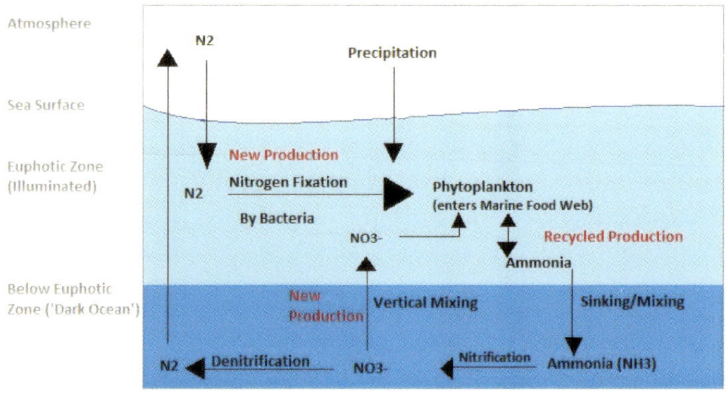

A schematic representing the Marine Nitrogen Cycle

Ammonium is thought to be the preferred source of fixed nitrogen for phytoplankton because its assimilation does not involve a redox reaction and therefore requires little energy. Nitrate requires a redox reaction for assimilation but is more abundant so most phytoplankton have adapted to have the enzymes necessary to undertake this reduction (nitrate reductase). There are a few notable and well-known exceptions that include Prochlorococcus and some Synechococcus. These species can only take up nitrogen as ammonium.

The nutrients in the ocean are not uniformly distributed. Areas of upwelling provide supplies of nitrogen from below the euphotic zone. Coastal zones provide nitrogen from runoff and upwelling occurs readily along the coast. However, the rate at which nitrogen can be taken up by phytoplankton is decreased in oligotrophic waters year-round and temperate water in the summer resulting in lower primary production. The distribution of the different forms of nitrogen varies throughout the oceans as well.

Nitrate is depleted in near-surface water except in upwelling regions. Coastal upwelling regions usually have high nitrate and chlorophyll levels as a result of the increased production. However, there are regions of high surface nitrate but low chlorophyll that are referred to as HNLC (high nitrogen, low chlorophyll) regions. The best explanation for HNLC regions relates to iron scarcity in the ocean, which may play an important part in ocean dynamics and nutrient cycles. The input of iron varies by region and is delivered to the ocean by dust (from dust storms) and leached out of rocks. Iron is under consideration as the true limiting element to ecosystem productivity in the ocean.

Ammonium and nitrite show a maximum concentration at 50–80 m (lower end of the euphotic zone) with decreasing concentration below that depth. This distribution can be accounted for by the fact that nitrite and ammonium are intermediate species. They are both rapidly produced and consumed through the water column. The amount of ammonium in the ocean is about 3 orders of magnitude less than nitrate. Between ammonium, nitrite, and nitrate, nitrite has the fastest turnover rate. It can be produced during nitrate assimilation, nitrification, and denitrification; however, it is immediately consumed again.

New vs. Regenerated Nitrogen

Nitrogen entering the euphotic zone is referred to as new nitrogen because it is newly arrived from outside the productive layer. The new nitrogen can come from below the euphotic zone or from outside sources. Outside sources are upwelling from deep water and nitrogen fixation. If the organic matter is eaten, respired, delivered to the water as ammonia, and re-incorporated into organic matter by phytoplankton it is considered recycled/regenerated production.

New production is an important component of the marine environment. One reason is that only continual input of new nitrogen can determine the total capacity of the ocean to produce a sustainable fish harvest. Harvesting fish from regenerated nitrogen areas will lead to a decrease in nitrogen and therefore a decrease in primary production. This will have a negative effect on the system. However, if fish are harvested from areas of new nitrogen the nitrogen will be replenished.

Oxygen

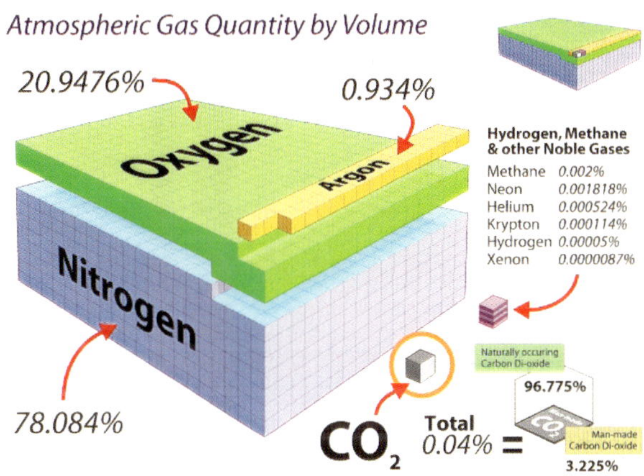

Atmospheric Gases and their Proportions

It is the second largest gas of the atmosphere constituting 20.9476 percent of the total dry atmospheric gases. It is very essential for the survival of many of the living organisms of this planet. It is chemically very active gas. It is combined with several other elements and forms varied compounds. Oxygen is vital for combustion of fuels. When anything burns, oxygen is consumed and helps in burning that substance. Though oxygen is found beyond 100 km but it is reasonably in good proportion within 16 km of height. With increasing height, the amount of oxygen decreases very rapidly. On mountain slope, the available oxygen for breath is very scanty and the mountaineers are supposed to carry oxygen for them.

Oxygen Cycle

The oxygen cycle is the biogeochemical cycle of oxygen within its four main reservoirs: the atmosphere (air), the total content of biological matter within the biosphere (the global sum of all ecosystems), the hydrosphere (the combined mass of water found on, under, and over the surface of planet Earth), and the lithosphere/Earth's crust. The main driving factor of the oxygen cycle is photosynthesis, which is responsible for the modern Earth's atmosphere and life on Earth.

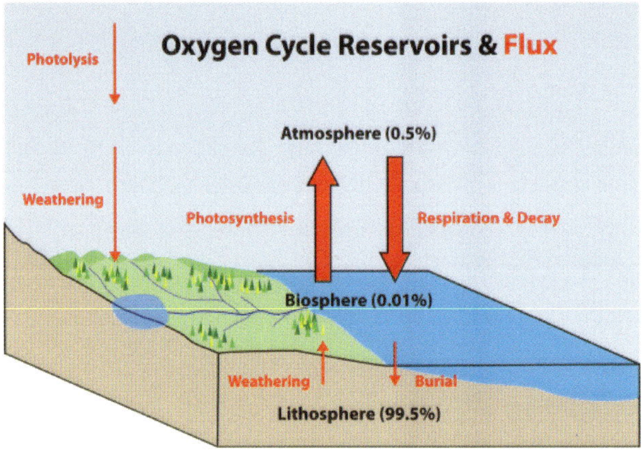

The oxygen cycle

Reservoirs

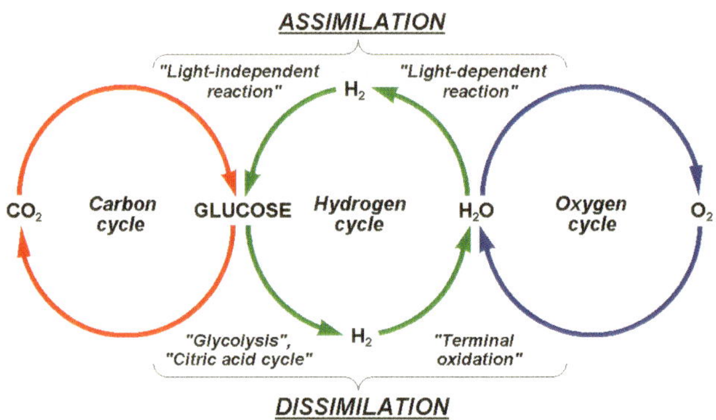

Interconnection between carbon, hydrogen and oxygen cycle in metabolism of photosynthesizing plants

By far the largest reservoir of Earth's oxygen is within the silicate and oxide minerals of the crust and mantle (99.5% by weight). The Earth's atmosphere, hydrosphere and biosphere together weigh less than 0.05% of the Earth's total mass. Oxygen is one of the most abundant elements on Earth and represents a large portion of each main reservoir:

- Atmosphere is 21% oxygen by volume present mainly as free oxygen molecules (O_2) with other oxygen-containing molecules including ozone (O_3), carbon dioxide (CO_2), water vapor (H_2O), and sulfur and nitrogen oxides (SO_2, NO, N_2O, etc.)
- Biosphere is 22% oxygen by volume present mainly as a component of organic molecules ($C_xH_xN_xO_x$) and water molecules
- Hydrosphere is 33% oxygen by volume present mainly as a component of water molecules with dissolved molecules including free oxygen and carbonic acids (H_xCO_3)
- Lithosphere is 94% oxygen by volume present mainly as silica minerals (SiO_2) and other oxide minerals

Movement of oxygen between the reservoirs is facilitated in large part by the presence of atmospheric free oxygen. The main source of atmospheric free oxygen is photosynthesis, which produces sugars and free oxygen from carbon dioxide and water:

$$6\,CO_2 + 6H_2O + \text{energy} \rightarrow C_6H_{12}O_6 + 6\,O_2$$

Photosynthesizing organisms include the plant life of the land areas as well as the phytoplankton of the oceans. The tiny marine cyanobacterium Prochlorococcus was discovered in 1986 and accounts for more than half of the photosynthesis of the open ocean.

An additional source of atmospheric free oxygen comes from photolysis, whereby high-energy ultraviolet radiation breaks down atmospheric water and nitrous oxide into component atoms. The free H and N atoms escape into space, leaving O_2 in the atmosphere:

$$2\,H_2O + \text{energy} \rightarrow 4\,H + O_2$$

$$2\,N_2O + \text{energy} \rightarrow 4\,N + O_2$$

The main way free oxygen is lost from the atmosphere is via respiration and decay, mechanisms in which animal life and bacteria consume oxygen and release carbon dioxide.

The lithosphere also consumes atmospheric free oxygen by chemical weathering and surface reactions. An example of surface weathering chemistry is formation of iron oxides (rust):

$$4\,FeO + O_2 \rightarrow 2\,Fe_2O_3$$

Oxygen is also cycled between the biosphere and lithosphere. Marine organisms in the biosphere create calcium carbonate shell material ($CaCO_3$) that is rich in oxygen. When the organism dies, its shell is deposited on the shallow sea floor and buried over time to create the limestone sedimentary rock of the lithosphere. Weathering processes initiated by organisms can also free oxygen from the lithosphere. Plants and animals extract nutrient minerals from rocks and release oxygen in the process.

Capacities and Fluxes

The following tables offer estimates of oxygen cycle reservoir capacities and fluxes. These numbers are based primarily on estimates from (Walker, J. C. G.):

Table: Major reservoirs involved in the oxygen cycle

Reservoir	Capacity ($kg\ O_2$)	Flux in/out ($kg\ O_2$ per year)	Residence time (years)
Atmosphere	1.4×10^{18}	3×10^{14}	4500
Biosphere	1.6×10^{16}	3×10^{14}	50
Lithosphere	2.9×10^{20}	6×10^{11}	500000000

Table: Annual gain and loss of atmospheric oxygen (Units of 10^{10} kg O_2 per year)

Photosynthesis (land)	16,500
Photosynthesis (ocean)	13,500
Photolysis of N_2O	1.3
Photolysis of H_2O	0.03
Total gains	~ 30,000

Losses - respiration and decay	
Aerobic respiration Microbial oxidation Combustion of fossil fuel (anthropogenic) Photochemical oxidation Fixation of N_2 by lightning Fixation of N_2 by industry (anthropogenic) Oxidation of volcanic gases	23,000 5,100 1,200 600 12 10 5
Losses - weathering	
Chemical weathering Surface reaction of O_3	50 12
Total losses	~ 30,000

Argon

In terms of percentage, argon is the third largest gas in the atmosphere constituting 0.934 percent of total dry atmosphere. It is an inert gas and chemically it is inactive. It is also found in the earth's crust and sea water. It is used in electric bulb and fluorescent lights.

Nearly all of the argon in the Earth's atmosphere is radiogenic argon-40, derived from the decay of potassium-40 in the Earth's crust. In the universe, argon-36 is by far the most common argon isotope, as it is the most easily produced by stellar nucleosynthesis in supernovas.

The name "argon" is derived from the Greek word, meaning "lazy" or "inactive", as a reference to the fact that the element undergoes almost no chemical reactions. The complete octet (eight electrons) in the outer atomic shell makes argon stable and resistant to bonding with other elements. Its triple point temperature of 83.8058 K is a defining fixed point in the International Temperature Scale of 1990.

Argon is produced industrially by the fractional distillation of liquid air. Argon is mostly used as an inert shielding gas in welding and other high-temperature industrial processes where ordinarily unreactive substances become reactive; for example, an argon atmosphere is used in graphite electric furnaces to prevent the graphite from burning. Argon is also used in incandescent, fluorescent lighting, and other gas-discharge tubes. Argon makes a distinctive blue-green gas laser. Argon is also used in fluorescent glow starters.

Occurrence

Argon constitutes 0.934% by volume and 1.288% by mass of the Earth's atmosphere, and air is the primary industrial source of purified argon products. Argon is isolated from air by fractionation, most commonly by cryogenic fractional distillation, a process that also produces purified nitrogen, oxygen, neon, krypton and xenon. The Earth's crust and seawater contain 1.2 ppm and 0.45 ppm of argon, respectively.

Isotopes

The main isotopes of argon found on Earth are ^{40}Ar (99.6%), ^{36}Ar (0.34%), and ^{38}Ar (0.06%).

Naturally occurring ^{40}K, with a half-life of 1.25×10^9 years, decays to stable ^{40}Ar (11.2%) by electron capture or positron emission, and also to stable ^{40}Ca (88.8%) by beta decay. These properties and ratios are used to determine the age of rocks by K–Ar dating.

In the Earth's atmosphere, ^{39}Ar is made by cosmic ray activity, primarily by neutron capture of ^{40}Ar followed by two-neutron emission. In the subsurface environment, it is also produced through neutron capture by ^{39}K, followed by proton emission. ^{37}Ar is created from the neutron capture by ^{40}Ca followed by an alpha particle emission as a result of subsurface nuclear explosions. It has a half-life of 35 days.

Between locations in the Solar System, the isotopic composition of argon varies greatly. Where the major source of argon is the decay of ^{40}K in rocks, ^{40}Ar will be the dominant isotope, as it is on Earth. Argon produced directly by stellar nucleosynthesis, is dominated by the alpha-process nuclide ^{36}Ar. Correspondingly, solar argon contains 84.6% ^{36}Ar (according to solar wind measurements), and the ratio of the three isotopes $^{36}Ar : ^{38}Ar : ^{40}Ar$ in the atmospheres of the outer planets is 8400 : 1600 : 1. This contrasts with the low abundance of primordial ^{36}Ar in Earth's atmosphere, which is only 31.5 ppmv (= 9340 ppmv × 0.337%), comparable with that of neon (18.18 ppmv) on Earth and with interplanetary gasses, measured by probes.

The atmospheres of Mars, Mercury and Titan (the largest moon of Saturn) contain argon, predominantly as ^{40}Ar, and its content may be as high as 1.93% (Mars).

The predominance of radiogenic ^{40}Ar is the reason the standard atomic weight of terrestrial argon is greater than that of the next element, potassium, a fact that was puzzling when argon was discovered. Mendeleev positioned the elements on his periodic table in order of atomic weight, but the inertness of argon suggested a placement *before* the reactive alkali metal. Henry Moseley later solved this problem by showing that the periodic table is actually arranged in order of atomic number.

Production

Industrial

Argon is produced industrially by the fractional distillation of liquid air in a cryogenic air separation unit; a process that separates liquid nitrogen, which boils at 77.3 K, from argon, which boils at 87.3 K, and liquid oxygen, which boils at 90.2 K. About 700,000 tonnes of argon are produced worldwide every year.

In Radioactive Decays

^{40}Ar, the most abundant isotope of argon, is produced by the decay of ^{40}K with a half-life of 1.25×10^9 years by electron capture or positron emission. Because of this, it is used in potassium–argon dating to determine the age of rocks.

Safety

Although argon is non-toxic, it is 38% denser than air and therefore considered a dangerous asphyxiant in closed areas. It is difficult to detect because it is colorless, odorless, and tasteless. A

1994 incident, in which a man was asphyxiated after entering an argon-filled section of oil pipe under construction in Alaska, highlights the dangers of argon tank leakage in confined spaces and emphasizes the need for proper use, storage and handling.

Carbon Dioxide

It is the fourth abundant gas of the atmosphere. It is densest gas and found in lower parts. It is found upto a height of about 30 km, it is concentrated in the lower strata. Its percentage is very low, i.e., 0.04 percent but it is most vital for the growth of vegetative life of biosphere. It is transparent to the incoming solar radiation but does not allow to escape the same. And hence, it is called as greenhouse gas. It plays a very crucial role in increasing the global temperature.

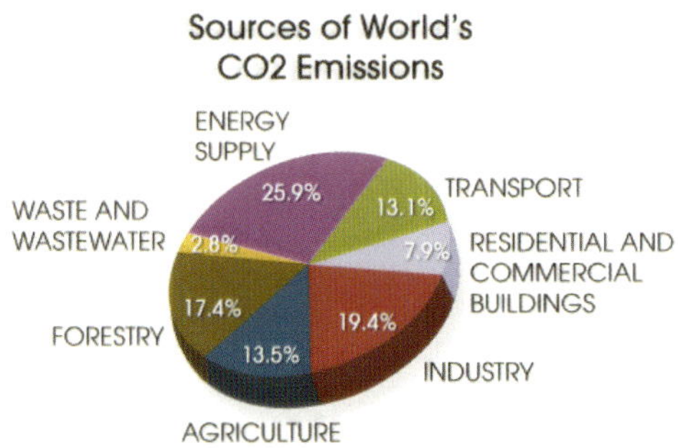

Different Sources of Carbon Dioxide

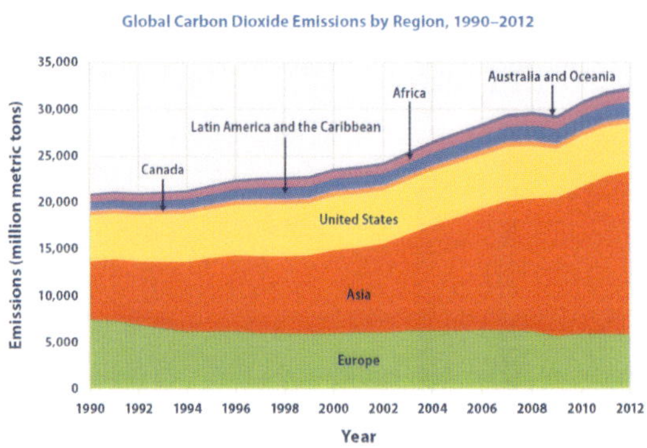

Global Carbon Dioxide Emission by Region: 1990-2012

It is also known as variable gas as its amount is dependent upon the combustion, human activities and vegetative cover of the planet. The carbon dioxide is reaching to the atmosphere due to several human activities like energy utilization, transport, industry, agriculture, waste generation etc. Apart from these human induced sources, some natural sources are like plant respiration and release to air from stored carbon in the rocks through natural process of denudation. All these are leading to increase in the atmosphere. This gas is on rise with the advancement in economic development of the society. Economically developed countries/regions are generating big amount of carbon dioxide.

Reconstructions show that concentrations of CO_2 in the atmosphere have varied from as high as 7,000 parts per million (ppm) during the Cambrian period about 500 million years ago to as low as 180 ppm during the Quaternary glaciation of the last two million years. Global annual mean CO_2 concentration has increased by more than 45% since the start of the Industrial Revolution from 280 ppm during the 10,000 years up to the mid-18th century, to 407 ppm as of mid-2017, the highest concentration in at least 800,000 and possibly 20 million years. The increase has been caused by human activities, particularly the burning of fossil fuels and deforestation. This increase of CO_2 and other greenhouse gases in Earth's atmosphere, of which water vapor is by far the most abundant, has produced the current episode of global warming. About 30–40% of the CO_2 released by humans into the atmosphere dissolves into oceans, rivers and lakes, which has produced ocean acidification.

Current Concentration

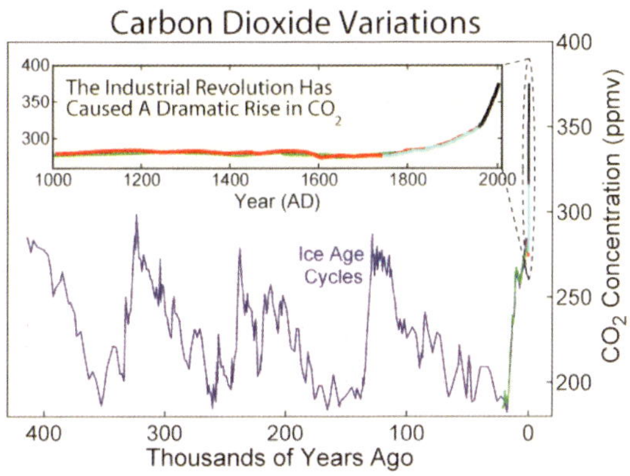

CO_2 concentrations over the last 400,000 years

Over the past 400,000 years, CO_2 concentrations have shown several cycles of variation from about 180 parts per million during the deep glaciations of the Holocene and Pleistocene to 280 parts per million during the interglacial periods. Following the start of the Industrial Revolution, atmospheric CO_2 concentration has increased to over 400 parts per million and continues to increase, causing the phenomenon of global warming. The daily average concentration of atmospheric CO_2 at Mauna Loa Observatory first exceeded 400 ppm on 10 May 2013 although this concentration had already been reached in the Arctic in June 2012. It currently constitutes about 0.041% by volume of the atmosphere, (equal to 410 ppm) which corresponds to approximately 3200 gigatons of CO_2, containing approximately 870 gigatons of carbon. Each part per million by volume of CO_2 in the atmosphere thus represents approximately 2.13 gigatonnes of carbon. The global mean CO_2 concentration is currently rising at a rate of approximately 2 ppm/year and accelerating. There is an annual fluctuation of about 3–9 ppm which is negatively correlated with the Northern Hemisphere's growing season. The Northern Hemisphere dominates the annual cycle of CO_2 concentration because it has much greater land area and plant biomass than the Southern Hemisphere. Concentrations reach a peak in May as the Northern Hemisphere spring greenup begins, and decline to a minimum in October, near the end of the growing season.

Since global warming is attributed to increasing atmospheric concentrations of greenhouse gases such as CO_2, scientists closely monitor atmospheric CO_2 concentrations and their impact on the present-day biosphere. The *National Geographic* wrote that the concentration of carbon dioxide in the atmosphere is this high "for the first time in 55 years of measurement—and probably more than 3 million years of Earth history." The current concentration may be the highest in the last 20 million years.

As of April 2018, the average monthly level of CO_2 in Earth's atmosphere exceeded 410 parts per million.

Past Concentration

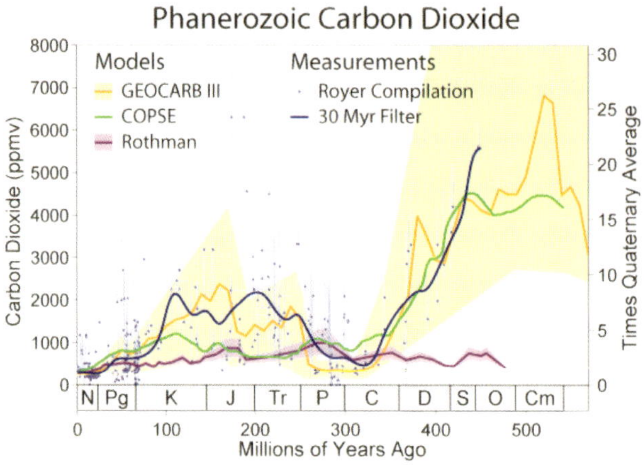

Changes in carbon dioxide during the Phanerozoic (the last 542 million years)

The recent period is located on the left side of the plot. This figure illustrates a range of events over the last 550 million years during which CO_2 played a role in global climate. The graph begins (on the right) with an era predating terrestrial plant life, during which solar output was more than 4% lower than today. Land plants only became widespread after 400Ma, during the Devonian (D) period, and their diversification (along with the evolution of leaves) may have been partially driven by a decrease in CO_2 concentration. Toward the left side of the graph the sun gradually approaches modern levels of solar output, while vegetation spreads, removing large amounts of CO_2 from the atmosphere. The last 200 million years includes periods of extreme warmth, and sea levels so high that 200 metre-deep shallow seas formed on continental land masses (for example, at 100Ma during the Cretaceous (K) Greenhouse). At the far left of the graph, we see modern CO_2 levels and the appearance of the climate under which human species and human civilization developed.

Carbon dioxide concentrations have varied widely over the Earth's 4.54 billion year history. Carbon dioxide is believed to have been present in Earth's first atmosphere, shortly after Earth's formation. Earth's second atmosphere emerged after the lighter gases, hydrogen and helium, escaped to space or like oxygen were bound up in molecules and is thought to have consisted largely of nitrogen, carbon dioxide and inert gases produced by outgassing from volcanism, supplemented by gases produced during the late heavy bombardment of Earth by asteroids. The production of free oxygen by cyanobacterial photosynthesis eventually led to the oxygen catastrophe that ended Earth's second atmosphere and brought about the Earth's third atmosphere (the modern atmosphere) 2.4 billion years before the present. Carbon dioxide concentrations dropped from 7,000 parts per million during the Cambrian period about 500 million years ago to as low as 180 parts per million during the Quaternary glaciation of the last two million years.

Drivers of Ancient-Earth Carbon Dioxide Concentration

On long timescales, atmospheric CO_2 concentration is determined by the balance among geochemical processes including organic carbon burial in sediments, silicate rock weathering, and volcanism. The net effect of slight imbalances in the carbon cycle over tens to hundreds of millions of years has been to reduce atmospheric CO_2. On a timescale of billions of years, such downward trend appears bound to continue indefinitely as occasional massive historical releases of buried carbon due to volcanism will become less frequent (as earth mantle cooling and progressive exhaustion of internal radioactive heat proceeds further). The rates of these processes are extremely slow; hence they are of no relevance to the atmospheric CO_2 concentration over the next hundreds or thousands of years.

In billion-year timescales, it is predicted that plant, and therefore animal, life on land will die off altogether, since by that time most of the remaining carbon in the atmosphere will be sequestered underground, and natural releases of CO_2 by radioactivity-driven tectonic activity will have continued to slow down. The loss of plant life would also result in the eventual loss of oxygen. Some microbes are capable of photosynthesis at concentrations of CO_2 of a few parts per million and so the last life forms would probably disappear finally due to the rising temperatures and loss of the atmosphere when the sun becomes a red giant some four billion years from now.

Measuring Ancient-Earth Carbon Dioxide Concentration

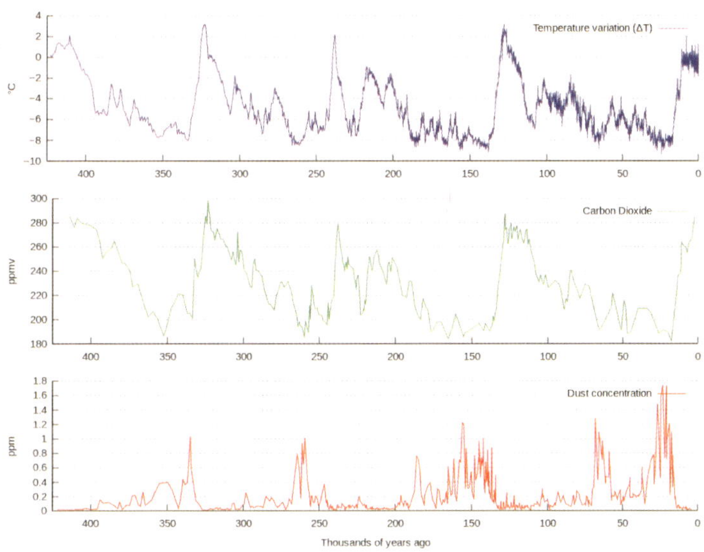

Graph of CO_2 (green), reconstructed temperature (blue) and dust (red) from the Vostok ice core for the past 420,000 years

The most direct method for measuring atmospheric carbon dioxide concentrations for periods before instrumental sampling is to measure bubbles of air (fluid or gas inclusions) trapped in the Antarctic or Greenland ice sheets. The most widely accepted of such studies come from a variety of Antarctic cores and indicate that atmospheric CO_2 concentrations were about 260–280 ppmv immediately before industrial emissions began and did not vary much from this level during the preceding 10,000 years. The longest ice core record comes from East Antarctica, where ice has been sampled to an age of 800,000 years. During this time, the atmospheric carbon dioxide concentration has varied between 180–210 ppm during ice ages, increasing to 280–300 ppm during warmer interglacials. The beginning of human agriculture during the current Holocene epoch may have been strongly connected to the atmospheric CO_2 increase after the last ice age ended, a fertilization effect raising plant biomass growth and reducing stomatal conductance requirements for CO_2 intake, consequently reducing transpiration water losses and increasing water usage efficiency.

Various proxy measurements have been used to attempt to determine atmospheric carbon dioxide concentrations millions of years in the past. These include boron and carbon isotope ratios in certain types of marine sediments, and the number of stomata observed on fossil plant leaves. While these measurements give much less precise estimates of carbon dioxide concentration than ice cores, there is evidence for very high CO_2 volume concentrations between 200 and 150 million years ago of over 3,000 ppm, and between 600 and 400 million years ago of over 6,000 ppm. In more recent times, atmospheric CO_2 concentration continued to fall after about 60 million years ago. About 34 million years ago, the time of the Eocene–Oligocene extinction event and when the Antarctic ice sheet started to take its current form, CO_2 is found to have been about 760 ppm, and there is geochemical evidence that concentrations were less than 300 ppm by about 20 million years ago. Carbon dioxide decrease, with a tipping point of 600 ppm, was the primary agent forcing Antarctic glaciation. Low CO_2 concentrations may have been the stimulus that favored the evolution of C4 plants, which increased greatly in abundance between 7 and 5 million years ago. Based on an analysis of fossil leaves, Wagner et al. argued that atmospheric CO_2 concentrations during the last 7,000–10,000 year period were significantly higher than 300 ppm and contained substantial variations that may be correlated to climate variations. Others have disputed such claims, suggesting they are more likely to reflect calibration problems than actual changes in CO_2. Relevant to this dispute is the observation that Greenland ice cores often report higher and more variable CO_2 values than similar measurements in Antarctica. However, the groups responsible for such measurements (e.g. H. J Smith et al.) believe the variations in Greenland cores result from *in situ* decomposition of calcium carbonate dust found in the ice. When dust concentrations in Greenland cores are low, as they nearly always are in Antarctic cores, the researchers report good agreement between measurements of Antarctic and Greenland CO_2 concentrations.

Atmospheric Carbon Dioxide and the Greenhouse Effect

Earth's natural greenhouse effect makes life as we know it possible and carbon dioxide plays a significant role in providing for the relatively warm temperature that the planet enjoys. The greenhouse effect is a process by which thermal radiation from a planetary atmosphere warms the planet's surface beyond the temperature it would have in the absence of its atmosphere. Without the greenhouse effect, the Earth's temperature would be about –18 °C (-0.4 °F) compared to Earth's actual surface temperature of approximately 14 °C (57.2 °F).

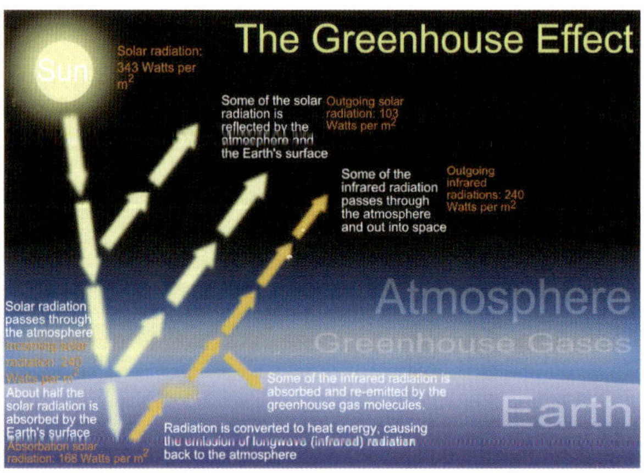

A pictogram of the greenhouse effect

Carbon dioxide is believed to have played an important effect in regulating Earth's temperature throughout its 4.7 billion year history. Early in the Earth's life, scientists have found evidence of liquid water indicating a warm world even though the Sun's output is believed to have only been 70% of what it is today. It has been suggested by scientists that higher carbon dioxide concentrations in the early Earth atmosphere might help explain this faint young sun paradox. When Earth first formed, Earth's atmosphere may have contained more greenhouse gases and CO_2 concentrations may have been higher, with estimated partial pressure as large as 1,000 kPa (10 bar), because there was no bacterial photosynthesis to reduce the gas to carbon compounds and oxygen. Methane, a very active greenhouse gas which reacts with oxygen to produce CO_2 and water vapor, may have been more prevalent as well, with a mixing ratio of 10^{-4} (100 parts per million by volume).

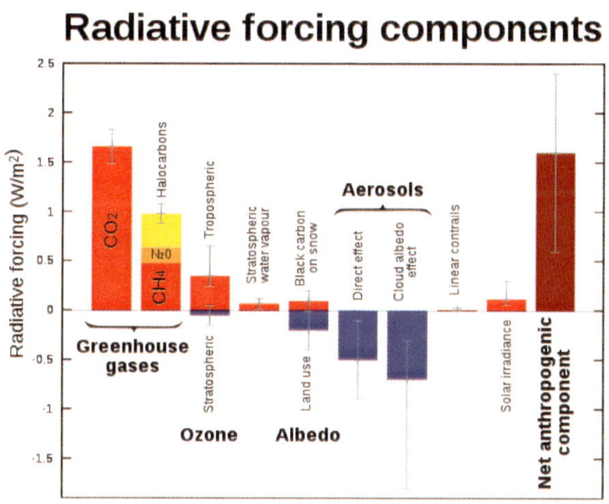

Contribution of natural factors and human activities to radiative forcing (RF) of climate change

RF values are for year 2005, relative to pre-industrial (1750). The contribution of solar irradiance to RF is 5% the value of the combined RF due to increases in the atmospheric concentrations of carbon dioxide, methane and nitrous oxide.

Though water is responsible for most (about 36-70%) of the total greenhouse effect, the role of water vapor as a greenhouse gas depends on temperature. On earth, carbon dioxide is the most relevant direct anthropologically influenced greenhouse gas. Carbon dioxide is often mentioned in the context of its increased influence as a greenhouse gas since the pre-industrial (1750) era. In the IPCC Fifth Assessment Report the increase in CO_2 was estimated to be responsible for 1.82 $W{\cdot}m^2$ of the 2.63 $W{\cdot}m^2$ change in radiative forcing on earth (about 70%).

The concept of atmospheric CO_2 increasing ground temperature was first published by Svante Arrhenius in 1896. The increased radiative forcing due to increased CO_2 in the earth's atmosphere is based on the physical properties of CO_2 and the non-saturated absorption windows where CO_2 absorbs outgoing long-wave energy.

Atmospheric Carbon Dioxide and the Carbon Cycle

Atmospheric carbon dioxide plays an integral role in the Earth's carbon cycle whereby carbon dioxide is removed from the atmosphere by some natural processes such as photosynthesis and deposition of carbonates, to form limestones for example, and added back to the atmosphere by other natural processes such as respiration and the acid dissolution of carbonate deposits. There are two broad carbon cycles on earth: the fast carbon cycle and the slow carbon cycle. The fast carbon cycle refers to movements of carbon between the environment and living things in the biosphere whereas the slow carbon cycle involves the movement of carbon between the atmosphere, oceans, soil, rocks and volcanism. Both carbon cycles are intrinsically interconnected and atmospheric gaseous carbon dioxide facilitates the carbon cycle.

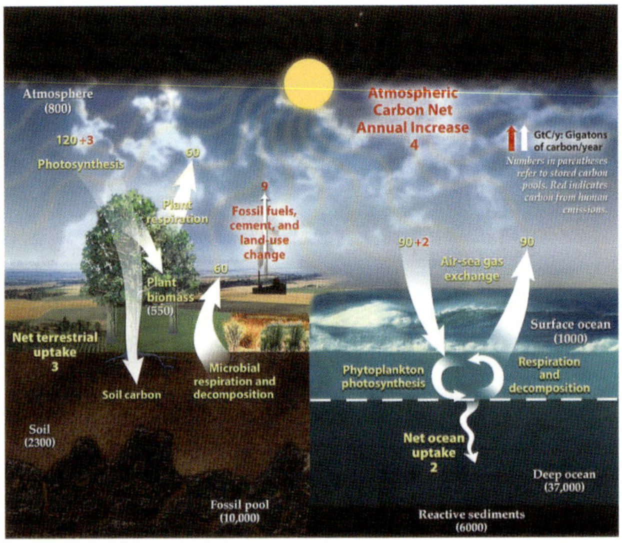

Fast carbon cycle shows the movement of carbon

This diagram of the fast carbon cycle shows the movement of carbon between land, atmosphere, and oceans in billions of tons of carbon per year. Yellow numbers are natural fluxes, red are human contributions in billions of tons of carbon per year. White numbers indicate stored carbon.

Natural sources of atmospheric carbon dioxide include volcanic outgassing, the combustion of organic matter, wildfires and the respiration processes of living aerobic organisms. Man-made sources of carbon dioxide include the burning of fossil fuels for heating, power generation and transport, as well as some industrial processes such as cement making. It is also produced by various microorganisms from fermentation and cellular respiration. Plants, algae and cyanobacteria convert carbon dioxide to carbohydrates by a process called photosynthesis. They gain the energy needed for this reaction from absorption of sunlight by chlorophyll and other pigments. Oxygen, produced as a by-product of photosynthesis, is released into the atmosphere and subsequently used for respiration by heterotrophic organisms and other plants, forming a cycle.

Most sources of CO_2 emissions are natural, and are balanced to various degrees by natural CO_2 sinks. For example, the natural decay of organic material in forests and grasslands and the action of forest fires results in the release of about 439 gigatonnes of carbon dioxide every year, while new growth entirely counteracts this effect, absorbing 450 gigatonnes per year. Although the initial carbon dioxide in the atmosphere of the young Earth was produced by volcanic activity, modern volcanic activity releases only 130 to 230 megatonnes of carbon dioxide each year. These natural sources are nearly balanced by natural sinks, physical and biological processes which remove carbon dioxide from the atmosphere. For example, some is directly removed from the atmosphere by land plants for photosynthesis and it is soluble in water forming carbonic acid. There is a large natural flux of CO_2 into and out of the biosphere and oceans. In the pre-industrial era these fluxes were largely in balance. Currently about 57% of human-emitted CO_2 is removed by the biosphere and oceans. From pre-industrial era to 2010, the terrestrial biosphere represented a net source of atmospheric CO_2 prior to 1940, switching subsequently to a net sink. The ratio of the increase in atmospheric CO_2 to emitted CO_2 is known as the *airborne fraction*; this varies for short-term averages and is typically about 45% over longer (5 year) periods. Estimated carbon in global terrestrial vegetation increased from approximately 740 billion tons in 1910 to 780 billion tons in 1990.

Atmospheric Carbon Dioxide and Photosynthesis

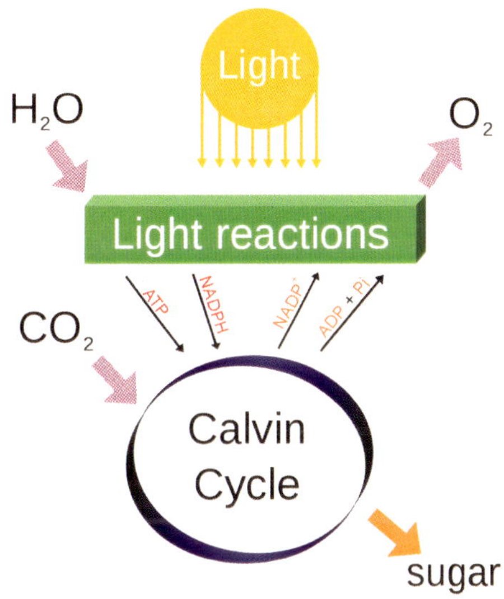

Photosynthesis changes sunlight into chemical energy, splits water to liberate O_2, and fixes CO_2 into sugar

Carbon dioxide in the Earth's atmosphere is essential to life and to the present planetary biosphere. Over the course of Earth's geologic history CO_2 concentrations have played a role in biological evolution. The first photosynthetic organisms probably evolved early in the evolutionary history of life and most likely used reducing agents such as hydrogen or hydrogen sulfide as sources of electrons, rather than water. Cyanobacteria appeared later, and the excess oxygen they produced contributed to the oxygen catastrophe, which rendered the evolution of complex life possible. In recent geologic times, low CO_2 concentrations below 600 parts per million might have been the stimulus that favored the evolution of C4 plants which increased greatly in abundance between 7 and 5 million years ago over plants that use the less efficient C3 metabolic pathway. At current atmospheric pressures photosynthesis shuts down when atmospheric CO_2 concentrations fall below 150 ppm and 200 ppm although some microbes can extract carbon from the air at much lower concentrations. Today, the average rate of energy capture by photosynthesis globally is approximately 130 terawatts, which is about six times larger than the current power consumption of human civilization. Photosynthetic organisms also convert around 100–115 thousand million metric tonnes of carbon into biomass per year.

Photosynthetic organisms are photoautotrophs, which means that they are able to synthesize food directly from CO_2 and water using energy from light. However, not all organisms that use light as a source of energy carry out photosynthesis, since *photoheterotrophs* use organic compounds, rather than CO_2, as a source of carbon. In plants, algae and cyanobacteria, photosynthesis releases oxygen. This is called *oxygenic photosynthesis*. Although there are some differences between oxygenic photosynthesis in plants, algae, and cyanobacteria, the overall process is quite similar in these organisms. However, there are some types of bacteria that carry out anoxygenic photosynthesis, which consumes CO_2 but does not release oxygen.

Carbon dioxide is converted into sugars in a process called carbon fixation. Carbon fixation is an endothermic redox reaction, so photosynthesis needs to supply both a source of energy to drive this process, and the electrons needed to convert CO_2 into a carbohydrate. This addition of the electrons is a reduction reaction. In general outline and in effect, photosynthesis is the opposite of cellular respiration, in which glucose and other compounds are oxidized to produce CO_2 and water, and to release exothermic chemical energy to drive the organism's metabolism. However, the two processes take place through a different sequence of chemical reactions and in different cellular compartments.

Most organisms that utilize photosynthesis to produce oxygen use visible light to do so, although at least three use shortwave infrared or, more specifically, far-red radiation.

Effects of Increased CO_2 on Plants and Crops

A 1993 review of scientific greenhouse studies found that a doubling of CO_2 concentration would stimulate the growth of 156 different plant species by an average of 37%. Response varied significantly by species, with some showing much greater gains and a few showing a loss. For example, a 1979 greenhouse study found that with doubled CO_2 concentration the dry weight of 40-day-old cotton plants doubled, but the dry weight of 30-day-old maize plants increased by only 20%.

In addition to greenhouse studies, field and satellite measurements attempt to understand the effect of increased CO_2 in more natural environments. In free-air carbon dioxide enrichment (FACE) experiments plants are grown in field plots and the CO_2 concentration of the surrounding air is artificially elevated. These experiments generally use lower CO_2 levels than the greenhouse studies. They show lower gains in growth than greenhouse studies, with the gains depending heavily on the species under study. A 2005 review of 12 experiments at 475-600 ppm showed an average gain of 17% in crop yield, with legumes typically showing a greater response than other species and C4 plants generally showing less. The review also stated that the experiments have their own limitations. The studied CO_2 levels were lower, and most of the experiments were carried out in temperate regions. Satellite measurements found increasing leaf area index for 25% to 50% of Earth's vegetated area Earth over the past 35 years, providing evidence for a positive CO_2 fertilization effect.

A 2017 article states that increased CO_2 levels have a negative impact on the nutritional quality of various human food crops, by increasing the levels of carbohydrates, such as glucose, while decreasing the levels of important nutrients such as protein, iron, and zinc. Crops experiencing a decrease in protein include rice, wheat, barley and potatoes.

Atmospheric Carbon Dioxide and the Oceanic Carbon Cycle

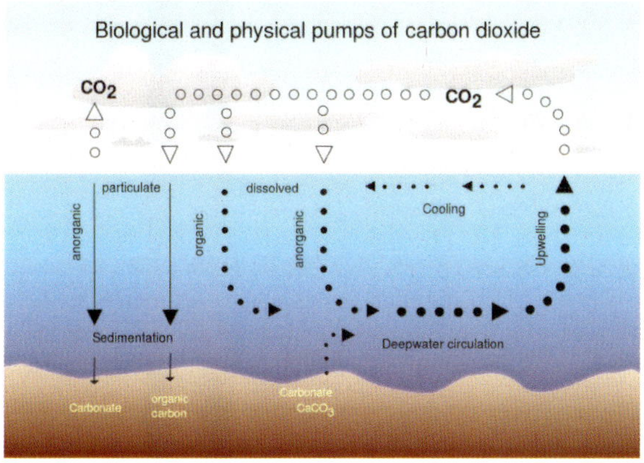

Air-sea exchange of CO_2

The Earth's oceans contain a large amount of CO_2 in the form of bicarbonate and carbonate ions — much more than the amount in the atmosphere. The bicarbonate is produced in reactions between rock, water, and carbon dioxide. One example is the dissolution of calcium carbonate:

$$CaCO_3 + CO_2 + H_2O \rightleftharpoons Ca^{2+} + 2\ HCO^-_3$$

Reactions like this tend to buffer changes in atmospheric CO_2. Since the right side of the reaction produces an acidic compound, adding CO_2 on the left side decreases the pH of sea water, a process which has been termed ocean acidification (pH of the ocean becomes more acidic although the pH value remains in the alkaline range). Reactions between CO_2 and non-carbonate rocks also add bicarbonate to the seas. This can later undergo the reverse of the above reaction to form carbonate rocks, releasing half of the bicarbonate as CO_2. Over hundreds of millions of years, this has produced huge quantities of carbonate rocks.

Ultimately, most of the CO_2 emitted by human activities will dissolve in the ocean; however, the rate at which the ocean will take it up in the future is less certain. Even if equilibrium is reached, including dissolution of carbonate minerals, the increased concentration of bicarbonate and decreased or unchanged concentration of carbonate ion will give rise to a higher concentration of un-ionized carbonic acid and dissolved CO_2. This, along with higher temperatures, would mean a higher equilibrium concentration of CO_2 in the air.

Anthropogenic CO_2 Emissions

While CO_2 absorption and release is always happening as a result of natural processes, the recent rise in CO_2 levels in the atmosphere is known to be mainly due to human (anthropogenic) activity. There are 4 ways human activity, especially fossil fuel burning, is known to have caused the rapid increase in atmospheric CO_2 over the last few centuries. 1) Various national statistics accounting for fossil fuel consumption, combined with knowledge of how much atmospheric CO_2 is produced per unit of fossil fuel (e.g. liter of gasoline). 2) By examining the ratio of various carbon isotopes in the atmosphere. The burning of long-buried fossil fuels releases CO_2 containing carbon of different isotopic ratios to those of living plants, enabling distinction between natural and human-caused contributions to CO_2 concentration. 3) Higher atmospheric CO_2 concentrations in the Northern Hemisphere, where most of the world's population lives (and emissions originate from), compared to the southern hemisphere. This difference has increased as anthropogenic emissions have increased. 4) Atmospheric O_2 levels are decreasing in earth's atmosphere as it reacts with the carbon in fossil fuels to form CO_2.

Burning fossil fuels such as coal, petroleum, and natural gas is the leading cause of increased anthropogenic CO_2; deforestation is the second major cause. In 2010, 9.14 gigatonnes of carbon (GtC, equivalent to 33.5 gigatonnes of CO_2 or about 4.3 ppm in earth's atmosphere) were released from fossil fuels and cement production worldwide, compared to 6.15 GtC in 1990. In addition, land use change contributed 0.87 GtC in 2010, compared to 1.45 GtC in 1990. In 1997, human-caused Indonesian peat fires were estimated to have released between 13% and 40% of the average carbon emissions caused by the burning of fossil fuels around the world in a single year. In the period 1751 to 1900, about 12 GtC were released as CO_2 to the atmosphere from burning of fossil fuels, whereas from 1901 to 2013 the figure was about 380 GtC.

CO_2 in Earth's atmosphere if *half* of anthropogenic CO_2 emissions are *not* absorbed

Anthropogenic carbon emissions exceed the amount that can be taken up or balanced out by natural sinks. As a result, carbon dioxide has gradually accumulated in the atmosphere, and as of 2013, its concentration is almost 43% above pre-industrial levels. Various techniques have been proposed for removing excess carbon dioxide from the atmosphere in carbon dioxide sinks. Currently about half of the carbon dioxide released from the burning of fossil fuels is not absorbed by vegetation and the oceans and remains in the atmosphere.

Excess CO_2 emitted since the pre-industrial era is projected to remain in the atmosphere for centuries to millennia, even after emissions stop. Even if human carbon dioxide emissions were to completely cease, atmospheric temperatures are not expected to decrease significantly for thousands of years.

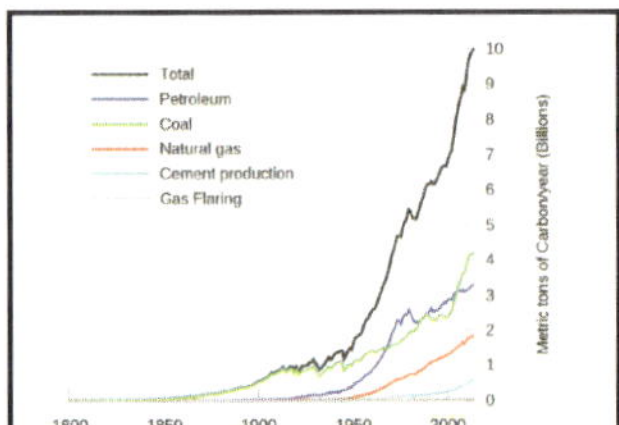	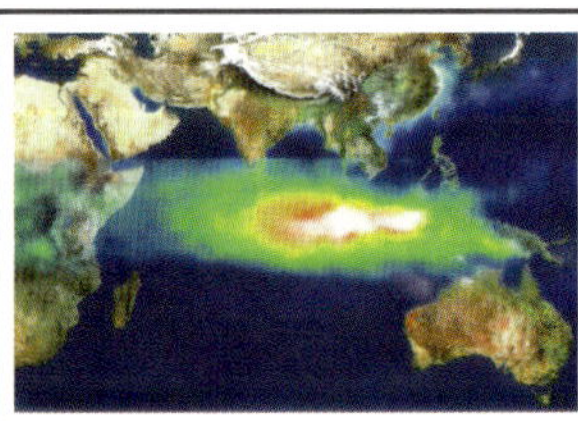	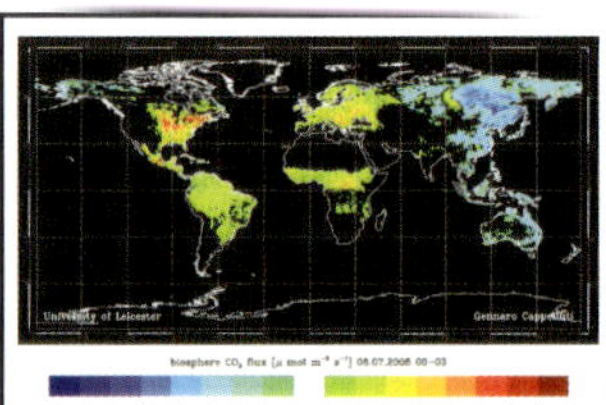	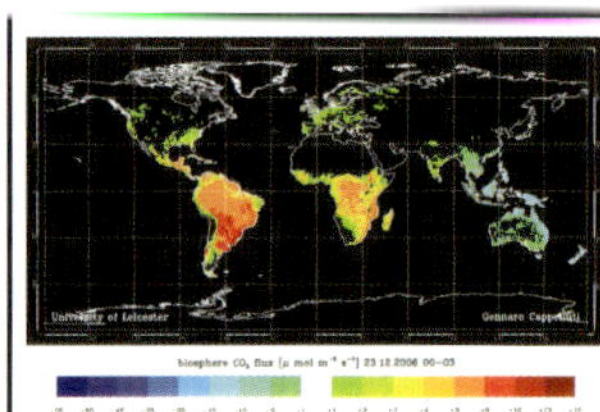
Global fossil carbon emissions 1800–2007	False-color image of smoke and ozone pollution from Indonesian fires, 1997	Biosphere CO_2 flux in the northern hemisphere summer (NOAA Carbon Tracker)	Biosphere CO_2 flux in the northern hemisphere winter (NOAA Carbon Tracker)

Ongoing Measurements of Atmospheric CO_2

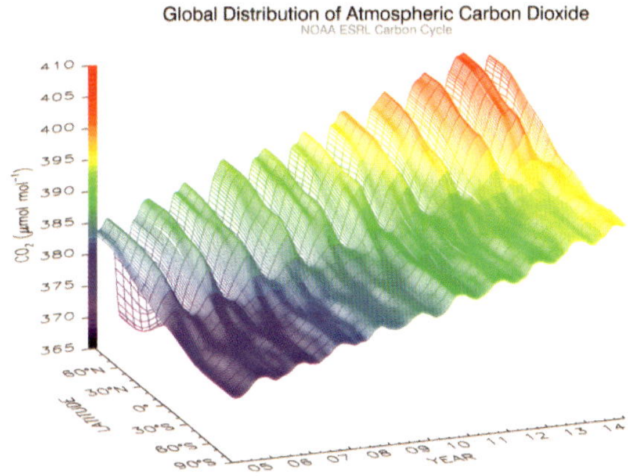

Carbon Dioxide observations from 2005 to 2014 showing the seasonal variations and the difference between northern and southern hemispheres

The first reproducibly accurate measurements of atmospheric CO_2 were from flask sample measurements made by Dave Keeling at Caltech in the 1950s. A few years later in March 1958 the first ongoing measurements were started by Keeling at Mauna Loa. Measurements at Mauna Loa have been ongoing since then. Now measurements are made at many sites globally. Additional measurement techniques are also used as well. Many measurement sites are part of larger global networks. Global network data are often made publicly available on the conditions of proper acknowledgement according to the respective data user policies.

There are several surface measurement (including flasks and continuous in situ) networks including NOAA/ERSL, WDCGG, and RAMCES. The NOAA/ESRL Baseline Observatory Network, and the Scripps Institution of Oceanography Network data are hosted at the CDIAC at ORNL. The World Data Centre for Greenhouse Gases (WDCGG), part of GAW, data are hosted by the JMA. The Reseau Atmospherique de Mesure des Composes an Effet de Serre database (RAMCES) is part of IPSL.

From these measurements, further products are made which integrate data from the various sources. These products also address issues such as data discontinuity and sparseness. GLOBALVIEW-CO2 is one of these products.

Ongoing ground-based total column measurements began more recently. Column measurements typically refer to an averaged column amount denoted X_{CO2}, rather than a surface only measurement. These measurements are made by the TCCON. These data are also hosted on the CDIAC, and made publicly available according to the data use policy.

Satellite measurements are also a recent addition to atmospheric X_{CO2} measurements. SCIAMACHY aboard ESA's ENVISAT made global column X_{CO2} measurements from 2002-2012. AIRS aboard NASA's Aqua satellite makes global X_{CO2} measurements and was launched shortly after ENVISAT in 2012. More recent satellites have significantly improved the data density and precision of global measurements. Newer missions have higher spectral and spatial resolutions. JAXA's GOSAT was the first dedicated GHG monitoring satellite to successfully achieve orbit in 2009. NASA's OCO-2 launched in 2014 was the second. Various other satellites missions to measure atmospheric X_{CO2} are planned.

Methane

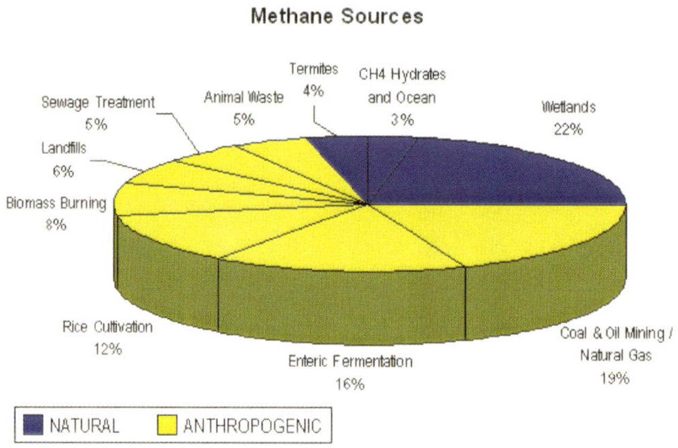

Different Sources of Methane

Methane is the also a greenhouse gas which absorbs the radiation and cause more temperature of the air. Paddy cultivation also generates methane in the air. It is also produced from the wetlands and waterlogged soils and released in the atmosphere. Fossil fuel is also a source to release of methane in the atmosphere. Its amount in the atmosphere is variable.

Atmospheric methane is the methane present in earth's atmosphere. Atmospheric methane concentrations are of interest because it is one of the most potent greenhouse gases in earth's atmosphere. The 100-year global warming potential of methane is 28. That is, over a 100-year period, it traps 28 times more heat per mass unit than carbon dioxide and 32 times the effect when accounting for aerosol interactions. Global methane concentrations had risen from 722 parts per billion (ppb) in pre-industrial times to 1800 ppb by 2011, an increase by a factor of 2.5 and the highest value in at least 800,000 years. Its concentration is higher in the Northern Hemisphere since most sources (both natural and human) are located on land and the Northern Hemisphere has more land mass. The concentrations vary seasonally, with, for example, a minimum in the northern tropics during April–May mainly due to removal by the hydroxyl radical.

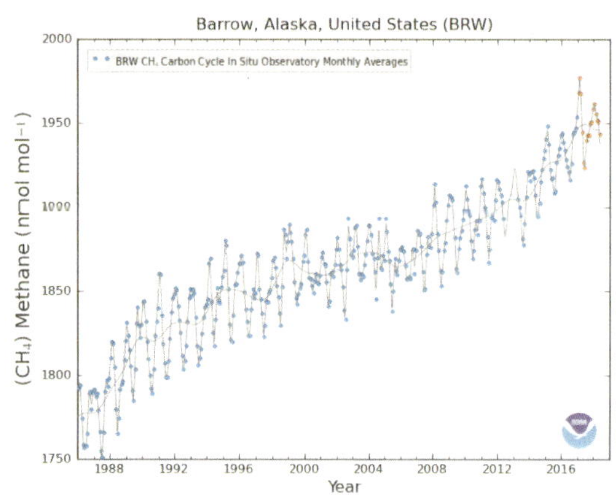

Methane concentrations up to May 2018: A monthly peak of 1887.65 ppb was reached in November 2017

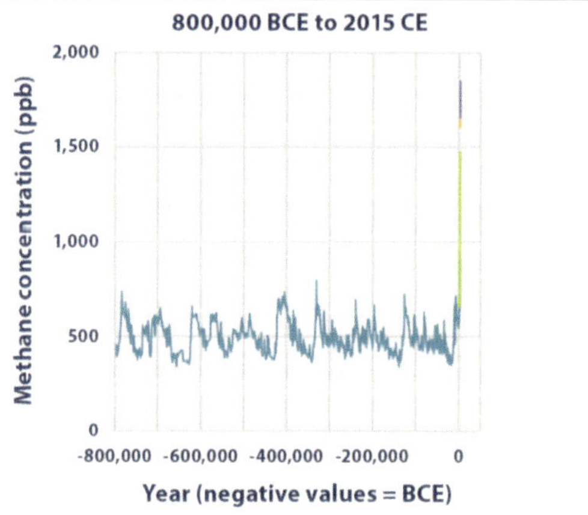

Compilation of paleo-climatology data of methane

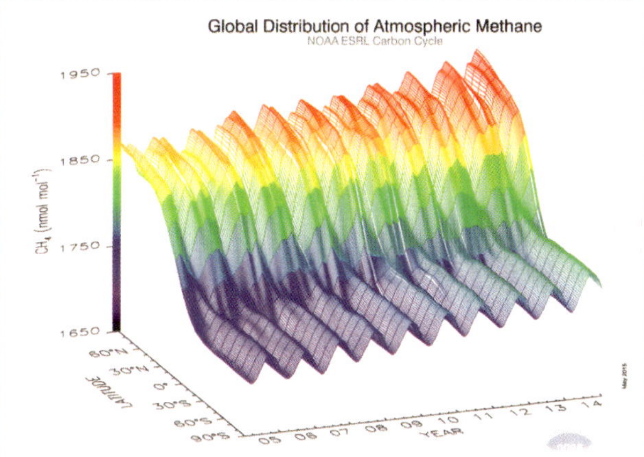

Methane observations from 2005 to 2014 showing the seasonal variations and the difference between northern and southern hemispheres

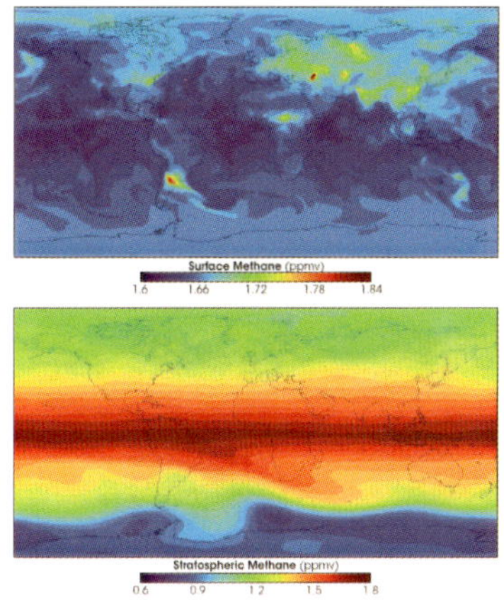

Computer models showing the amount of methane (parts per million by volume) at the surface (top) and in the stratosphere (bottom)

Early in the earth's history carbon dioxide and methane likely produced a greenhouse effect. The carbon dioxide would have been produced by volcanoes and the methane by early microbes. During this time, earth's earliest life appeared. These first, ancient bacteria added to the methane concentration by converting hydrogen and carbon dioxide into methane and water. Oxygen did not become a major part of the atmosphere until photosynthetic organisms evolved later in earth's history. With no oxygen, methane stayed in the atmosphere longer and at higher concentrations than it does today.

Methane forms near the Earth's surface, and it is carried into the stratosphere by rising air. Uncontrolled build-up of methane in earth's atmosphere is naturally checked—although human influence can upset this natural regulation—by methane's reaction with hydroxyl radicals formed from singlet oxygen atoms and with water vapor.

Methane as a Greenhouse Gas

Methane in the earth's atmosphere is a strong greenhouse gas with a global warming potential (GWP) 104 times greater than CO_2 in a 20-year time frame; methane is not as persistent a gas as CO_2 and tails off to about GWP of 28 for a 100-year time frame. This means that a methane emission will have 28 times the impact on temperature of a carbon dioxide emission of the same mass over the following 100 years. Methane has a large effect but for a relatively brief period, having an estimated lifetime of 8.9±0.6 years in the atmosphere, whereas carbon dioxide has a small effect for a long period, having an estimated lifetime of over 100 years.

One of the most significant consequences of using a GWP 100 year time horizon for all GHGs regardless of their lifetime and warming potential is in sequestration. Here it leads to a serious underestimation, by a factor of approximately 3, for the volumes of CO_2 sequestration which would be required to counter the warming effects of a given volume of methane. On the other hand, the benefits of rapid direct reductions of methane are also thought to be substantially obscured by this choice of warming time horizon for methane.

The concentration of methane in earth's atmosphere has increased by about 150 percent since

1750, and it now accounts for 20% of the total radiative forcing from all of the long-lived and globally mixed greenhouse gases. According to NOAA the atmospheric methane concentration was measured at 1890 ppb in Ireland (Northern Hemisphere) and 1760 ppb (Tasmania, Southern Hemisphere) in 2013 – arctic methane release from melting methane clathrates has played a role, but other recent emissions include peat fires in Southeast Asia. In all of the 420,000 year time-period before the industrial era, the atmospheric methane concentration has been less than half the current level, including the period since the Last Glacial Maximum.

Global Methane Cycle

This simple diagram depicts the flow of methane from sources into the atmosphere as well as the sinks that consume methane

1. Permafrost, glaciers, and ice cores – A source that slowly releases methane trapped in frozen environments as global temperatures rise.
2. Wetlands – Warm temperatures and moist environments are ideal for methane production.
3. Forest fire – Mass burning of organic matter releases methane into the atmosphere.
4. Rice paddies – The warmer and moister the rice field, the more methane is produced.
5. Animals – Microorganisms breaking down difficult to digest material in the guts of ruminant livestock and termites produce methane that is then released during defecation.
6. Plants – While methane can be consumed in soil before being released into the atmosphere, plants allow for direct travel of methane up through the roots and leaves and into the atmosphere. Plants may also be direct producers of methane.
7. Landfills – Decaying organic matter and anaerobic conditions cause landfills to be a significant source of methane.
8. Waste water treatment facilities – Anaerobic treatment of organic compounds in the water results in the production of methane.
9. Hydroxyl radical – OH in the atmosphere is the largest sink for atmospheric methane as well as one of the most significant sources of water vapor in the upper atmosphere.
10. Chlorine radical – Free chlorine in the atmosphere also reacts with methane.

Emissions Accounting of Methane

The balance between sources and sinks of methane is not yet fully understood. There are "large uncertainties in the current bottom-up estimates of components of the global source", and the balance between sources and sinks is not yet well known. The most important sink in the methane cycle is reaction with the hydroxyl radical, which is produced photochemically in the atmosphere. Production of this radical is not fully understood and has a large effect on atmospheric concentrations. This uncertainty is exemplified by observations that have shown between the year 2000 and 2006 increases in atmospheric concentration of methane ceased, for reasons still being investigated.

Various sources give the following values for methane emissions:

Estimates of the global methane budget (in Tg(CH_4)/yr)					
Reference:	Fung et al. (1991)	Hein et al. (1997)	Lelieveld et al. (1998)	Houweling et al. (1999)	Bousquet et al. (2006)
Base year:	1980s	–	1992	-	-
Natural emissions					
Wetlands	115	237	225[nb 1]	145	147±15
Termites	20	–	20	20	23±4
Ocean	10	–	15	15	19±6
Hydrates	5	–	10	–	–
Anthropogenic emissions					
Energy	75	97	110	89	110±13
Landfills	40	35	40	73	55±11[nb 2]
Ruminants (livestock)	80	90[nb 3]	115	93	
Waste treatment	–	[nb 3]	25	–	[nb 2]
Rice agriculture	100	88	[nb 1]	–	31±5
Biomass burning	55	40	40	–	50±8
Other	–	–	–	20	90±14[nb 4]
Total source	500	587	600		525±8
Sinks					
Soils	10				
30	40	21±3			
Tropospheric OH	450	489	510		448±1
Stratospheric loss					
46	40		37±1		
Total sink	460	535	580		506

Natural Sources of Atmospheric Methane

Any process that results in the production of methane and its release into the atmosphere can be considered a "source." The two main processes that are responsible for methane production occur as a result of microorganisms anaerobically converting organic compounds into methane.

Methanogenesis

Most ecological emissions of methane relate directly to methanogens generating methane in warm, moist soils as well as in the digestive tracts of certain animals. Methanogens are methane producing microorganisms. In order to produce energy, they use an anaerobic process called methanogenesis. This process is used in lieu of aerobic, or with oxygen, processes because methanogens are unable to metabolise in the presence of even small concentrations of oxygen. When acetate is broken down in methanogenesis, the result is the release of methane into the surrounding environment.

Methanogenesis, the scientific term for methane production, occurs primarily in anaerobic conditions because of the lack of availability of other oxidants. In these conditions, microscopic organisms called archaea use acetate and hydrogen to break down essential resources in a process called fermentation.

Acetoclastic methanogenesis – certain archaea cleave acetate produced during anaerobic fermentation to yield methane and carbon dioxide.

$$H_3C\text{-}COOH \rightarrow CH_4 + CO_2$$

Hydrogenotrophic methanogenesis – archaea oxidize hydrogen with carbon dioxide to yield methane and water.

$$4H_2 + CO_2 \rightarrow CH_4 + 2H_2O$$

While acetoclastic methanogenesis and hydrogenotrophic methanogenesis are the two major source reactions for atmospheric methane, other minor biological methane source reactions also occur. For example, it has been discovered that leaf surface wax exposed to UV radiation in the presence of oxygen is an aerobic source of methane.

Wetlands

Wetlands account for approximately 20 percent of atmospheric methane through emissions from soils and plants. Wetlands counteract the sinking action that normally occurs with soil because of the high water table. The level of the water table represents the boundary between anaerobic methane production and aerobic methane consumption. When the water table is low, the methane generated within the wetland soil has to come up through the soil and get past a deeper layer of methanotrophic bacteria, thereby reducing emission. Methane transport by vascular plants can bypass this aerobic layer, thus increasing emission.

Animals

Ruminant animals, particularly cows and sheep, contain bacteria in their gastrointestinal systems that help to break down plant material. Some of these microorganisms use the acetate from the plant material to produce methane, and because these bacteria live in the stomachs and intestines of ruminants, whenever the animal "burps" or defecates, it emits methane as well. Based upon a study in the Snowy Mountains region, the amount of methane emitted by one cow is equivalent to the amount of methane that around 3.4 hectares of methanotrophic bacteria can consume.

Termites also contain methanogenic microorganisms in their gut. However, some of these microorganisms are so unique that they live nowhere else in the world except in the third gut of termites. These microorganisms also break down biotic components to produce ethanol, as well as methane byproduct. However, unlike ruminants who lose 20 percent of the energy from the plants they eat, termites only lose 2 percent of their energy in the process. Thus comparatively, termites do not have to eat as much food as ruminants to obtain the same amount of energy, and give off proportionally less methane.

Plants

Living plants (e.g. forests) have recently been identified as a potentially important source of methane, possibly being responsible for approximately 10 to 30 percent of atmospheric methane. A 2006 paper calculated emissions of 62–236 Tg a^{-1}, and "this newly identified source may have important implications". However the authors stress "our findings are preliminary with regard to the methane emission strength".

These findings have been called into question in a 2007 paper which found "there is no evidence for substantial aerobic methane emission by terrestrial plants, maximally 0.3% of the previously published values".

While the details of plant methane emissions have yet to be confirmed, plants as a significant methane source would help fill in the gaps of previous global methane budgets as well as explain large plumes of methane that have been observed over the tropics.

In wetlands, where the rate of methane production is high, plants help methane travel into the atmosphere—acting like inverted lightning rods as they direct the gas up through the soil and into the air. They are also suspected to produce methane themselves, but because the plants would have to use aerobic conditions to produce methane, the process itself is still unidentified.

Methane Gas from Methane Clathrates

At high pressures, such as are found on the bottom of the ocean, methane forms a solid clathrate with water, known as methane hydrate. An unknown, but possibly very large quantity of methane is trapped in this form in ocean sediments. The release of large volumes of methane gas from such sediments into the atmosphere has been suggested as a possible cause for rapid global warming events in the Earth's distant past, such as the Paleocene–Eocene Thermal Maximum of 55 million years ago, and the Great Dying.

Theories suggest that should global warming cause them to heat up sufficiently, all of this methane gas could again be released into the atmosphere. Since methane gas is twenty-five times stronger (for a given weight, averaged over 100 years) than CO_2 as a greenhouse gas; this would immensely magnify the greenhouse effect. However, most of this reservoir of hydrates appears isolated from changes to the surface climate, so any such release is likely to happen over geological timescales of a millennium or more.

Permafrost

Methane that gets frozen in permafrost – land that is frozen for several years at a time – is slowly released from bogs as the permafrost melts. With rising global temperatures, the amount of permafrost melting and releasing methane continues to increase.

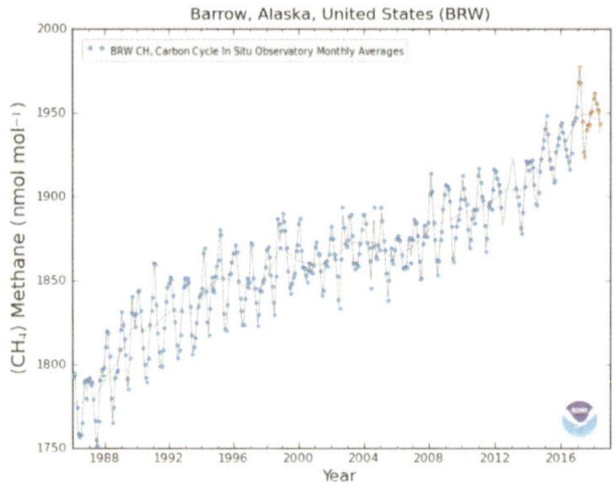

Arctic methane concentrations up to May 2018

Although records of permafrost are limited, recent years (1999 to 2007) have seen record thawing of permafrost in Alaska and Siberia. Measurements during 2006 in Siberia show that the methane released is five times greater than previously estimated. Melting yedoma, a type of permafrost, is a significant source of atmospheric methane (about 4 Tg of CH_4 per year).

The Woods Hole Research Center, citing two 2015 studies on permafrost carbon says there may be a self-reinforcing tipping point where an estimated equivalent of 205 gigatons of carbon dioxide in the form of methane could cause up to 0.5 °C (up to 0.9 °F) warming by the end of the century, which would trigger more warming. Permafrost contains almost twice as much carbon as is present in the atmosphere. The Intergovernmental Panel on Climate Change does not adequately account for arctic methane in permafrost.

Anthropogenic Sources of Atmospheric Methane

Slightly over half of the total emission is due to human activity. Since the Industrial Revolution humans have had a major impact on concentrations of atmospheric methane, increasing atmospheric concentrations roughly 250%.

Ecological Conversion

Conversion of forests and natural environments into agricultural plots increases the amount of nitrogen in the soil, which inhibits methane oxidation, weakening the ability of the methanotrophic bacteria in the soil to act as sinks. Additionally, by changing the level of the water table, humans can directly affect the soil's ability to act as a source or sink.

Farm Animals

A 2006 UN FAO report reported that livestock generate more greenhouse gases as measured in CO2 equivalents than the entire transportation sector. Livestock accounts for 9 percent of anthropogenic CO2, 65 percent of anthropogenic nitrous oxide and 37 percent of anthropogenic methane. A senior UN official and co-author of the report, Henning Steinfeld, said "Livestock are one of the most significant contributors to today's most serious environmental problems."

Recent NASA research has confirmed the vital role of enteric fermentation in livestock on global warming. "We understand that other greenhouse gases apart from carbon dioxide are important for climate change today," said Gavin Schmidt, the lead author of the study and a researcher at NASA's Goddard Institute for Space Studies in New York, NY and Columbia University's Center for Climate Systems Research. Other recent peer reviewed NASA research published in the journal Science has also indicated that the contribution of methane to global warming has been underestimated.

Nicholas Stern, the author of the 2006 Stern Review on climate change has stated "people will need to turn vegetarian if the world is to conquer climate change". President of the National Academy of Sciences Ralph Cicerone (an atmospheric scientist), has indicated the contribution of methane by livestock flatulence and eructation to global warming is a "serious topic." Cicerone states "Methane is the second-most-important greenhouse gas in the atmosphere now. The population of beef cattle and dairy cattle has grown so much that methane from cows now is big. This is not a trivial issue."

Approximately 5% of the methane is released via the flatus, whereas the other 95% is released via eructation. Vaccines are under development to reduce the amount introduced through eructation.

Rice Agriculture

Due to a continuously growing world population, rice agriculture has become one of the most significant anthropogenic sources of methane. With warm weather and water-logged soil, rice paddies act like wetlands, but are generated by humans for the purpose of food production. Due to the swamp-like environment of rice fields, these paddies yield 50-100 million metric tons of methane emission each year. This means that rice agriculture is responsible for approximately 15 to 20 percent of anthropogenic methane emissions. An article written by William F. Ruddiman explores the possibility that methane emissions began to rise as a result of anthropogenic activity 5000 years ago when ancient cultures started to settle and use agriculture, rice irrigation in particular, as a primary food source.

Landfills

Due to the large collections of organic matter and availability of anaerobic conditions, landfills are the third largest source of atmospheric methane in the United States, accounting for roughly 18.2% of methane emissions globally in 2014. When waste is first added to a landfill, oxygen is abundant and thus undergoes aerobic decomposition; during which time very little methane is produced. However, generally within a year oxygen levels are depleted and anaerobic conditions dominate the landfill allowing methanogens to takeover the decomposition process. These methanogens emit methane into the atmosphere and even after the landfill is closed, the mass amount of decaying matter allows the methanogens to continue producing methane for years.

Waste Water Treatment

Waste water treatment facilities act to remove organic matter, solids, pathogens, and chemical hazards as a result of human contamination. Methane emission in waste treatment facilities occurs as a result of anaerobic treatments of organic compounds and anaerobic biodegradation of sludge.

Biomass Burning

Incomplete burning of both living and dead organic matter results in the emission of methane. While natural wildfires can contribute to methane emissions, the bulk majority of biomass burning occurs as a result of humans – including everything from accidental burnings by civilians to deliberate burnings used to clear out land to biomass burnings occurring as a result of destroying waste.

Natural Gas Distribution

Methane is a primary component of natural gas, and thus during the production, processing, storage, transmission, and distribution of natural gas, a significant amount of methane is lost into the atmosphere.

According to the EPA Inventory of U.S Greenhouse Gases and Sinks: 1990-2009 report, dated April 2011, methane emissions from oil and natural gas systems make up 3.8 percent of total greenhouse gas emissions in the United States. Methane emissions occur in all sectors of the natural gas industry, from drilling and production, through gathering and processing and transmission, to distribution. These emissions occur through normal operation, routine maintenance, fugitive leaks, system upsets, and venting of equipment. In the oil industry, some underground crude contains natural gas that is entrained in the oil at high reservoir pressures. When oil is removed from the reservoir, associated natural gas is produced.

Additionally, the inventory shows that the oil and natural gas industry emitted 624 billion cubic feet (Bcf) of methane in 2009. Of this amount, 63% was from production operations, 7% from processing, 18% from transmission and storage systems, and 12% from distribution systems. Oil and natural gas systems are the largest human-made source of methane emissions (37 percent) in the United States.

Similarly, according to the EPA Global Anthropogenic Non CO_2 Greenhouse Gas Emissions: 1990-2020 report, dated June 2006, oil and natural gas operations are a significant source of global methane emissions, constituting approximately 18 percent of total human-made methane emissions. The report also indicates that in 2005, global oil and natural gas methane emissions totaled approximately 82 billion cubic meters (Bcm), or 2,896 billion cubic feet (Bcf), equivalent to about 1,165 million tonnes carbon dioxide equivalent (MtCO2e).

Recent estimates of natural gas losses during distribution are as high as 9%.

Gas flares are a tool the energy industry can use to reduce its environmental impact.

Methane Slip from Gas Engines

The use of natural gas and biogas in ICE (Internal combustion engine) for such applications as electricity production / cogeneration / CHP (Combined Heat and Power) and heavy vehicles or marine vessels such as LNG carriers using the boil off gas for propulsion, emits a certain percentage of UHC, unburned hydrocarbon of which 85% is methane. The climate issues of using gas to fuel ICE may offset or even cancel out the advantages of less CO2 and particle emissions is described in this 2016 EU Issue Paper on methane slip from marine engines: "Emissions of unburnt methane (known as the 'methane slip') were around 7g per kg LNG at higher engine loads, rising to 23–36g at lower loads. This increase could be due to slow combustion at lower temperatures, which allows small quantities of gas to avoid the combustion process". Road vehicles run more on low load than marine engines causing relatively higher methane slip.

Coal Mining

In 2014 NASA researchers reported the discovery of a 2,500 square miles (6,500 km2) methane cloud floating over the Four Corners region of the south-west United States. The discovery was based on data from the European Space Agency's Scanning Imaging Absorption Spectrometer for Atmospheric Chartography instrument from 2002 to 2012.

The report concluded that "the source is likely from established gas, coal, and coalbed methane mining and processing." The region emitted 590,000 metric tons of methane every year between 2002 and 2012—almost 3.5 times the widely used estimates in the European Union's Emissions Database for Global Atmospheric Research.

Removal Processes

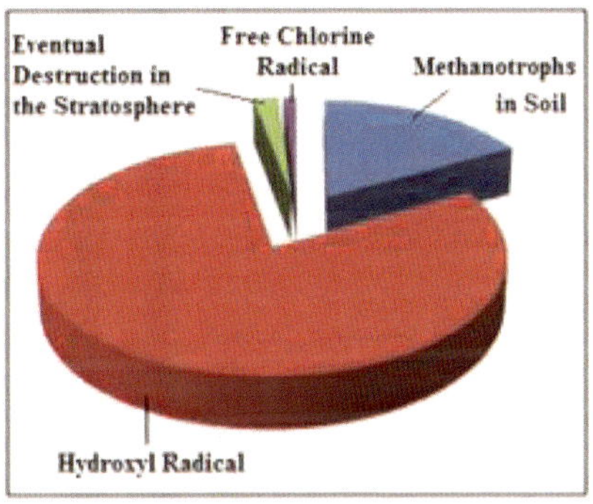

A pie chart demonstrating the relative effects of various sinks of atmospheric methane

Any process that consumes methane from the atmosphere can be considered a "sink" of atmospheric methane. The most prominent of these processes occur as a result of methane either being destroyed in the atmosphere or broken down in soil. Humans have yet to act as any significant sink of atmospheric methane.

Reaction with the hydroxyl radical – The major removal mechanism of methane from the atmosphere involves radical chemistry; it reacts with the hydroxyl radical (·OH) in the troposphere or stratosphere to create the ·CH3 radical and water vapor. In addition to being the largest known sink for atmospheric methane, this reaction is one of the most important sources of water vapor in the upper atmosphere.

$$CH_4 + \cdot OH \rightarrow \cdot CH_3 + H_2O$$

This reaction in the troposphere gives a methane lifetime of 9.6 years. Two more minor sinks are soil sinks (160 year lifetime) and stratospheric loss by reaction with ·OH, ·Cl and $\cdot O^1D$ in the stratosphere (120 year lifetime), giving a net lifetime of 8.4 years. Oxidation of methane is the main source of water vapor in the upper stratosphere (beginning at pressure levels around 10 kPa).

The methyl radical formed in the above reaction will, during normal daytime conditions in the troposphere, usually react with another hydroxyl radical to form formaldehyde. Note that this is not strictly oxidative pyrolysis as described previously. Formaldehyde can react again with a hydroxyl radical to form carbon dioxide and more water vapor. Sidechains in these reactions may interact with nitrogen compounds that will likely produce ozone, thus supplanting radicals required in the initial reaction.

Natural Sinks of Atmospheric Methane

Most natural sinks occur as a result of chemical reactions in the atmosphere as well as oxidation by methane consuming bacteria in Earth's soils.

Methanotrophs in Soils

Soils act as a major sink for atmospheric methane through the methanotrophic bacteria that reside within them. This occurs with two different types of bacteria. "High capacity-low affinity" methanotrophic bacteria grow in areas of high methane concentration, such as waterlogged soils in wetlands and other moist environments. And in areas of low methane concentration, "low capacity-high affinity" methanotrophic bacteria make use of the methane in the atmosphere to grow, rather than relying on methane in their immediate environment.

Forest soils act as good sinks for atmospheric methane because soils are optimally moist for methanotroph activity, and the movement of gases between soil and atmosphere (soil diffusivity) is high. With a lower water table, any methane in the soil has to make it past the methanotrophic bacteria before it can reach the atmosphere.

Wetland soils, however, are often sources of atmospheric methane rather than sinks because the water table is much higher, and the methane can be diffused fairly easily into the air without having to compete with the soil's methanotrophs.

Methanotrophic bacteria in soils – Methanotrophic bacteria that reside within soil use methane as a source of carbon in methane oxidation. Methane oxidation allows methanotrophic bacteria to use methane as a source of energy, reacting methane with oxygen and as a result producing carbon dioxide and water.

$$CH_4 + 2O_2 \rightarrow CO_2 + 2H_2O$$

Troposphere

The most effective sink of atmospheric methane is the hydroxyl radical in the troposphere, or the lowest portion of Earth's atmosphere. As methane rises into the air, it reacts with the hydroxyl radical to create water vapor and carbon dioxide. The lifespan of methane in the atmosphere was estimated at 9.6 years as of 2001; however, increasing emissions of methane over time reduce the concentration of the hydroxyl radical in the atmosphere. With less OH° to react with, the lifespan of methane could also increase, resulting in greater concentrations of atmospheric methane.

Stratosphere

If it is not destroyed in the troposphere, methane will last approximately 120 years before it is eventually destroyed in Earth's next atmospheric layer: the stratosphere. Destruction in the stratosphere occurs the same way that it does in the troposphere: methane is oxidized to produce carbon dioxide and water vapor. Based on balloon-borne measurements since 1978, the abundance of stratospheric methane has increased by 13.4%±3.6% between 1978 and 2003.

Reaction with Free Chlorine

Methane also reacts with natural chlorine gas in the atmosphere to produce chloromethane and hydrochloric acid (HCl). This process is known as free radical halogenations.

$$CH_4 + Cl_2 \rightarrow CH_3Cl + HCl$$

The HCl produced in this reaction leads to catalytic ozone destruction in the stratosphere.

Patterns of Methane Change over Time

Since the 1800s, atmospheric methane concentrations have increased annually at a rate of about 0.9%. Long term atmospheric measurements of methane by NOAA show that the build up of methane has slowed dramatically over the decade leading to 2006, after nearly tripling since pre-industrial times. Although scientists have yet to pinpoint the exact reasons for this sudden drop in growth rates, it is thought that this reduction is due to reduced industrial emissions and drought in wetland areas.

The only exceptions to this drop in growth rate occurred in 1991 and 1998 when growth rates increased suddenly to 14-15 nmol/mol per year for those years, nearly double the growth rates of the years before.

The 1991 spike is believed to have occurred due to the volcanic eruption of Mt. Pinatubo in June of that year. Volcanoes affect atmospheric methane emissions when they erupt, releasing ash and sulfur dioxide into the air. As a result, photochemistry of plants is affected and the removal of methane via the tropospheric hydroxyl radical is reduced. However, growth rates quickly fell due to lower temperatures and global reduction in rainfall.

The cause of the 1998 spike is unresolved, but scientists are currently attributing it to a combination of increased wetland and rice field emissions as well as an increased amount of biomass burning. 1998 was also the warmest year since surface temperatures were first recorded, suggesting that anomalously high temperatures can induce elevated methane emission.

Data from 2007 suggested methane concentrations were beginning to rise again. This was confirmed in 2010 when a study showed methane levels were on the rise for the 3 years 2007 to 2009. After a decade of near-zero growth in methane levels, "globally averaged atmospheric methane increased by approximately 7 nmol/mol per year during 2007 and 2008. During the first half of 2009, globally averaged atmospheric CH_4 was approx. 7 nmol/mol greater than it was in 2008, suggesting that the increase will continue in 2009."

Methane emissions levels vary greatly depending on the local geography. For both natural and anthropogenic sources, higher temperatures and higher water levels result in the anaerobic environment that is necessary for methane production.

Natural Methane Cycles

Emissions of methane into the atmosphere are directly related to temperature and moisture. Thus, the natural environmental changes that occur during seasonal change act as a major control of methane emission. Additionally, even changes in temperature during the day can affect the amount of methane that is produced and consumed.

For example, plants that produce methane can emit as much as two to four times more methane during the day than during the night. This is directly related to the fact that plants tend to rely on solar energy to enact chemical processes.

Additionally, methane emissions are affected by the level of water sources. Seasonal flooding during the spring and summer naturally increases the amount of methane released into the air.

Changes in Anthropogenic Sources

The most clearly identified rise in atmospheric methane as a result of human activity occurred in the 1700s during the industrial revolution. As technology increased at a considerable rate, humans began to build factories and plants, burn fossil fuels for energy, and clear out forests and other vegetation for the purpose of building and agriculture. This growth continued to rise at a rate of almost 1 percent per year until around 1990 when growth rates dropped to almost zero.

A 2003 article from William F. Ruddiman, however, indicates that the anthropogenic change in methane may have started 5000 years prior to the industrial revolution. The methane insolation cycles of the ice core remained stable and predictable until 5000 years ago, most likely due to some anthropogenic effect. Ruddiman suggests that the transition of humans from hunter gatherers into agricultural farming was the first instance of humans affecting methane concentration in the atmosphere. Ruddiman's hypothesis is supported by the fact that early rice irrigation occurred approximately 5000 years ago—the same time the ice core cycles lost their predictability. Due to the inefficiency of humans first learning how to grow rice, extensive rice paddies would have been needed to feed even a small population. These, over-flooded and filled with weeds, would have resulted in huge methane emitting wetlands.

Another source of methane emissions has been identified in Russia. Near Yamburg and Urengoy exist gas fields with a methane concentration of 97 percent. The gas obtained from these fields is taken and exported to Western and Central Europe through an extensive pipeline system known as the Trans-Siberian natural gas pipeline system. In accordance with the IPCC and other natural gas emissions control groups, measurements had to be taken throughout the pipeline to measure methane emissions from technological discharges and leaks at the pipeline fittings and vents. Although the majority of the natural gas leaks were carbon dioxide, a significant amount of methane was also being consistently released from the pipeline as a result of leaks and breakdowns. In 2001, natural gas emissions from the pipeline and natural gas transportation system accounted for 1 percent of the natural gas produced. Fortunately, between 2001 and 2005, this number reduced to 0.7 percent, and even the 2001 value is still significantly less than that of 1996.

However that may be, transportation is only part of the problem. Howarth *et al.* have argued that:

> We believe the preponderance of evidence indicates shale gas has a larger GHG (green house gas) footprint than conventional gas, considered over any time scale. The GHG footprint of shale gas also exceeds that of oil or coal when considered at decadal time scales.

For subsequent works confirming these results see Howarth's *"A bridge to nowhere: methane emissions and the greenhouse gas footprint of natural gas,"* and *"Methane emissions and climatic warming risk from hydraulic fracturing and shale gas development: implications for policy."* A 2013 study by Miller *et al.* indicates that current greenhouse gas reduction policies in the US are based on what appear to be significant underestimates of anthropogenic methane emissions. The authors state:

> We find greenhouse gas emissions from agriculture and fossil fuel extraction and processing (*i.e.,* oil and/or natural gas) are likely a factor of two or greater than cited in existing studies.

Impacts

The direct radiative greenhouse gas forcing effect has been estimated at 0.5 W/m^2.

Methane is a strong GHG with a global warming potential 84 times greater than CO_2 in a 20-year time frame; Methane is not as persistent a gas and tails off to about 28 for a 100-year time frame.

In addition to the direct heating effect and the normal feedbacks, the methane breaks down to carbon dioxide and water. This water is often above the tropopause where little water usually reaches. Ramanathan (1988) notes that both water and ice clouds, when formed at cold lower stratospheric temperatures, are extremely efficient in enhancing the atmospheric greenhouse effect. He also notes that there is a distinct possibility that large increases in future methane may lead to a surface warming that increases nonlinearly with the methane concentration.

Ozone Layer

Methane also affects the degradation of the ozone layer, when methane is transformed into water in the stratosphere. This process is enhanced by global warming, because warmer air holds more water vapor than colder air, so the amount of water vapor in the atmosphere increases as it is warmed by the greenhouse effect. Climate models also indicate that greenhouse gases such as carbon dioxide and methane may enhance the transport of water into the stratosphere; though this is not fully understood.

Methane Management Techniques

In an effort to mitigate climate change, humans have started to develop alternative methods and medicines.

For example, in order to counteract the amount of methane that ruminants give off, a type of drug called monensin has been developed. This drug is classified as an ionophore, which is an antibiotic that is naturally produced by a harmless bacteria strain. This drug not only improves feed efficiency but also reduces the amount of methane gas emitted from the animal and its manure.

In addition to medicine, specific manure management techniques have been developed to counteract emissions from livestock manure. Educational resources have begun to be provided for small farms. Management techniques include daily pickup and storage of manure in a completely closed off storage facility that will prevent runoff from making it into bodies of water. The manure can then be kept in storage until it is either reused for fertilizer or taken away and stored in an offsite compost. Nutrient levels of various animal manures are provided for optimal use as compost for gardens and agriculture.

In order to reduce effects on methane oxidation in soil, several steps can be taken. Controlling the usage of nitrogen enhancing fertilizer and reducing the amount of nitrogen pollution into the air can both lower inhibition of methane oxidation. Additionally, using drier growing conditions for crops such as rice and selecting strains of crops that produce more food per unit area can reduce the amount of land with ideal conditions for methanogenesis. Careful selection of areas of land conversion (for example, plowing down forests to create agricultural fields) can also reduce the destruction of major areas of methane oxidation.

To counteract methane emissions from landfills, on March 12, 1996, the EPA (Environmental Protection Agency) added the "Landfill Rule" to the Clean Air Act. This rule requires large landfills that have ever accepted municipal solid waste, have been used as of November 8, 1987, can hold at least 2.5 million metric tons of waste with a volume greater than 2.5 million cubic meters, and/or have nonmethane organic compound (NMOC) emissions of at least 50 metric tons per year to collect and combust emitted landfill gas. This set of requirements excludes 96% of the landfills in the USA. While the direct result of this is landfills reducing emission of non-methane compounds that form smog, the indirect result is reduction of methane emissions as well.

Furthermore, in an attempt to absorb the methane that is already being produced from landfills, experiments in which nutrients were added to the soil to allow methanotrophs to thrive have been conducted. These nutrient supplemented landfills have been shown to act as a small scale methane sink, allowing the abundance of methanotrophs to sponge the methane from the air to use as energy, effectively reducing the landfill's emissions.

To reduce emissions from the natural gas industries, the EPA developed the Natural Gas STAR Program, also known as Gas STAR.

Another program was also developed by the EPA to reduce emissions from coal mining. The Coalbed Methane Outreach Program (CMOP) helps and encourages the mining industry to find ways to use or sell methane that would otherwise be released from the coal mine into the atmosphere.

A portable methane detector has been developed which, mounted in a vehicle, can detect excess levels of methane in the ambient atmosphere and differentiate between natural methane from rotting vegetation or manure and gas leaks. As of 2013 the technology was being deployed by Pacific Gas & Electric.

Ozone

Ozone's concentration lays in a belt between the heights of 15 to 50 km of atmosphere. Instead of normal two atoms of oxygen, ozone has three atoms of oxygen formed together denoted by O3. It is formed when atmospheric oxygen molecules are broken by ultraviolet solar radiation. It may even be formed at the time of electrical discharge during thunderstorms. This gas is also termed as variable as its formation and disintegration is dependent upon numerous activities. Though ozone is very less in quantity (0.00006 percent), this thin layer is very significant for the survival of living world as it absorbs the dangerous ultraviolet rays and protects the earth. Neon, helium, krypton, hydrogen, xenon are other minor gases. Some gases are still extremely less in quantity, they are termed as trace gases. Important among them are ammonia, carbon monoxide, sulphur dioxide, nitrogen dioxide, nitrous oxide and sulphur hexafluoride etc.

Ozone in Earth's Atmosphere

The standard way to express total ozone levels (the amount of ozone in a given vertical column) in the atmosphere is by using Dobson units. Point measurements are reported as mole fractions in nmol/mol (parts per billion, ppb) or as concentrations in μg/m³. The study of ozone concentration in the atmosphere started in the 1920s.

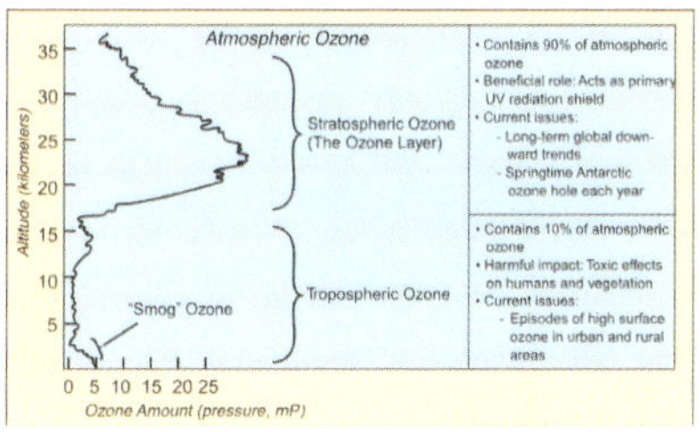

The distribution of atmospheric ozone in partial pressure as a function of altitude

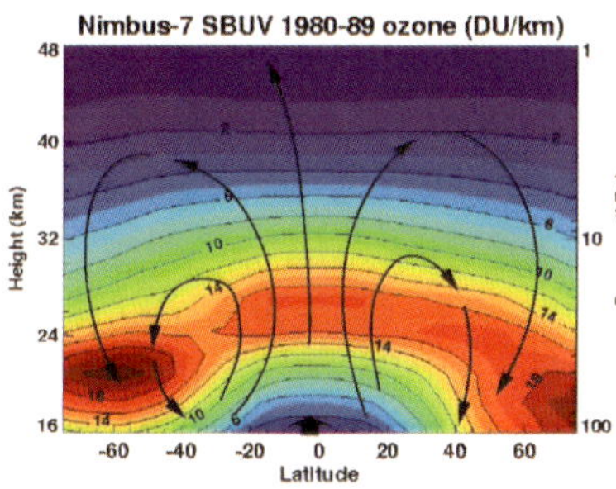

Concentration of ozone as measured by the Nimbus-7 satellite

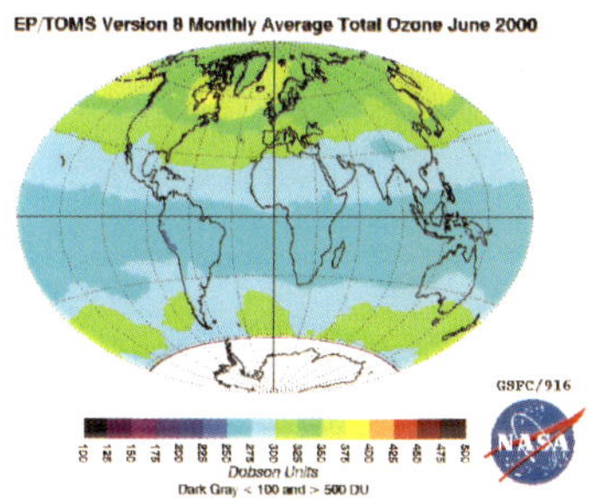

Total ozone concentration in June 2000 as measured by EP-TOMS satellite instrument

Ozone Layer

Location and Production

The highest levels of ozone in the atmosphere are in the stratosphere, in a region also known as the ozone layer between about 10 km and 50 km above the surface (or between about 6 and 31 miles). However, even in this "layer", the ozone concentrations are only two to eight parts per million, so most of the oxygen there is dioxygen, O_2, at about 210,000 parts per million by volume.

Ozone in the stratosphere is mostly produced from short-wave ultraviolet rays between 240 and 160 nm. Oxygen starts to absorb weakly at 240 nm in the Herzberg bands, but most of the oxygen is dissociated by absorption in the strong Schumann–Runge bands between 200 and 160 nm where ozone does not absorb. While shorter wavelength light, extending to even the X-Ray limit, is energetic enough to dissociate molecular oxygen, there is relatively little of it, and, the strong solar emission at Lyman-alpha, 121 nm, falls at a point where molecular oxygen absorption is a minimum.

The process of ozone creation and destruction is called the Chapman cycle and starts with the photolysis of molecular oxygen

$$O_2 + \text{photon (radiation } \lambda < 240 \text{ nm)} \rightarrow 2\,O$$

followed by reaction of the oxygen atom with another molecule of oxygen to form ozone.

$$O + O_2 + M \rightarrow O_3 + M$$

where "M" denotes the third body that carries off the excess energy of the reaction. The ozone molecule can then absorb a UV-C photon and dissociate

$$O_3 \rightarrow O + O_2 + \text{kinetic energy}$$

The excess kinetic energy heats the stratosphere when the O atoms and the molecular oxygen fly apart and collide with other molecules. This conversion of UV light into kinetic energy warms the stratosphere. The oxygen atoms produced in the photolysis of ozone then react back with other oxygen molecule as in the previous step to form more ozone. In the clear atmosphere, with only nitrogen and oxygen, ozone can react with the atomic oxygen to form two molecules of O_2

$$O_3 + O \rightarrow 2\,O_2$$

An estimate of the rate of this termination step to the cycling of atomic oxygen back to ozone can be found simply by taking the ratios of the concentration of O_2 to O_3. The termination reaction is catalysed by the presence of certain free radicals, of which the most important are hydroxyl (OH), nitric oxide (NO) and atomic chlorine (Cl) and bromine (Br). In the second half of the 20th Century the amount of ozone in the stratosphere was discovered to be declining, mostly because of increasing concentrations of chlorofluorocarbons (CFC) and similar chlorinated and brominated organic molecules. The concern over the health effects of the decline led to the 1987 Montreal Protocol, the ban on the production of many ozone depleting chemicals and in the first and second decade of the 21st Century the beginning of the recovery of stratospheric ozone concentrations.

Importance to Surface-dwelling Life on Earth

Ozone in the ozone layer filters out sunlight wavelengths from about 200 nm UV rays to 315 nm, with ozone peak absorption at about 250 nm. This ozone UV absorption is important to life, since it extends the absorption of UV by ordinary oxygen and nitrogen in air (which absorb all wavelengths < 200 nm) through the lower UV-C (200–280 nm) and the entire UV-B band (280–315 nm). The small unabsorbed part that remains of UV-B after passage through ozone causes sunburn in humans, and direct DNA damage in living tissues in both plants and animals. Ozone's effect on mid-range UV-B rays is illustrated by its effect on UV-B at 290 nm, which has a radiation intensity 350 million times as powerful at the top of the atmosphere as at the surface. Nevertheless, enough of UV-B radiation at similar frequency reaches the ground to cause some sunburn, and these same wavelengths are also among those responsible for the production of vitamin D in humans.

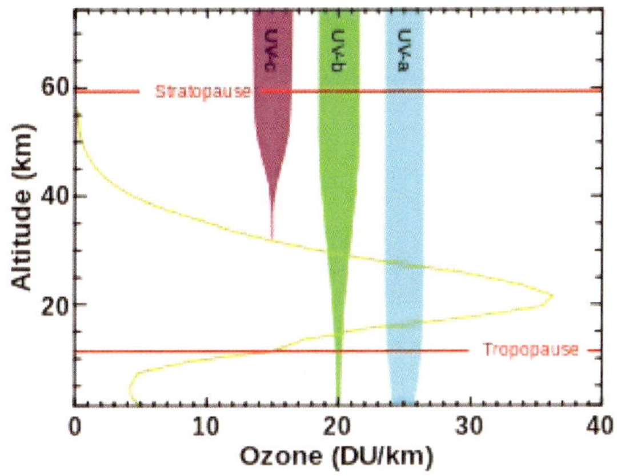

Levels of ozone at various altitudes and blocking of different bands of ultraviolet radiation. Essentially all UVC (100–280 nm) is blocked by dioxygen (at 100–200 nm) or by ozone (at 200–280 nm) in the atmosphere

The shorter portion of this band and even more energetic UV causes the formation of the ozone layer, when single oxygen atoms produced by UV photolysis of dioxygen (below 240 nm) react with more dioxygen. The ozone layer itself then blocks most, but not quite all, sunburn-producing UVB (280–315 nm). The band of UV closest to visible light, UVA (315–400 nm), is hardly affected by ozone, and most of it reaches the ground.

The ozone layer has little effect on the longer UV wavelengths called UV-A (315–400 nm), but this radiation does not cause sunburn or direct DNA damage, and while it probably does cause long-term skin damage in certain humans, it is not as dangerous to plants and to the health of surface-dwelling organisms on Earth in general.

Low Level Ozone

Low level ozone (or tropospheric ozone) is an atmospheric pollutant. It is not emitted directly by car engines or by industrial operations, but formed by the reaction of sunlight on air containing hydrocarbons and nitrogen oxides that react to form ozone directly at the source of the pollution or many kilometers down wind.

Ozone reacts directly with some hydrocarbons such as aldehydes and thus begins their removal from the air, but the products are themselves key components of smog. Ozone photolysis by UV light leads to production of the hydroxyl radical HO• and this plays a part in the removal of hydrocarbons from the air, but is also the first step in the creation of components of smog such as peroxyacyl nitrates, which can be powerful eye irritants. The atmospheric lifetime of tropospheric ozone is about 22 days; its main removal mechanisms are being deposited to the ground, the above-mentioned reaction giving HO•, and by reactions with OH and the peroxy radical HO_2•.

There is evidence of significant reduction in agricultural yields because of increased ground-level ozone and pollution which interferes with photosynthesis and stunts overall growth of some plant species. The United States Environmental Protection Agency is proposing a secondary regulation to reduce crop damage, in addition to the primary regulation designed for the protection of human health.

Certain examples of cities with elevated ozone readings are Houston, Texas, and Mexico City, Mexico. Houston has a reading of around 41 nmol/mol, while Mexico City is far more hazardous, with a reading of about 125 nmol/mol.

Ozone Cracking

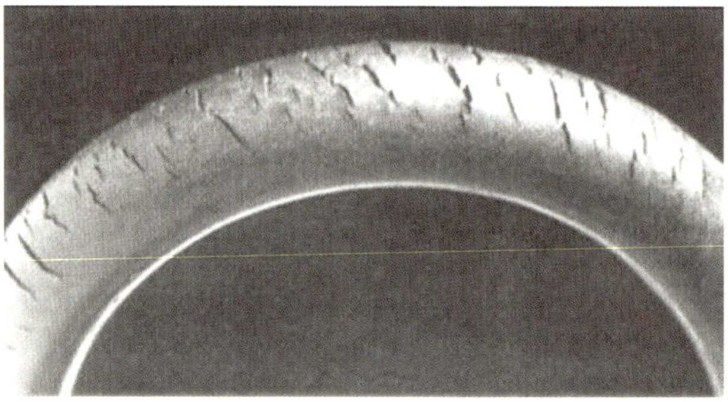

Ozone cracking in natural rubber tubing

Ozone gas attacks any polymer possessing olefinic or double bonds within its chain structure, such as natural rubber, nitrile rubber, and styrene-butadiene rubber. Products made using these polymers are especially susceptible to attack, which causes cracks to grow longer and deeper with time, the rate of crack growth depending on the load carried by the rubber component and the concentration of ozone in the atmosphere. Such materials can be protected by adding antiozonants, such as waxes, which bond to the surface to create a protective film or blend with the material and provide long term protection. Ozone cracking used to be a serious problem in car tires for example, but the problem is now seen only in very old tires. On the other hand, many critical products, like gaskets and O-rings, may be attacked by ozone produced within compressed air systems. Fuel lines made of reinforced rubber are also susceptible to attack, especially within the engine compartment, where some ozone is produced by electrical components. Storing rubber products in close proximity to a DC electric motor can accelerate ozone cracking. The commutator of the motor generates sparks which in turn produce ozone.

Ozone as a Greenhouse Gas

Although ozone was present at ground level before the Industrial Revolution, peak concentrations are now far higher than the pre-industrial levels, and even background concentrations well away from sources of pollution are substantially higher. Ozone acts as a greenhouse gas, absorbing some of the infrared energy emitted by the earth. Quantifying the greenhouse gas potency of ozone is difficult because it is not present in uniform concentrations across the globe. However, the most widely accepted scientific assessments relating to climate change (e.g. the Intergovernmental Panel on Climate Change Third Assessment Report) suggest that the radiative forcing of tropospheric ozone is about 25% that of carbon dioxide.

The annual global warming potential of tropospheric ozone is between 918–1022 tons carbon dioxide equivalent/tons tropospheric ozone. This means on a per-molecule basis, ozone in the troposphere has a radiative forcing effect roughly 1,000 times as strong as carbon dioxide. However, tropospheric ozone is a short-lived greenhouse gas, which decays in the atmosphere much more quickly than carbon dioxide. This means that over a 20-year span, the global warming potential of tropospheric ozone is much less, roughly 62 to 69 tons carbon dioxide equivalent/ton tropospheric ozone.

Because of its short-lived nature, tropospheric ozone does not have strong global effects, but has very strong radiative forcing effects on regional scales. In fact, there are regions of the world where tropospheric ozone has a radiative forcing up to 150% of carbon dioxide.

Water Vapour

Water vapour is small in amount but it is one of the most important part of atmosphere with respect to the distribution of vegetation and life. Water vapour exists all the time in the atmosphere but with varying degree of amount depending upon the season (temperature condition) and the supply of water for evaporation and evapotranspiration. Air is hardly completely dry. In summer, the water holding capacity of the air is large as the temperature is high while in winter it is low.

Availability of sufficient amount water on the earth surface or water body in an area witness greater vapour while less availability of avoidance of the same shows low vapour. Examples 8 may be taken as equatorial region rich in water bodies (high vapour) and subtropical hot desert region with less to no water availability (low vapour). Though, vapour and air both are in gaseous form, their mixing and movement are quite natural but the same homogeneity it not seen.

At any particular point of time, the amount of vapour is not more than four present of the total volume of atmosphere. It is found in the troposphere only and its concentration is in the lower level. About 90 percent of the total vapour lays below six km. It is estimated that the about 50 percent is within two km of height. Water vapour plays a vital role in keeping the earth warm as it has greenhouse characteristics.

In Earth's Atmosphere

Gaseous water represents a small but environmentally significant constituent of the atmosphere. The percentage water vapor in surface air varies from 0.01% at -42 °C (-44 °F) to 4.24% when the dew point is 30 °C (86 °F). Approximately 99.13% of it is contained in the troposphere. The condensation of water vapor to the liquid or ice phase is responsible for clouds, rain, snow, and other precipitation, all of which count among the most significant elements of what we experience as weather. Less obviously, the latent heat of vaporization, which is released to the atmosphere whenever condensation occurs, is one of the most important terms in the atmospheric energy budget on both local and global scales. For example, latent heat release in atmospheric convection is directly responsible for powering destructive storms such as tropical cyclones and severe thunderstorms. Water vapor is the most potent greenhouse gas owing to the presence of the hydroxyl bond which strongly absorbs in the infra-red region of the light spectrum.

Water in Earth's atmosphere is not merely below its boiling point (100 °C), but at altitude it goes below its freezing point (0 °C), due to water's highly polar attraction. When combined with its quantity, water vapor then has a relevant dew point and frost point, unlike e. g., carbon dioxide and methane. Water vapor thus has a scale height a fraction of that of the bulk atmosphere, as the water condenses and exits, primarily in the troposphere, the lowest layer of the atmosphere. Carbon dioxide (CO_2) and methane, being non-polar, rise above water vapor. The absorption and emission of both compounds contribute to Earth's emission to space, and thus the planetary greenhouse effect. This greenhouse forcing is directly observable, via distinct spectral features versus water vapor, and observed to be rising with rising CO_2 levels. Conversely, adding water vapor at high altitudes has a disproportionate impact, which is why methane (rising, then oxidizing to CO_2 and two water molecules) and jet traffic have disproportionately high warming effects.

It is less clear how cloudiness would respond to a warming climate; depending on the nature of the response, clouds could either further amplify or partly mitigate warming from long-lived greenhouse gases.

In the absence of other greenhouse gases, Earth's water vapor would condense to the surface; this has likely happened, possibly more than once. Scientists thus distinguish between non-condensable (driving) and condensable (driven) greenhouse gases- i. e., the above water vapor feedback.

Fog and clouds form through condensation around cloud condensation nuclei. In the absence of nuclei, condensation will only occur at much lower temperatures. Under persistent condensation or deposition, cloud droplets or snowflakes form, which precipitate when they reach a critical mass.

The water content of the atmosphere as a whole is constantly depleted by precipitation. At the same time it is constantly replenished by evaporation, most prominently from seas, lakes, rivers, and moist earth. Other sources of atmospheric water include combustion, respiration, volcanic eruptions, the transpiration of plants, and various other biological and geological processes. The mean global content of water vapor in the atmosphere is roughly sufficient to cover the surface of the planet with a layer of liquid water about 25 mm deep. The mean annual precipitation for the planet is about 1 meter, which implies a rapid turnover of water in the air – on average, the residence time of a water molecule in the troposphere is about 9 to 10 days.

Episodes of surface geothermal activity, such as volcanic eruptions and geysers, release variable amounts of water vapor into the atmosphere. Such eruptions may be large in human terms, and major explosive eruptions may inject exceptionally large masses of water exceptionally high into the atmosphere, but as a percentage of total atmospheric water, the role of such processes is minor. The relative concentrations of the various gases emitted by volcanoes varies considerably according to the site and according to the particular event at any one site. However, water vapor is consistently the commonest volcanic gas; as a rule, it comprises more than 60% of total emissions during a subaerial eruption.

Atmospheric water vapor content is expressed using various measures. These include vapor pressure, specific humidity, mixing ratio, dew point temperature, and relative humidity.

Radar and Satellite Imaging

Because water molecules absorb microwaves and other radio wave frequencies, water in the atmosphere attenuates radar signals. In addition, atmospheric water will reflect and refract signals to an extent that depends on whether it is vapor, liquid or solid.

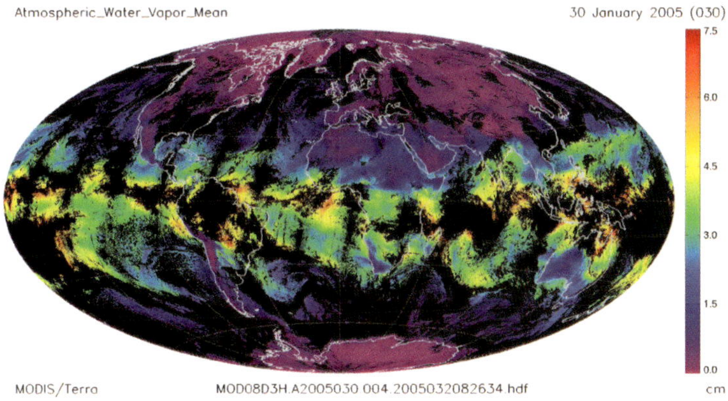

MODIS/Terra global mean atmospheric water vapor

Generally, radar signals lose strength progressively the farther they travel through the troposphere. Different frequencies attenuate at different rates, such that some components of air are opaque to some frequencies and transparent to others. Radio waves used for broadcasting and other communication experience the same effect.

Water vapor reflects radar to a lesser extent than do water's other two phases. In the form of drops and ice crystals, water acts as a prism, which it does not do as an individual molecule; however, the existence of water vapor in the atmosphere causes the atmosphere to act as a giant prism.

A comparison of GOES-12 satellite images shows the distribution of atmospheric water vapor relative to the oceans, clouds and continents of the Earth. Vapor surrounds the planet but is unevenly distributed. The image loop above shows monthly average of water vapor content with the units are given in centimeters, which is the precipitable water or equivalent amount of water that could be produced if all the water vapor in the column were to condense. The lowest amounts of water vapor (0 centimeters) appear in yellow, and the highest amounts (6 centimeters) appear in dark blue. Areas of missing data appear in shades of gray. The maps are based on data collected by the Moderate Resolution Imaging Spectroradiometer (MODIS) sensor on NASA's Aqua satellite. The most noticeable pattern in the time series is the influence of seasonal temperature changes and incoming sunlight on water vapor. In the tropics, a band of extremely humid air wobbles north and south of the equator as the seasons change. This band of humidity is part of the Intertropical Convergence Zone, where the easterly trade winds from each hemisphere converge and produce near-daily thunderstorms and clouds. Farther from the equator, water vapor concentrations are high in the hemisphere experiencing summer and low in the one experiencing winter. Another pattern that shows up in the time series is that water vapor amounts over land areas decrease more in winter months than adjacent ocean areas do. This is largely because air temperatures over land drop more in the winter than temperatures over the ocean. Water vapor condenses more rapidly in colder air.

As water vapour absorbs light in the visible spectral range, its absorption can be used in spectroscopic applications (such as DOAS) to determine the amount of water vapor in the atmosphere. This is done operationally, e.g. from the GOME spectrometers on ERS and MetOp. The weaker water vapor absorption lines in the blue spectral range and further into the UV up to its dissociation limit around 243 nm are mostly based on quantum mechanical calculations and are only partly confirmed by experiments.

Lightning Generation

Water vapor plays a key role in lightning production in the atmosphere. From cloud physics, usually clouds are the real generators of static charge as found in Earth's atmosphere. The ability of clouds to hold massive amounts of electrical energy is directly related to the amount of water vapor present in the local system.

The amount of water vapor directly controls the permittivity of the air. During times of low humidity, static discharge is quick and easy. During times of higher humidity, fewer static discharges occur. Permittivity and capacitance work hand in hand to produce the megawatt outputs of lightning.

After a cloud, for instance, has started its way to becoming a lightning generator, atmospheric water vapor acts as a substance (or insulator) that decreases the ability of the cloud to discharge its electrical energy. Over a certain amount of time, if the cloud continues to generate and store more static electricity, the barrier that was created by the atmospheric water vapor will ultimately break down from the stored electrical potential energy. This energy will be released to a locally, oppositely charged region in the form of lightning. The strength of each discharge is directly related to the atmospheric permittivity, capacitance, and the source's charge generating ability.

Dust Particles

Huge amount of dust particles are available in the lower layer of atmosphere in a suspended form. These dust particles are solid substances generated from various sources and being carried by winds. Greater velocity wind have greater amount of dust particles. It is not only a matter of suspended solid particles, they are transported to great distances as well. They varies widely in sizes. They may be big sized suspended only when the drafting ability of the wind is great, but settles swiftly when the carrying capacity of the wind is reduced. You must have observed yourselves too during gales/storms.

Huge number of microscopic dust particles are suspended even in completely calm air. Dust particles are variable and it is more during dry seasons as the soils are loose and easily carried by winds, but reverse is the case when it is rainy season when they are settled and compacted. Over the globe, it is less in equatorial and polar areas while more in subtropical hot desert areas. Minute dust particles are found several km above the surface while the coarse sized are abundant near the surface. Microscopic particles are nuclei for condensation and precipitation and they have a great importance in this respect.

Atmospheric or wind-borne fugitive dust, also known as *aeolian dust*, comes from arid and dry regions where high velocity winds are able to remove mostly silt-sized material, deflating susceptible surfaces. This includes areas where grazing, ploughing, vehicle use, and other human activities have further destabilized the land, though not all source areas have been largely affected by anthropogenic impacts. One-third of the global land area is covered by dust-producing surfaces, made up of hyper-arid regions like the Sahara which covers 0.9 billion hectares, and drylands which occupy 5.2 billion hectares.

Dust in the atmosphere is produced by saltation and sandblasting of sand-sized grains, and it is transported through the troposphere. This airborne dust is considered an aerosol and once in the atmosphere, it can produce strong local radiative forcing. Saharan dust in particular can be transported and deposited as far as the Caribbean and the Amazon basin, and may affect air temperatures, cause ocean cooling, and alter rainfall amounts.

Middle East

Dust in the Middle East has been a historic phenomenon. Recently, because of climate change and the escalating process of desertification, the problem has worsened dramatically. As a multi-factor phenomenon, there is not yet a clear consensus on the sources or potential solutions to the problem.

In Iran, the dust is already affecting more than 5 million people directly, and has emerged as a serious government issue in recent years. In the province of Khuzestan it has led to the severe reduction of air quality. The amount of pollutants in the air has surpassed more than 50 times the normal level several times in a year. Recently, initiatives such as Project-Dust have been established to directly study the Middle Eastern dust.

Roads

Dust kicked up by vehicles traveling on roads may make up 33% of air pollution. Road dust consists of deposits of vehicle exhausts and industrial exhausts, particles from tire and brake wear, dust from paved roads or potholes, and dust from construction sites. Road dust is a significant source contributing to the generation and release of particulate matter into the atmosphere. Control of road dust is a significant challenge in urban areas, and also in other locations with high levels of vehicular traffic upon unsealed roads, such as mines and landfill dumps.

Road dust may be suppressed by mechanical methods like street sweeper vehicles equipped with vacuum cleaners, vegetable oil sprays, or with water sprayers. Improvements in automotive engineering have reduced the amount of PM10s produced by road traffic; the proportion representing re-suspension of existing particulates has increased as a result.

Aerosols

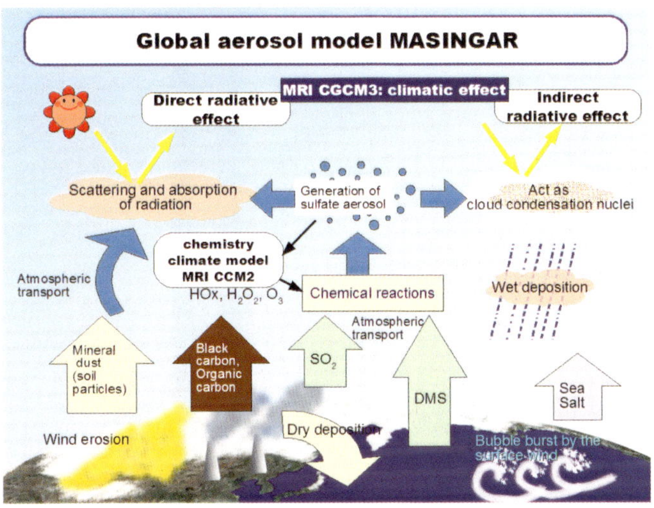

Different Sources of Aerosols

Aerosols are extremely fine-sized solid particles or liquid droplets which continue to be in suspended form in gas for very-very long time. They could be seen when their concentration is more otherwise they are invisible. Aerosols themselves are non-gaseous microscopic substance released in the atmosphere from various sources – natural and human created. They could be pollen, minute earthly dust, sea salt, carbon soot from burning fuels, volcanic dust etc. Human activities also help the aerosols to enter the atmosphere. Their concentration is more over the industrial and urban areas. Burning of fossil fuels and generation of smoke also pump the aerosols in the air. Therefore, the source of aerosols are both natural as well as human generated. They are grouped into two – hygroscopic (moisture absorbing and retaining) and non-hygroscopic (moisture non-absorbing). Hygroscopic aerosols form the nuclei for condensation and in this way, they help in precipitation.

From the above description, it is fairly obvious that our atmosphere is made up of innumerable minute molecules of several gases about which a discussion has been presented above. Apart from that several non-gaseous substances are also available in the air which are part and parcel of the atmosphere. They have their own significance and play very essential role for the earth to be a liveable planet.

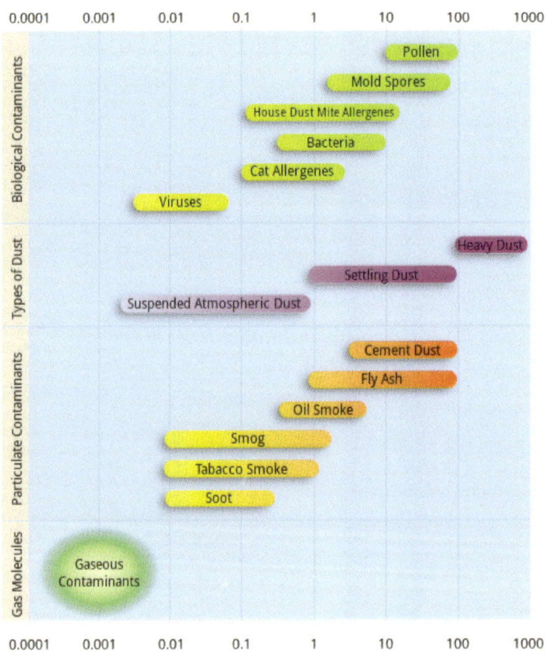

This diagram shows types, and size distribution in micrometres, of atmospheric particulate matter

Atmospheric aerosol particles, also known as atmospheric particulate matter, particulate matter (PM), particulates, or suspended particulate matter (SPM) are microscopic solid or liquid matter suspended in Earth's atmosphere. The term aerosol commonly refers to the particulate/air mixture, as opposed to the particulate matter alone. Sources of particulate matter can be natural or anthropogenic. They have impacts on climate and precipitation that adversely affect human health.

Subtypes of atmospheric particles include suspended particulate matter (SPM), thoracic and respirable particles, inhalable coarse particles, which are coarse particles with a diameter between

2.5 and 10 micrometers (μm) (PM_{10}), fine particles with a diameter of 2.5 μm or less ($PM_{2.5}$), ultrafine particles, and soot.

The IARC and WHO designate airborne particulates a Group 1 carcinogen. Particulates are the deadliest form of air pollution due to their ability to penetrate deep into the lungs and blood streams unfiltered, causing permanent DNA mutations, heart attacks, and premature death. In 2013, a study involving 312,944 people in nine European countries revealed that there was no safe level of particulates and that for every increase of 10 μg/m^3 in PM_{10}, the lung cancer rate rose 22%. The smaller $PM_{2.5}$ were particularly deadly, with a 36% increase in lung cancer per 10 μg/m^3 as it can penetrate deeper into the lungs.

Sources of Atmospheric Particulate Matter

Some particulates occur naturally, originating from volcanoes, dust storms, forest and grassland fires, living vegetation, and sea spray. Human activities, such as the burning of fossil fuels in vehicles, power plants and various industrial processes, also generate significant amounts of particulates. Coal combustion in developing countries is the primary method for heating homes and supplying energy. Because salt spray over the oceans is the overwhelmingly most common form of particulate in the atmosphere, *anthropogenic* aerosols—those made by human activities—currently account for about 10 percent of the total mass of aerosols in our atmosphere.

Composition

The composition of aerosols and particles depends on their source. Wind-blown mineral dust tends to be made of mineral oxides and other material blown from the Earth's crust; this particulate is light-absorbing. Sea salt is considered the second-largest contributor in the global aerosol budget, and consists mainly of sodium chloride originated from sea spray; other constituents of atmospheric sea salt reflect the composition of sea water, and thus include magnesium, sulfate, calcium, potassium, etc. In addition, sea spray aerosols may contain organic compounds, which influence their chemistry.

Secondary particles derive from the oxidation of primary gases such as sulfur and nitrogen oxides into sulfuric acid (liquid) and nitric acid (gaseous). The precursors for these aerosols—i.e. the gases from which they originate—may have an anthropogenic origin (from fossil fuel or coal combustion) and a natural biogenic origin. In the presence of ammonia, secondary aerosols often take the form of ammonium salts; i.e. ammonium sulfate and ammonium nitrate (both can be dry or in aqueous solution); in the absence of ammonia, secondary compounds take an acidic form as sulfuric acid (liquid aerosol droplets) and nitric acid (atmospheric gas), all of which may contribute to the health effects of particulates.

Secondary sulfate and nitrate aerosols are strong light-scatterers. This is mainly because the presence of sulfate and nitrate causes the aerosols to increase to a size that scatters light effectively.

Organic matter (OM) can be either primary or secondary, the latter part deriving from the oxidation of VOCs; organic material in the atmosphere may either be biogenic or anthropogenic.

Organic matter influences the atmospheric radiation field by both scattering and absorption. Another important aerosol type is elemental carbon (EC, also known as *black carbon*, BC): this aerosol type includes strongly light-absorbing material and is thought to yield large positive radiative forcing. Organic matter and elemental carbon together constitute the carbonaceous fraction of aerosols. Secondary organic aerosols (SOAs), tiny "tar balls" resulting from combustion products of internal combustion engines, have been identified as a danger to health.

The chemical composition of the aerosol directly affects how it interacts with solar radiation. The chemical constituents within the aerosol change the overall refractive index. The refractive index will determine how much light is scattered and absorbed.

The composition of particulate matter that generally causes visual effects such as smog consists of sulfur dioxide, nitrogen oxides, carbon monoxide, mineral dust, organic matter, and elemental carbon also known as black carbon or soot. The particles are hygroscopic due to the presence of sulfur, and SO_2 is converted to sulfate when high humidity and low temperatures are present. This causes the reduced visibility and yellow color.

Size Distribution of Particulates

Aerosol particles of natural origin (such as windblown dust) tend to have a larger radius than human-produced aerosols such as particle pollution.

Among the most obvious patterns that the size distribution time series shows is that in the planet's most southerly latitudes, nearly all the aerosols are large, but in the high northern latitudes, smaller aerosols are very abundant. Most of the Southern Hemisphere is covered by ocean, where the largest source of aerosols is natural sea salt from dried sea spray. Because the land is concentrated in the Northern Hemisphere, the amount of small aerosols from fires and human activities is greater there than in the Southern Hemisphere. Over land, patches of large-radius aerosols appear over deserts and arid regions, most prominently, the Sahara Desert in North Africa and the Arabian Peninsula, where dust storms are common. Places where human-triggered or natural fire activity is common (land-clearing fires in the Amazon from August–October, for example, or lightning-triggered fires in the forests of northern Canada in Northern Hemisphere summer) are dominated by smaller aerosols. Human-produced (fossil fuel) pollution is largely responsible for the areas of small aerosols over developed areas such as the eastern United States and Europe, especially in their summer.

Satellite measurements of aerosols, called aerosol optical thickness, are based on the fact that the particles change the way the atmosphere reflects and absorbs visible and infrared light.

Deposition Processes

In general, the smaller and lighter a particle is, the longer it will stay in the air. Larger particles (greater than 10 micrometers in diameter) tend to settle to the ground by gravity in a matter of hours whereas the smallest particles (less than 1 micrometer) can stay in the atmosphere for weeks and are mostly removed by precipitation. Diesel particulate matter is highest near the source of emission. Any info regarding DPM and the atmosphere, flora, height, and distance from major sources would be useful to determine health effects.

Climate Effects

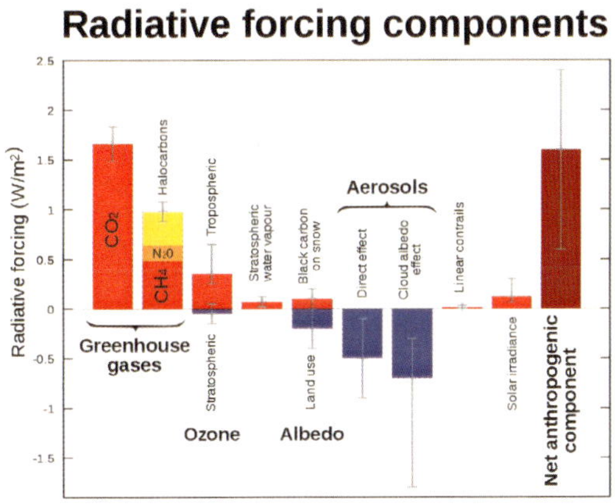

Radiative forcings and uncertainties as estimated by the IPCC

Atmospheric aerosols affect the climate of the earth by changing the amount of incoming solar radiation and outgoing terrestrial longwave radiation retained in the earth's system. This occurs through several distinct mechanisms which are split into direct, indirect and semi-direct aerosol effects. The aerosol climate effects are the biggest source of uncertainty in future climate predictions. The Intergovernmental Panel on Climate Change, Third Assessment Report, says: *While the radiative forcing due to greenhouse gases may be determined to a reasonably high degree of accuracy... the uncertainties relating to aerosol radiative forcings remain large, and rely to a large extent on the estimates from global modelling studies that are difficult to verify at the present time.*

Aerosol Radiative Effects

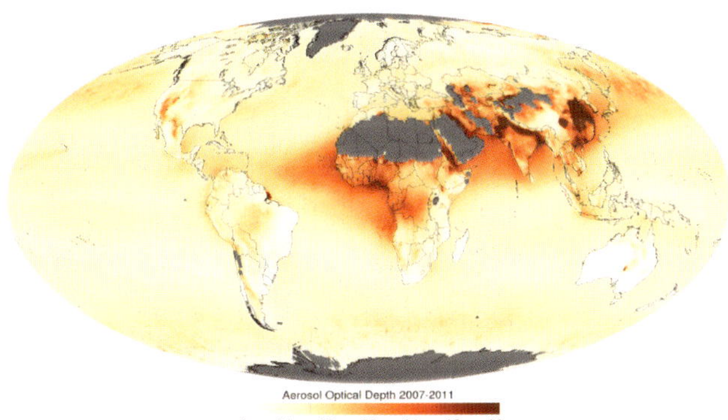

Global aerosol optical thickness. The aerosol scale (yellow to dark reddish-brown) indicates the relative amount of particles that absorb sunlight

Direct Effect

The direct aerosol effect consists of any direct interaction of radiation with atmospheric aerosols, such as absorption or scattering. It affects both short and longwave radiation to produce a net negative radiative forcing. The magnitude of the resultant radiative forcing due to the direct effect of an aerosol is dependent on the albedo of the underlying surface, as this affects the net amount of radiation absorbed or scattered to space. e.g. if a highly scattering aerosol is above a surface of low albedo it has a greater radiative forcing than if it was above a surface of high albedo. The converse is true of absorbing aerosol, with the greatest radiative forcing arising from a highly absorbing aerosol over a surface of high albedo. The direct aerosol effect is a first order effect and is therefore classified as a radiative forcing by the IPCC. The interaction of an aerosol with radiation is quantified by the single-scattering albedo (SSA), the ratio of scattering alone to scattering plus absorption (*extinction*) of radiation by a particle. The SSA tends to unity if scattering dominates, with relatively little absorption, and decreases as absorption increases, becoming zero for infinite absorption. For example, the sea-salt aerosol has an SSA of 1, as a sea-salt particle only scatters, whereas soot has an SSA of 0.23, showing that it is a major atmospheric aerosol absorber.

Particulates in the air causing shades of grey and pink in Mumbai during sunset

Indirect Effect

The Indirect aerosol effect consists of any change to the earth's radiative budget due to the modification of clouds by atmospheric aerosols, and consists of several distinct effects. Cloud droplets form onto pre-existing aerosol particles, known as cloud condensation nuclei (CCN).

For any given meteorological conditions, an increase in CCN leads to an increase in the number of cloud droplets. This leads to more scattering of shortwave radiation i.e. an increase in the albedo of the cloud, known as the Cloud albedo effect, First indirect effect or Twomey effect. Evidence supporting the cloud albedo effect has been observed from the effects of ship exhaust plumes and biomass burning on cloud albedo compared to ambient clouds. The Cloud albedo aerosol effect is a first order effect and therefore classified as a radiative forcing by the IPCC.

An increase in cloud droplet number due to the introduction of aerosol acts to reduce the cloud droplet size, as the same amount of water is divided into more droplets. This has the effect of suppressing precipitation, increasing the cloud lifetime, known as the cloud lifetime aerosol effect, second indirect effect or Albrecht effect. This has been observed as the suppression of drizzle in ship exhaust plume compared to ambient clouds, and inhibited precipitation in biomass burning plumes. This cloud lifetime effect is classified as a climate feedback (rather than a radiative forcing) by the IPCC due to the interdependence between it and the hydrological cycle. However, it has previously been classified as a negative radiative forcing.

Semi-direct Effect

The Semi-direct effect concerns any radiative effect caused by absorbing atmospheric aerosol such as soot, apart from direct scattering and absorption, which is classified as the direct effect. It encompasses many individual mechanisms, and in general is more poorly defined and understood than the direct and indirect aerosol effects. For instance, if absorbing aerosols are present in a layer aloft in the atmosphere, they can heat surrounding air which inhibits the condensation of water vapour, resulting in less cloud formation. Additionally, heating a layer of the atmosphere relative to the surface results in a more stable atmosphere due to the inhibition of atmospheric convection. This inhibits the convective uplift of moisture, which in turn reduces cloud formation. The heating of the atmosphere aloft also leads to a cooling of the surface, resulting in less evaporation of surface water. The effects described here all lead to a reduction in cloud cover i.e. an increase in planetary albedo. The semi-direct effect classified as a climate feedback) by the IPCC due to the interdependence between it and the hydrological cycle. However, it has previously been classified as a negative radiative forcing.

Roles of Different Aerosol Species

Sulfate Aerosol

Sulfate aerosol has two main effects, direct and indirect. The direct effect, via albedo, is a cooling effect that slows the overall rate of global warming: the IPCC's best estimate of the radiative forcing is −0.4 watts per square meter with a range of −0.2 to −0.8 W/m^2 but there are substantial uncertainties. The effect varies strongly geographically, with most cooling believed to be at and downwind of major industrial centres. Modern climate models addressing the attribution of recent climate change take into account sulfate forcing, which appears to account (at least partly) for the slight drop in global temperature in the middle of the 20th century. The indirect effect (via the aerosol acting as cloud condensation nuclei, CCN, and thereby modifying the cloud properties -albedo and lifetime-) is more uncertain but is believed to be a cooling.

Black Carbon

Black carbon (BC), or carbon black, or elemental carbon (EC), often called soot, is composed of pure carbon clusters, skeleton balls and buckyballs, and is one of the most important absorbing aerosol species in the atmosphere. It should be distinguished from organic carbon (OC): clustered or aggregated organic molecules on their own or permeating an EC buckyball. BC from fossil fuels is estimated by the IPCC in the Fourth Assessment Report of the IPCC, 4AR, to contribute a global mean radiative forcing of +0.2 W/m^2 (was +0.1 W/m^2 in the Second Assessment Report of the IPCC, SAR), with a range +0.1 to +0.4 W/m^2. Bond et al., however, states that "the best estimate for the industrial-era (1750 to 2005) direct radiative forcing of atmospheric black carbon is +0.71 W/m^2 with 90% uncertainty bounds of (+0.08, +1.27) W/m^2" with "total direct forcing by all black carbon sources, without subtracting the preindustrial background, is estimated as +0.88 (+0.17, +1.48) W/m^2".

Instances of Aerosol Affecting Climate

Volcanoes are a large natural source of aerosol and have been linked to changes in the earth's climate often with consequences for the human population. Eruptions linked to changes in climate include the 1600 eruption of Huaynaputina which was linked to the Russian famine of 1601 - 1603, leading to the deaths of two million, and the 1991 eruption of Mount Pinatubo which caused a global cooling of approximately 0.5 °C lasting several years. Research tracking the effect of light-scattering aerosols in the stratosphere during 2000 and 2010 and comparing its pattern to volcanic activity show a close correlation. Simulations of the effect of anthropogenic particles showed little influence at present levels.

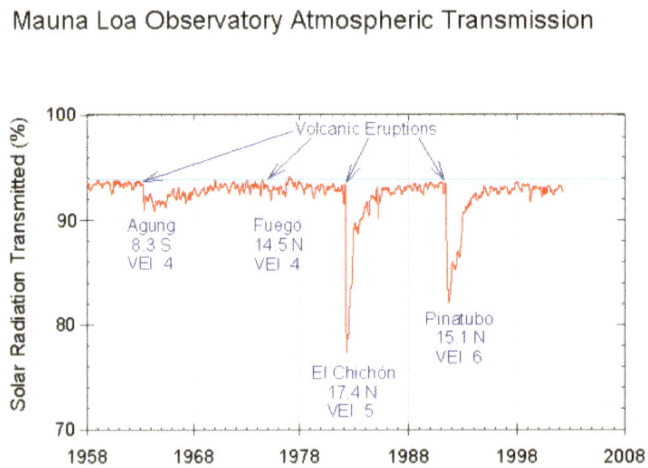

Solar radiation reduction due to volcanic eruptions

Aerosols are also thought to affect weather and climate on a regional scale. The failure of the Indian Monsoon has been linked to the suppression of evaporation of water from the Indian Ocean due to the semi-direct effect of anthropogenic aerosol.

Recent studies of the Sahel drought and major increases since 1967 in rainfall over the Northern Territory, Kimberley, Pilbara and around the Nullarbor Plain have led some scientists to conclude that the aerosol haze over South and East Asia has been steadily shifting tropical rainfall in both hemispheres southward.

The latest studies of severe rainfall decline over southern Australia since 1997 have led climatologists there to consider the possibility that these Asian aerosols have shifted not only tropical but also midlatitude systems southward.

Structure of Atmosphere

Structure means the arrangement of different part into one. In another words, it is the skeleton or organization or anatomy of a whole by looking at the relationships with its parts. According this background, the study of different parts of the atmosphere and the relationship with its parts is said to be the structure of the atmosphere. Vertically, the atmosphere is divided into different layers/parts. Therefore, the study of different layers is known as structure of atmosphere.

Based on chemical composition, the atmosphere is classified into two. They are homosphere and heterosphere.

Homosphere and Heterosphere

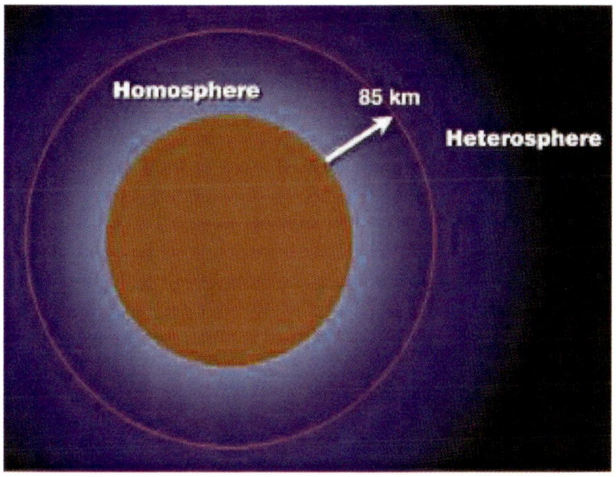

Homosphere and Heterosphere

Homosphere is that part of atmosphere where the chemical composition of the air is uniform or similar. It is the lowest layer in terms of chemical composition. It extends from the earth's/ocean surface to about 85 km. The density of the air changes very rapidly with increasing altitude but the proportion of the major gases found there remain alike throughout this layer with the exception of water vapour, pollutants, ozone and some trace/very minor gases.

On the basis of the changes in temperature, the atmosphere is divided into five layers out of them, three lower layers falls under homosphere (i.e. within 85 km of altitude). They are troposphere, stratosphere and mesosphere.

The atmosphere laying beyond the homosphere is termed as heterosphere. The term itself is self-explanatory and it is used for that part of atmosphere where the air is not uniform. In this part of atmosphere, the air is rare and the molecules are wide apart. Relatively heavier gas molecules are concentrated in the lower part whereas the lighter are forced to be above.

Beyond 85 km height, the composition of the atmosphere with increasing altitude vary significantly. The mixing of the gases are not possible as the turbulence is not happening there. Different layers of prominently different gases are nitrogen layer, oxygen layer, helium layer and hydrogen layer are differentiated. However, the heterosphere, is divided into two main spheres – thermosphere and exosphere.

Troposphere

It is the lowest and densest layer of the atmosphere. It extends till a height of about 8 km over pole but over equator, it is 18 km. About 80 percent of the total mass of the atmosphere lays in this layer. With increase in height, the temperature keeps on declining till the limit of this layer. On an average, the decrease in temperature with height is 60Celsius par km. The upper boundary is known as troposphere laying between 8 and 18 km. At this level, the average temperature reaches to minus 500 to minus 600 Celsius. Water vapour is found in this layer in abundance and about 99 percent of the total atmospheric water vapour is concentrated here but wide variation is seen in terms of height and longitudes. Vapour plays very vital role in regulating the temperature of the earth by creating greenhouse effect.

All weather phenomena are occurring in this layer only. Troposphere is the home of all types of clouds, atmospheric turbulence and mixing of the air. Both horizontal and vertical mixing is quite prominent here. In fact, the term troposphere is derived from the Greek word 'tropos'means 'turn'. Sphere is signifying 'ball' or a structure which is round in shape attaining a three-dimensional space. Therefore, the troposphere is a three-dimensional object with turning or mixing characteristics.Every sort of living life is confined to the biosphere which include land water and air. The upper limit of troposphere is tropopause which is a transition zone another upper layer known as stratosphere.

The word *troposphere* derives from the Greek: "tropos" for "turn, turn toward, trope" and "-sphere" (as in, the Earth), reflecting the fact that rotational turbulent mixing plays an important role in the troposphere's structure and behaviour. Most of the phenomena associated with day-to-day weather occur in the troposphere.

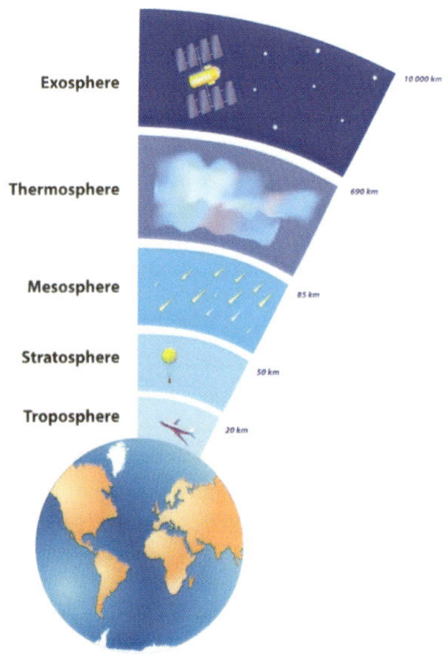

Different Layers of Atmosphere

Pressure and Temperature Structure

A view of Earth's troposphere from an airplane

Composition

By volume, dry air contains 78.08% nitrogen, 20.95% oxygen, 0.93% argon, 0.04% carbon dioxide, and small amounts of other gases. Air also contains a variable amount of water vapor. Except for the water vapor content, the composition of the troposphere is essentially uniform. The source of water vapor is at the Earth's surface through the process of evaporation. The temperature of the troposphere decreases with altitude. And, saturation vapor pressure decreases strongly as temperature drops. Hence, the amount of water vapor that can exist in the atmosphere decreases strongly with altitude and the proportion of water vapor is normally greatest near the surface of the Earth.

Pressure

The pressure of the atmosphere is maximum at sea level and decreases with altitude. This is because the atmosphere is very nearly in hydrostatic equilibrium so that the pressure is equal to the weight of air above a given point. The change in pressure with altitude can be equated to the density with the hydrostatic equation

$$\frac{dP}{dz} = -\rho g_n = -\frac{mPg_n}{RT}$$

where:

- g_n is the standard gravity
- ρ is the density
- z is the altitude
- P is the pressure
- R is the gas constant
- T is the thermodynamic (absolute) temperature
- m is the molar mass

Temperature

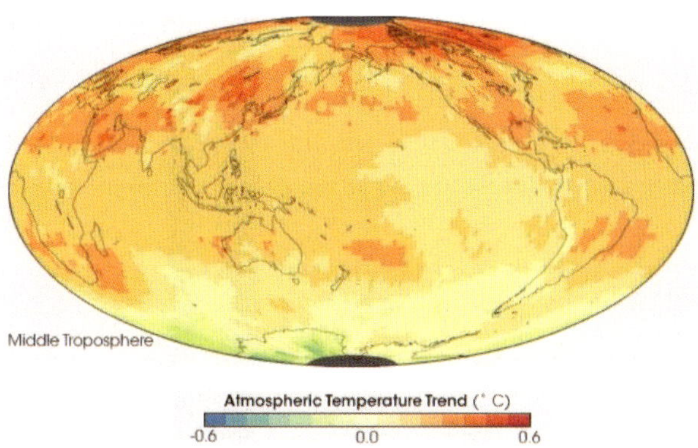

This image shows the temperature trend in the Middle Troposphere as measured by a series of satellite-based instruments between January 1979 and December 2005

The middle troposphere is centered around 5 kilometers above the surface. Oranges and yellows dominate the troposphere image, indicating that the air nearest the Earth's surface warmed during the period.

The temperature of the troposphere generally decreases as altitude increases. The rate at which the temperature decreases, $-dT / dz$, is called the environmental lapse rate (ELR). The ELR is nothing more than the difference in temperature between the surface and the tropopause divided by the height. The ELR assumes that the air is perfectly still, i.e. that there is no mixing of the layers of air from vertical convection, nor winds that would create turbulence and hence mixing of the layers of air. The reason for this temperature difference is that the ground absorbs most of the sun's energy, which then heats the lower levels of the atmosphere with which it is in contact. Meanwhile, the radiation of heat at the top of the atmosphere results in the cooling of that part of the atmosphere.

Environmental Lapse Rate (ELR)		
Altitude Region	**Lapse rate**	**Lapse Rate**
(m)	**(Kelvin/km)**	**(°F/1000 feet)**
0 – 11,000	−6.5	−3.57
11,000 – 20,000	0.0	0.0
20,000 – 32,000	1.0	0.55
32,000 – 47,000	2.8	1.54
47,000 – 51,000	0.0	0.0
51,000 – 71,000	−2.8	−1.54
71,000 – 85,000	−2.0	−1.09

The ELR assumes the atmosphere is still, but as air is heated it becomes buoyant and rises. The dry adiabatic lapse rate accounts for the effect of the expansion of dry air as it rises in the atmosphere and wet adiabatic lapse rates includes the effect of the condensation of water vapor on the lapse rate.

When a parcel of air rises, it expands, because the pressure is lower at higher altitudes. As the air parcel expands, it pushes the surrounding air outward, transferring energy in the form of work from that parcel to the atmosphere. As energy transfer to a parcel of air by way of heat is very slow, it is assumed to not exchange energy by way of heat with the environment. Such a process is called an adiabatic process (no energy transfer by way of heat). Since the rising parcel of air is losing energy as it does work on the surrounding atmosphere and no energy is transferred into it as heat from the atmosphere to make up for the loss, the parcel of air is losing energy, which manifests itself as a decrease in the temperature of the air parcel. The reverse, of course, will be true for a parcel of air that is sinking and is being compressed.

Since the process of compression and expansion of an air parcel can be considered reversible and no energy is transferred into or out of the parcel, such a process is considered isentropic, meaning that there is no change in entropy as the air parcel rises and falls, $dS = 0$. Since the heat exchanged $dQ = 0$ is related to the entropy change dS by $dQ = TdS$, the equation governing the temperature as a function of height for a thoroughly mixed atmosphere is

$$\frac{dS}{dz} = 0$$

where S is the entropy. The above equation states that the entropy of the atmosphere does not change with height. The rate at which temperature decreases with height under such conditions is called the adiabatic lapse rate.

For *dry* air, which is approximately an ideal gas, we can proceed further. The adiabatic equation for an ideal gas is

$$p(z)T(z)^{-\frac{\gamma}{\gamma-1}} = \text{constant}$$

where γ is the heat capacity ratio (γ =7/5, for air). Combining with the equation for the pressure, one arrives at the dry adiabatic lapse rate,

$$\frac{dT}{dz} = -\frac{mg}{R}\frac{\gamma-1}{\gamma} = -9.8^{\circ}\text{C} / \text{km}$$

If the air contains water vapor, then cooling of the air can cause the water to condense, and the behavior is no longer that of an ideal gas. If the air is at the saturated vapor pressure, then the rate at which temperature drops with height is called the saturated adiabatic lapse rate. More generally, the actual rate at which the temperature drops with altitude is called the environmental lapse rate. In the troposphere, the average environmental lapse rate is a drop of about 6.5 °C for every 1 km (1,000 meters) in increased height.

The environmental lapse rate (the actual rate at which temperature drops with height, dT / dz) is not usually equal to the adiabatic lapse rate (or correspondingly, $dS / dz \neq 0$). If the upper air is warmer than predicted by the adiabatic lapse rate ($dS / dz > 0$), then when a parcel of air rises and expands, it will arrive at the new height at a lower temperature than its surroundings. In this case, the air parcel is denser than its surroundings, so it sinks back to its original height, and the air is stable against being lifted. If, on the contrary, the upper air is cooler than predicted by the adiabatic lapse rate, then when the air parcel rises to its new height it will have a higher temperature and a lower density than its surroundings, and will continue to accelerate upward.

The troposphere is heated from below by latent heat, longwave radiation, and sensible heat. Surplus heating and vertical expansion of the troposphere occurs in the tropics. At middle latitudes, tropospheric temperatures decrease from an average of 15 °C at sea level to about −55 °C at the tropopause. At the poles, tropospheric temperature only decreases from an average of 0 °C at sea level to about −45 °C at the tropopause. At the equator, tropospheric temperatures decrease from an average of 20 °C at sea level to about −70 to −75 °C at the tropopause. The troposphere is thinner at the poles and thicker at the equator. The average thickness of the tropical tropopause is roughly 7 kilometers greater than the average tropopause thickness at the poles.

Tropopause

The tropopause is the boundary region between the troposphere and the stratosphere.

Measuring the temperature change with height through the troposphere and the stratosphere identifies the location of the tropopause. In the troposphere, temperature decreases with altitude. In the stratosphere, however, the temperature remains constant for a while and then increases with altitude. This coldest layer of the atmosphere, where the lapse rate changes from positive (in the troposphere) to negative (in the stratosphere), is defined as the tropopause. Thus, the tropopause is an inversion layer, and there is little mixing between the two layers of the atmosphere.

Atmospheric Flow

The flow of the atmosphere generally moves in a west to east direction. This, however, can often become interrupted, creating a more north to south or south to north flow. These scenarios are often described in meteorology as zonal or meridional. These terms, however, tend to be used in reference to localised areas of atmosphere (at a synoptic scale). A fuller explanation of the flow of atmosphere around the Earth as a whole can be found in the three-cell model.

Zonal Flow

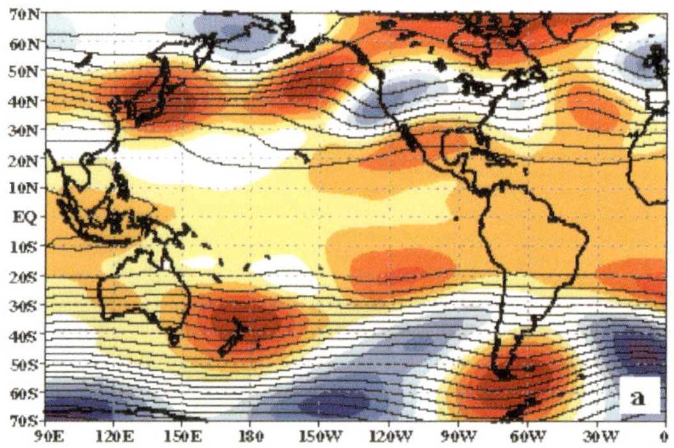

A zonal flow regime. Note the dominant west-to-east flow as shown in the 500 hPa height pattern

A zonal flow regime is the meteorological term meaning that the general flow pattern is west to east along the Earth's latitude lines, with weak shortwaves embedded in the flow. The use of the word "zone" refers to the flow being along the Earth's latitudinal "zones". This pattern can buckle and thus become a meridional flow.

Meridional Flow

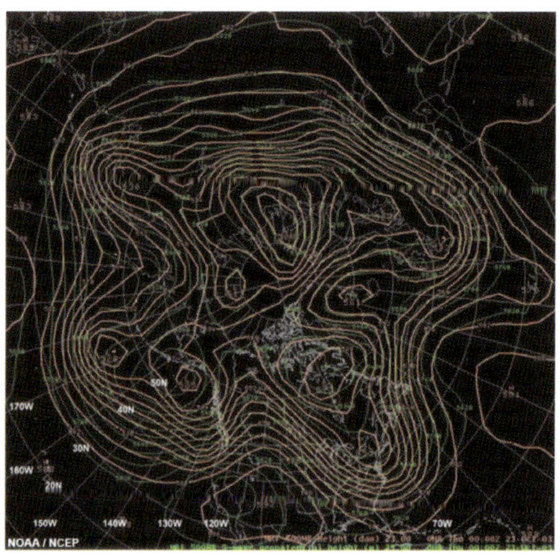

Meridional Flow pattern. Note the amplified troughs and ridges in this 500 hPa height pattern

When the zonal flow buckles, the atmosphere can flow in a more longitudinal (or meridional) direction, and thus the term "meridional flow" arises. Meridional flow patterns feature strong, amplified troughs of low pressure and ridges of high pressure, with more north-south flow in the general pattern than west-to-east flow.

Three-cell Model

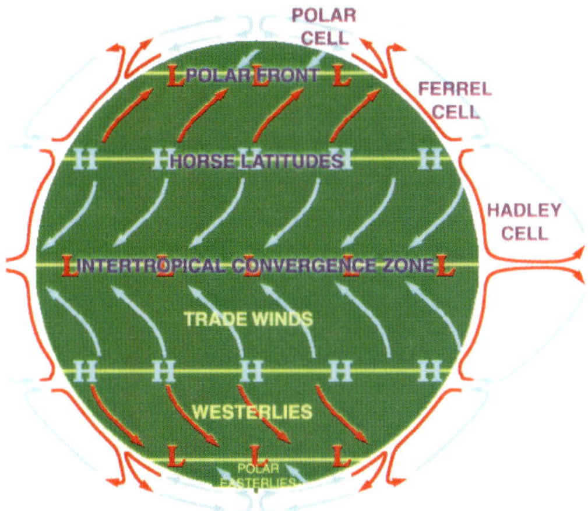

Atmospheric circulation shown with three large cells

The three cells model of the atmosphere attempts to describe the actual flow of the Earth's atmosphere as a whole. It divides the Earth into the tropical (Hadley cell), mid latitude (Ferrel cell), and polar (polar cell) regions, to describe energy flow and global atmospheric circulation (mass flow). Its fundamental principle is that of balance – the energy that the Earth absorbs from the sun each year is equal to that which it loses to space by radiation. This overall Earth energy balance, however, does not apply in each latitude due to the varying strength of the sun in each "cell" as a result of the tilt of the Earth's axis in relation to its orbit. The result is a circulation of the atmosphere that transports warm air poleward from the tropics and cold air equatorward from the poles. The effect of the three cells is the tendency to even out the heat and moisture in the Earth's atmosphere around the planet.

Synoptic Scale Observations and Concepts

Forcing

Forcing is a term used by meteorologists to describe the situation where a change or an event in one part of the atmosphere causes a strengthening change in another part of the atmosphere. It is usually used to describe connections between upper, middle or lower levels (such as upper-level divergence causing lower level convergence in cyclone formation), but also be to describe such connections over lateral distance rather than height alone. In some respects, teleconnections could be considered a type of forcing.

Divergence and Convergence

An area of convergence is one in which the total mass of air is increasing with time, resulting in an increase in pressure at locations below the convergence level (recall that atmospheric pressure is just the total weight of air above a given point). Divergence is the opposite of convergence – an area where the total mass of air is decreasing with time, resulting in falling pressure in regions below the area of divergence. Where divergence is occurring in the upper atmosphere, there will be air coming in to try to balance the net loss of mass (this is called the principle of mass conservation), and there is a resulting upward motion (positive vertical velocity). Another way to state this is to say that regions of upper air divergence are conducive to lower level convergence, cyclone formation, and positive vertical velocity. Therefore, identifying regions of upper air divergence is an important step in forecasting the formation of a surface low pressure area.

Stratosphere

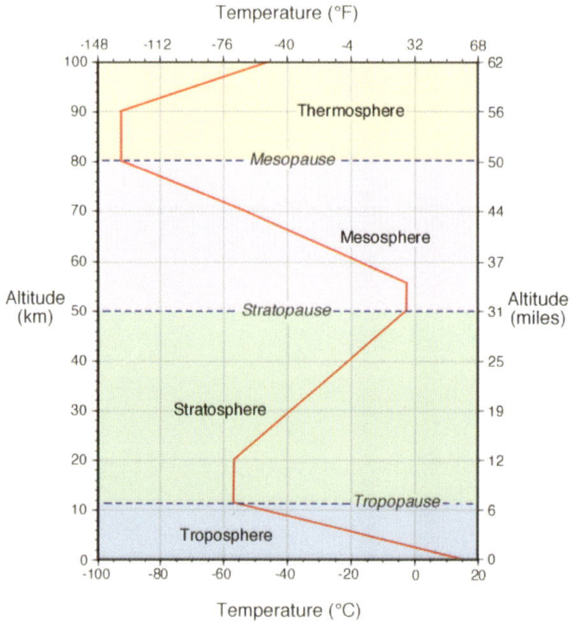

State of Temperature in Different Layers of Atmosphere

Stratosphere is the upward second layer as well as middle layer of the homosphere. It starts from tropopause to approximate height of 50 km. The temperature at the tropopause remains almost constant till the height of 20 km. After that, it starts increasing and continue the trend till the height of 50 km. At this level, the estimated temperature is about minus 100 to minus 150 Celsius. Though the temperature is on rise, but there is no atmospheric turbulence. This layer is completely free from clouds and other weather conditions. That is why, it has an advantage for flying long-distance supersonic jets/ aeroplanes through this layer. The increase in temperature in this layer is caused by absorption of solar radiation by ozone (O_3). Ozone is abundant in this layer and its 90 percent is concentrated (found between 15 km to 50 km) in this layer only. The upper limit is stratopausewhich is a very narrow strip of transition zone beyond which mesosphere is found.

Stratosphere is also termed as ozonosphere, a layer made up of ozone. This layer protects the living world of the planet from the harmful effects ultra-violet rays. The ozone depletion and hole over Antarctic was discovered in 1985. Since then, we have observed that this layer has reached to a dangerous level of depletion of ozone. The main ozone depleting substances are received from the release of chlorofluorocarbon (CFCs), hydro-chlorofluorocarbon (HCFCs) and carbon tetrachloride.

When the molecules of chlorine atom come into contact with ozone (O_3), it turns O3into O_2 which is normal oxygen. One chlorine atom can destroy more than a lakh atoms of ozone. Therefore, the chain of depletion is very serious. O_3 turned into O_2, an ordinary oxygen molecule is not capable to absorbing ultra-violet rays.

Ozone and Temperature

The mechanism describing the formation of the ozone layer was described by British mathematician Sydney Chapman in 1930. Molecular oxygen absorbs high energy sunlight in the UV-C region, at wavelengths shorter than about 240 nm. The oxygen atoms produced combine with molecular oxygen to form ozone. Ozone in turn is photolysed much more rapidly than molecular oxygen as it has a stronger absorption that occurs at longer wavelengths, where the solar emission is more intense. Ozone (O_3) photolysis produces O and O_2. The oxygen atom product combines with atmospheric molecular oxygen to reform O_3, releasing heat. The rapid photolysis and reformation of ozone heats the stratosphere resulting in a temperature inversion. This increase of temperature with altitude is characteristic of the stratosphere; its resistance to vertical mixing means that it is stratified. Within the stratosphere temperatures increase with altitude; the top of the stratosphere has a temperature of about 270 K (−3°C or 26.6°F). This vertical stratification, with warmer layers above and cooler layers below, makes the stratosphere dynamically stable: there is no regular convection and associated turbulence in this part of the atmosphere. However, exceptionally energetic convection processes, such as volcanic eruption columns and overshooting tops in severe supercell thunderstorms, may carry convection into the stratosphere on a very local and temporary basis. Overall the attenuation of solar UV at wavelengths that damage DNA by the ozone layer allows life to exist on the surface of the planet outside of the ocean. All air entering the stratosphere must pass through the tropopause, the temperature minimum that divides the troposphere and stratosphere. The rising air is literally freeze dried; the stratosphere is a very dry place. The top of the stratosphere is called the stratopause, above which the temperature decreases with height.

Sydney Chapman gave a correct description of the source of stratospheric ozone and its ability to generate heat within the stratosphere; he also wrote that ozone may be destroyed by reacting with atomic oxygen, making two molecules of molecular oxygen. We now know that there are additional ozone loss mechanisms, and that these mechanisms are catalytic meaning that a small amount of the catalyst can destroy a great number of ozone molecules. The first is due to the reaction of hydroxyl radicals OH• with ozone. OH is formed by the reaction of electronically excited oxygen atoms produced by ozone photolysis, with water vapor. While the stratosphere is dry, additional water vapor is produced in situ by the photochemical oxidation of methane (CH_4). The HO_2 radical produced by the reaction of OH with O_3 is recycled to OH by reaction with oxygen atoms or ozone. In addition, solar proton events can significantly effect ozone levels via Radiolysis with the subsequent formation of OH . Laughing gas or Nitrous oxide (N_2O) is produced by biological activity at the surface and is oxidised to NO in the stratosphere; the so-called NOx radical cycles also deplete stratospheric ozone. Finally, chlorofluorocarbon molecules are photolysed in the stratosphere releasing chlorine atoms that react with ozone giving ClO and O_2. The chlorine atoms are recycled when ClO reacts with O in the upper stratosphere, or when ClO reacts with itself in the chemistry of the Antarctic ozone hole. Paul J. Crutzen, Mario J. Molina and F. Sherwood Rowland were awarded the Nobel Prize in Chemistry in 1995 for their work describing the formation and decomposition of stratospheric ozone.

Aircraft Flight

Commercial airliners typically cruise at altitudes of 9–12 km (30,000–39,000 ft) which is in the lower reaches of the stratosphere in temperate latitudes. This optimizes fuel efficiency, mostly due to the low temperatures encountered near the tropopause and low air density, reducing parasitic drag on the airframe. Stated another way, it allows the airliner to fly faster while maintaining lift equal to the weight of the plane. (The fuel consumption depends on the drag, which is related to the lift by the lift-to-drag ratio.) It also allows them to stay above the turbulent weather of the troposphere.

The Concorde aircraft cruised at mach 2 at about 18,000 m (59,000 ft), and the SR-71 cruised at mach 3 at 26,000 m (85,000 ft), all within the stratosphere.

Because the temperature in the tropopause and lower stratosphere is largely constant with increasing altitude, very little convection and its resultant turbulence occurs there. Most turbulence at this altitude is caused by variations in the jet stream and other local wind shears, although areas of significant convective activity (thunderstorms) in the troposphere below may produce turbulence as a result of convective overshoot.

On October 24, 2014, Alan Eustace became the record holder for reaching the altitude record for a manned balloon at 135,890 ft (41,419 m). Dr Eustace also broke the world records for vertical speed skydiving, reached with a peak velocity of 1,321 km/h (822 mph) and total freefall distance of 123,414 ft (37,617 m) – lasting four minutes and 27 seconds.

Circulation and Mixing

The stratosphere is a region of intense interactions among radiative, dynamical, and chemical processes, in which the horizontal mixing of gaseous components proceeds much more rapidly than does vertical mixing. The overall circulation of the stratosphere is termed as Brewer-Dobson circulation, which is a single celled circulation, spanning from the tropics up to the poles, consisting of the tropical upwelling of air from the tropical troposphere and the extra-tropical downwelling of air. Stratospheric Circulation is a pre-dominantly wave-driven circulation in that the tropical upwelling is induced by the wave force by the westward propagating Rossby Waves, in a phenomenon called Rossby-Wave pumping.

An interesting feature of stratospheric circulation is the QBO in the tropical latitudes, which is driven by gravity waves that are convectively generated in the troposphere. The QBO induces a secondary circulation that is important for the global stratospheric transport of tracers, such as ozone or water vapor.

Another large-scale feature that significantly influences Stratospheric Circulation is the breaking planetary waves resulting in intense quasi-horizontal mixing in the midlatitudes. This breaking is much more pronounced in the winter hemisphere where this region is called the surf zone. This breaking is caused due to a highly non-linear interaction between the vertically propagating planetary waves and the isolated high potential vorticity region known as Polar Vortex. The resultant breaking causes large scale mixing of air and other trace gases throughout the midlatitude surf zone. The timescale of this rapid mixing is much smaller than the much slower timescales of upwelling in the tropics and downwelling in the extratropics.

During northern hemispheric winters, sudden stratospheric warmings, caused by the absorption of Rossby waves in the stratosphere, can be observed in approximately half of winters when easterly winds develop in the stratosphere. These events often precede unusual winter weather and may even be responsible for the cold European winters of the 1960s.

Stratospheric warming of the polar vortex results in its weakening. When the vortex is strong, it keeps the cold, high pressure air masses *contained* in the arctic; when the vortex weakens, air masses move equatorward, and results in rapid changes of weather in the mid latitudes.

Life

Bacteria

Bacterial life survives in the stratosphere, making it a part of the biosphere. In 2001 an Indian experiment, involving a high-altitude balloon, was carried out at a height of 41 kilometres and a sample of dust was collected with bacterial material inside.

Birds

Some bird species have been reported to fly at the lower levels of the stratosphere. On November 29, 1973, a Rüppell's vulture (*Gyps rueppelli*) was ingested into a jet engine 11,278 m (37,000 ft) above the Ivory Coast, and bar-headed geese (*Anser indicus*) reportedly overfly Mount Everest's summit, which is 8,848 m (29,029 ft).

Discovery

Léon Teisserenc de Bort from France and Richard Assmann from Germany, in separate publications and following years of observations, announced the discovery of an isothermal layer at around 11–14 km, which is the base of the lower stratosphere. This was based on temperature profiles from unmanned instrumented balloons.

Mesosphere

Noctilucent Clouds

Mesosphere is the third but the upper-most layer of the homosphere. After this layer, heterosphere starts. The literal meaning of mesosphere is the middle sphere. It is separated by tropopause below from troposphere and mesopause on the top from thermosphere. It is extended from

50 km to 85 km from the earth's surface. The air pressure is very low. It is 1 millibar at the lower limit whereas it is 0.01 millibar at the highest limit. This layer is characterised by decreasing temperature and the coldest/ lowest atmospheric temperature is recorded in this layer. The lowest temperature estimated near the mesosphere is around minus 130° Celsius. It is colder that the lowest temperature recorded over Antarctic.

Between 75 to 85 km from the earth, noctilucent clouds are normal affairs in the summer nights between 500 to 700 north and south latitudes. Its literal meaning is night shining. It is a deep twilightseen only when the sun is on horizon but the sunlight is still falling at that height. The seen clouds are made up of ice crystals. Meteoric dust particles work like nuclei for ice crystallization which are falling as well as produced due to burning of meteor caused by friction.

The stratosphere, mesosphere and lowest part of the thermosphere are collectively referred to as the "middle atmosphere", which spans heights from approximately 10 kilometres (33,000 ft; 6.2 mi) to 100 kilometres (62 mi; 330,000 ft). The mesopause, at an altitude of 80–90 km (50–56 mi), separates the mesosphere from the thermosphere—the second-outermost layer of the Earth's atmosphere. This is also around the same altitude as the turbopause, below which different chemical species are well mixed due to turbulent eddies. Above this level the atmosphere becomes non-uniform; the scale heights of different chemical species differ by their molecular masses.

The term near space is also sometimes used. This term does not have a technical definition, but typically refers the region of the atmosphere up to 100 km (65,000 and 328,000 feet), roughly between the Armstrong limit (above which humans need a pressure suit to survive) up to the Kármán line where astrodynamics must take over from aerodynamics in order to achieve flight. The definition of near space can vary depending on the source, but in general near space comprises the altitudes above where commercial airliners fly but below orbiting satellites. Some sources distinguish between the terms "near space" and "upper atmosphere," so that only the layers closest to the Karman line are called near space.

Temperature

Within the mesosphere, temperature decreases with increasing height, due to decreasing solar heating and increasing cooling by CO_2 radiative emission. The top of the mesosphere, called the mesopause, is the coldest part of Earth's atmosphere. Temperatures in the upper mesosphere fall as low as −101 °C (172 K; −150 °F), varying according to latitude and season.

Dynamic Features

The main dynamic features in this region are strong zonal (East-West) winds, atmospheric tides, internal atmospheric gravity waves (commonly called "gravity waves"), and planetary waves. Most of these tides and waves start in the troposphere and lower stratosphere, and propagate to the mesosphere. In the mesosphere, gravity-wave amplitudes can become so large that the waves become unstable and dissipate. This dissipation deposits momentum into the mesosphere and largely drives global circulation.

Noctilucent clouds are located in the mesosphere. The upper mesosphere is also the region of the ionosphere known as the *D layer*. The D layer is only present during the day when some ionization occurs with nitric oxide being ionized by Lyman series-alpha hydrogen radiation. The ionization is so weak that when night falls, and the source of ionization is removed, the free electron and ion form back into a neutral molecule. The mesosphere has been called the "ignorosphere" because it is poorly studied relative to the stratosphere (which can be accessed with high-altitude balloons) and the thermosphere (in which satellites can orbit).

A 5 km (3.1 mi; 16,000 ft) deep sodium layer is located between 80–105 km (50–65 mi; 262,000–344,000 ft). Made of unbound, non-ionized atoms of sodium, the sodium layer radiates weakly to contribute to the airglow. The sodium has an average concentration of 400,000 atoms per cubic centimetre. This band is regularly replenished by sodium sublimating from incoming meteors. Astronomers have begun utilizing this sodium band to create "guide stars" as part of the adaptive optical correction process used to produce ultra-sharp ground-based observations. Other metal layers, e.g. iron and potassium, exist in the upper mesosphere/lower thermosphere region as well.

Millions of meteors enter the Earth's atmosphere, averaging 40 tons per year. The ablated material, called meteoric smoke, is thought to serve as condensation nuclei for noctilucent clouds.

Exploration and uses

The mesosphere lies above altitude records for aircraft, while only the lowest few kilometers are accessible to balloons, for which the altitude record is 53.0 km. Meanwhile, the mesosphere is below the minimum altitude for orbital spacecraft due to high atmospheric drag. It has only been accessed through the use of sounding rockets, which are only capable of taking mesospheric measurements for a few minutes per mission. As a result, it is the least-understood part of the atmosphere, resulting in the humorous moniker *ignorosphere*. The presence of red sprites and blue jets (electrical discharges or lightning within the lower mesosphere), noctilucent clouds, and density shears within this poorly understood layer are of current scientific interest.

Near space was first explored in the 1930s. The early flights flew to the edge of space without computers, spacesuits, and with only crude life support systems. Notable people who flew in near space were Jean Piccard and his wife Jeannette, on the nearcraft The Century of Progress. Later exploration was mainly carried out by unmanned craft, although there have been skydiving attempts made from high-altitude balloons.

The area is of interest for military surveillance purposes, scientific study, as well as to commercial interests for communications, and tourism. Craft that fly in near space include high-altitude balloons, non-rigid airships, rockoons, sounding rockets, and the Lockheed U-2 aircraft. The region has been of interest to space travel. Early attempts used a craft known as a rockoon to reach extreme altitudes and orbit. These are still used today for sounding rockets.

High-altitude platform stations have been proposed for applications such as communications relays. There has been a resurgence of interest in near space to launch manned spacecraft by man. Groups like ARCASPACE, as well as the da Vinci Project are planning on launching manned suborbital space vehicles from high-altitude balloons. JP Aerospace has a proposal for a spaceport in near space, as part of their Airship to Orbit program.

High-altitude Balloons

Near space has long been used for scientific ballooning, for applications such as submillimetre astronomy. High-altitude balloons are often flown by students and by amateur groups to altitudes on the order of 100,000 ft (30,000 m), for both scientific and educational purposes.

The Earth from approximately 100,000 ft (30,000 m) above Oregon, United States

Phenomena in Mesosphere and Near Space

- Polar aurora
- Airglow
- Noctilucent cloud
- Meteors
- Upper atmospheric lightning a.k.a. Transient luminous event
- Nacreous clouds
- Ozone layer
- Atmospheric tides
- Ionosphere

Thermosphere

Aurora Borealis over Alaska, Poker Flat Research Range

This sphere extends from mesopouse i.e., 85 km to about 650 km from earth. The temperature is on rise in this layer due to absorption of solar radiation by small amount of oxygen molecules present. It is highly dependent upon the solar activities. The temperature reaches beyond 1200°C at an altitude of about 350 km but by 650 km it may even rise to 2000°C. This much high temperature is primarily defined by average speed with which molecules are moving. Because of this, the temperature may be high. The effectiveness of this temperature is not that great. Its exposure to astronaut, if they are coming out from the capsule, is not affecting at all.

The Aurora is conspicuous phenomena observed in this layer. It is a striking display of light and are maximum between 10pm to 2am in the magnetic polar region of northern and southern hemispheres. It is known as Aurora Borealis in the northern and Aurora Australis in southern hemisphere. They are referred as northern and southern lights.Charged particles coming from the solar flare reaches to the earth's upper atmosphere. These charged particles collide with nitrogen and oxygen molecules present there in rarity. The collusion results into innumerable little surges of lights. The scattering of these lights seen is very beautiful and magnificent.

The highly diluted gas in this layer can reach 2,500 °C (4,530 °F) during the day. Even though the temperature is so high, one would not feel warm in the thermosphere, because it is so near vacuum that there is not enough contact with the few atoms of gas to transfer much heat. A normal thermometer might indicate significantly below 0 °C (32 °F), at least at night, because the energy lost by thermal radiation would exceed the energy acquired from the atmospheric gas by direct contact. In the anacoustic zone above 160 kilometres (99 mi), the density is so low that molecular interactions are too infrequent to permit the transmission of sound.

The dynamics of the thermosphere are dominated by atmospheric tides, which are driven by the very significant diurnal heating. Atmospheric waves dissipate above this level because of collisions between the neutral gas and the ionospheric plasma.

The International Space Station orbits the Earth within the middle of the thermosphere, between 330 and 435 kilometres (205 and 270 mi).

Neutral Gas Constituents

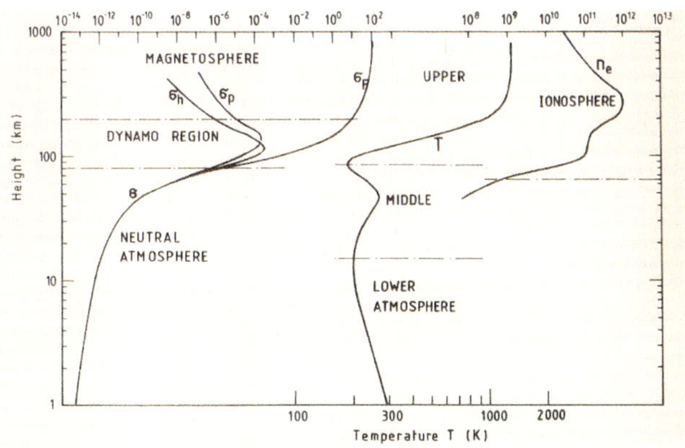

Nomenclature of atmospheric regions based on the profiles of electric conductivity (left), temperature (middle), and electron number density in m^{-3}(right)

It is convenient to separate the atmospheric regions according to the two temperature minima at about 12 km altitude (the tropopause) and at about 85 km (the mesopause). The thermosphere (or the upper atmosphere) is the height region above 85 km, while the region between the tropopause and the mesopause is the middle atmosphere (stratosphere and mesosphere) where absorption of solar UV radiation generates the temperature maximum near 45 km altitude and causes the ozone layer.

The density of the Earth's atmosphere decreases nearly exponentially with altitude. The total mass of the atmosphere is $M = \rho_A H \simeq 1$ kg/cm² within a column of one square centimeter above the ground (with $\rho_A = 1.29$ kg/m³ the atmospheric density on the ground at z = 0 m altitude, and $H \simeq 8$ km the average atmospheric scale height). 80% of that mass is concentrated within the troposphere. The mass of the thermosphere above about 85 km is only 0.002% of the total mass. Therefore, no significant energetic feedback from the thermosphere to the lower atmospheric regions can be expected.

Turbulence causes the air within the lower atmospheric regions below the turbopause at about 110 km to be a mixture of gases that does not change its composition. Its mean molecular weight is 29 g/mol with molecular oxygen (O_2) and nitrogen (N_2) as the two dominant constituents. Above the turbopause, however, diffusive separation of the various constituents is significant, so that each constituent follows its own barometric height structure with a scale height inversely proportional to its molecular weight. The lighter constituents atomic oxygen (O), helium (He), and hydrogen (H) successively dominate above about 200 km altitude and vary with geographic location, time, and solar activity. The ratio N_2/O which is a measure of the electron density at the ionospheric F region is highly affected by these variations. These changes follow from the diffusion of the minor constituents through the major gas component during dynamic processes.

The thermosphere contains an appreciable concentration of elemental sodium located in a 10-km thick band that occurs at the edge of the mesosphere, 80 to 100 km above Earth's surface. The sodium has an average concentration of 400,000 atoms per cubic centimeter. This band is regularly replenished by sodium sublimating from incoming meteors. Astronomers have begun utilizing this sodium band to create "guide stars" as part of the optical correction process in producing ultra-sharp ground-based observations.

Energy Input

Energy Budget

The thermospheric temperature can be determined from density observations as well as from direct satellite measurements. The temperature vs. altitude z can be simulated by the so-called Bates profile:

$$T = T_\infty - (T_\infty - T_0)\exp\{-s(z - z_0)\}$$

with T_∞ the exospheric temperature above about 400 km altitude, $T_0 = 355$ K, and $z_0 = 120$ km reference temperature and height, and s an empirical parameter depending on T_∞ and decreasing with T_∞. That formula is derived from a simple equation of heat conduction. One estimates a total heat input of $q_0 \simeq 0.8$ to 1.6 mW/m² above $z_0 = 120$ km altitude. In order to obtain equilibrium conditions, that heat input q_0 above z_0 is lost to the lower atmospheric regions by heat conduction.

The exospheric temperature T_∞ is a fair measurement of the solar XUV radiation. Since solar radio emission F at 10.7 cm wavelength is a good indicator of solar activity, one can apply the empirical formula for quiet magnetospheric conditions.

$$T_\infty \simeq 500 + 3.4F_0$$

with T_∞ in K, F_0 in 10^{-2} W m^{-2} Hz^{-1} (the Covington index) a value of F averaged over several solar cycles. The Covington index varies typically between 70 and 250 during a solar cycle, and never drops below about 50. Thus, T_∞ varies between about 740 and 1350 K. During very quiet magnetospheric conditions, the still continuously flowing magnetospheric energy input contributes by about 250 K to the residual temperature of 500 K in above equation. The rest of 250 K in above equation can be attributed to atmospheric waves generated within the troposphere and dissipated within the lower thermosphere.

Solar XUV Radiation

The solar X-ray and extreme ultraviolet radiation (XUV) at wavelengths < 170 nm is almost completely absorbed within the thermosphere. This radiation causes the various ionospheric layers as well as a temperature increase at these heights. While the solar visible light (380 to 780 nm) is nearly constant with a variability of not more than about 0.1% of the solar constant, the solar XUV radiation is highly variable in time and space. For instance, X-ray bursts associated with solar flares can dramatically increase their intensity over preflare levels by many orders of magnitude over a time span of tens of minutes. In the extreme ultraviolet, the Lyman α line at 121.6 nm represents an important source of ionization and dissociation at ionospheric D layer heights. During quiet periods of solar activity, it alone contains more energy than the rest of the XUV spectrum. Quasi-periodic changes of the order of 100% or greater, with periods with period of 27 days and 11 years, belong to the prominent variations of solar XUV radiation. However, irregular fluctuations over all time scales are present all the time. During low solar activity, about half of the total energy input into the thermosphere is thought to be solar XUV radiation. Evidently, that solar XUV energy input occurs only during daytime conditions, maximizing at the equator during equinox.

Solar Wind

A second source of energy input into the thermosphere is solar wind energy which is transferred to the magnetosphere by mechanisms that are not well understood. One possible way to transfer energy is via a hydrodynamic dynamo process. Solar wind particles penetrate into the polar regions of the magnetosphere where the geomagnetic field lines are essentially vertically directed. An electric field is generated, directed from dawn to dusk. Along the last closed geomagnetic field lines with their footpoints within the auroral zones, field aligned electric currents can flow into the ionospheric dynamo region where they are closed by electric Pedersen and Hall currents. Ohmic losses of the Pedersen currents heat the lower thermosphere (e.g., Magnetospheric electric convection field). In addition, penetration of high energetic particles from the magnetosphere into the auroral regions enhance drastically the electric conductivity, further increasing the electric currents and thus Joule heating. During quiet magnetospheric activity, the magnetosphere contributes perhaps by a quarter to the thermosphere's energy budget. This is about 250 K of the exospheric temperature in above equation. During very large activity, however, this heat input can increase substantially, by a factor of four or more. That solar wind input occurs mainly in the auroral regions during both day and night.

Atmospheric Waves

Two kinds of large-scale atmospheric waves within the lower atmosphere exist: internal waves with finite vertical wavelengths which can transport wave energy upward; and external waves with infinitely large wavelengths which cannot transport wave energy. Atmospheric gravity waves and most of the atmospheric tides generated within the troposphere belong to the internal waves. Their density amplitudes increase exponentially with height, so that at the mesopause these waves become turbulent and their energy is dissipated (similar to breaking of ocean waves at the coast), thus contributing to the heating of the thermosphere by about 250 K in above equation. On the other hand, the fundamental diurnal tide labelled (1, −2) which is most efficiently excited by solar irradiance is an external wave and plays only a marginal role within lower and middle atmosphere. However, at thermospheric altitudes, it becomes the predominant wave. It drives the electric Sq-current within the ionospheric dynamo region between about 100 and 200 km height.

Heating, predominately by tidal waves, occurs mainly at lower and middle latitudes. The variability of this heating depends on the meteorological conditions within troposphere and middle atmosphere, and may not exceed about 50%.

Dynamics

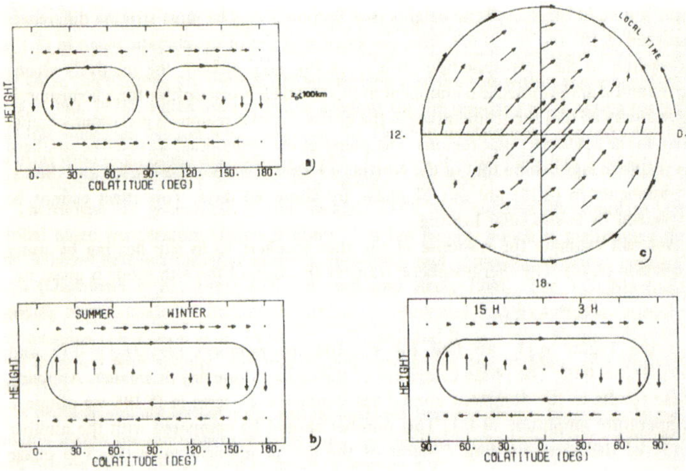

Schematic meridian-height cross-section of circulation of (a) symmetric wind component (P_2^0), (b) of antisymmetric wind component (P_1^0), and (d) of symmetric diurnal wind component (P_1^1) at 3 h and 15 h local time. Upper right panel (c) shows the horizontal wind vectors of the diurnal component in the northern hemisphere depending on local time

Within the thermosphere above about 150 km height, all atmospheric waves successively become external waves, and no significant vertical wave structure is visible. The atmospheric wave modes degenerate to the spherical functions P_n^m with m a meridional wave number and n the zonal wave number (m = 0: zonal mean flow; m = 1: diurnal tides; m = 2: semidiurnal tides; etc.). The thermosphere becomes a damped oscillator system with low-pass filter characteristics. This means that smaller-scale waves (greater numbers of (n,m)) and higher frequencies are suppressed in favor of large-scale waves and lower frequencies. If one considers very quiet magnetospheric disturbances and a constant mean exospheric temperature (averaged over the sphere), the observed temporal and spatial distribution of the exospheric temperature distribution can be described by a sum of spheric functions:

$$T(\varphi,\lambda,t) = T_{\infty}\{1 + \Delta T_2^0 P_2^0(\varphi) + \Delta T_1^0 P_1^0(\varphi)\cos[\omega_a(t - t_a)] + \Delta T_1^1 P_1^1(\varphi)\cos(\tau - \tau_d) + \cdots\}$$

Here, it is φ latitude, λ longitude, and t time, ω_a the angular frequency of one year, ω_d the angular frequency of one solar day, and $\tau = \omega_d t + \lambda$ the local time. t_a = June 21 is the date of northern summer solstice, and τ_d = 15:00 is the local time of maximum diurnal temperature.

The first term in on the right is the global mean of the exospheric temperature (of the order of 1000 K). The second term [with $P_2^o = 0.5(3 \sin^2(\varphi)-1)$] represents the heat surplus at lower latitudes and a corresponding heat deficit at higher latitudes (Fig. a). A thermal wind system develops with wind toward the poles in the upper level and wind away from the poles in the lower level. The coefficient $\Delta T_2^o \approx 0.004$ is small because Joule heating in the aurora regions compensates that heat surplus even during quiet magnetospheric conditions. During disturbed conditions, however, that term becomes dominant, changing sign so that now heat surplus is transported from the poles to the equator. The third term (with $P_1^o = \sin \varphi$) represents heat surplus on the summer hemisphere and is responsible for the transport of excess heat from the summer into the winter hemisphere (Fig. b). Its relative amplitude is of the order $\Delta T_1^o \simeq 0.13$. The fourth term (with $P_1^1(\varphi) = \cos \varphi$) is the dominant diurnal wave (the tidal mode (1,−2)). It is responsible for the transport of excess heat from the daytime hemisphere into the nighttime hemisphere (Fig. d). Its relative amplitude is $\Delta T_1^1 \simeq 0.15$, thus on the order of 150 K. Additional terms (e.g., semiannual, semidiurnal terms and higher order terms) must be added to the above equation. However, they are of minor importance. Corresponding sums can be developed for density, pressure, and the various gas constituents.

Thermospheric Storms

In contrast to solar XUV radiation, magnetospheric disturbances, indicated on the ground by geomagnetic variations, show an unpredictable impulsive character, from short periodic disturbances of the order of hours to long-standing giant storms of several days' duration. The reaction of the thermosphere to a large magnetospheric storm is called thermospheric storm. Since the heat input into the thermosphere occurs at high latitudes (mainly into the auroral regions), the heat transport represented by the term P_2^o in the above equation. is reversed. In addition, due to the impulsive form of the disturbance, higher-order terms are generated which, however, possess short decay times and thus quickly disappear. The sum of these modes determines the "travel time" of the disturbance to the lower latitudes, and thus the response time of the thermosphere with respect to the magnetospheric disturbance. Important for the development of an ionospheric storm is the increase of the ratio N_2/O during a thermospheric storm at middle and higher latitude. An increase of N_2 increases the loss process of the ionospheric plasma and causes therefore a decrease of the electron density within the ionospheric F-layer (negative ionospheric storm).

Exosphere

Exo means external. Therefore, exosphere the external or the outer most layer of the atmosphere. Its lower boundary starts from the thermopause (650 km) to the limit from where the void space begins. This limit is estimated to be about 10000 km. This much distance is little less than the diameter of the earth. It is really a very big size of the limit of the atmosphere. In exosphere, very light gases are traced and they are hydrogen and helium. Their molecules are spaced very widely. Beyond the upper limit of exosphere, the space is considered to be void.

Mercury and several large moons, such as the Moon and the Galilean satellites of Jupiter, have exospheres without a denser atmosphere underneath, referred to as a surface boundary exosphere. Here, molecules are ejected on elliptic trajectories until they collide with the surface. Smaller bodies such as asteroids, in which the molecules emitted from the surface escape to space, are not considered to have exospheres.

Earth's Exosphere

The most common molecules within Earth's exosphere are those of the lightest atmospheric gases. Hydrogen is present throughout the exosphere, with some helium, carbon dioxide, and atomic oxygen near its base. Because it can be hard to define the boundary between the exosphere and outer space, the exosphere may be considered a part of interplanetary or outer space.

Lower Boundary

The lower boundary of the exosphere is called the *exobase*. It is also called *exopause* and *'critical altitude'* as this is the altitude where barometric conditions no longer apply. Atmospheric temperature becomes nearly a constant above this altitude. On Earth, the altitude of the exobase ranges from about 500 to 1,000 kilometres (310 to 620 mi) depending on solar activity.

The exobase can be defined in one of two ways:

If we define the exobase as the height at which upward-traveling molecules experience one collision on average, then at this position the mean free path of a molecule is equal to one pressure scale height. This is shown in the following. Consider a volume of air, with horizontal area A and height equal to the mean free path l, at pressure p and temperature T. For an ideal gas, the number of molecules contained in it is:

$$n = \frac{pAl}{RT}$$

where R is the universal gas constant. From the requirement that each molecule traveling upward undergoes on average one collision, the pressure is:

$$p = \frac{m_A n g}{A}$$

where m_A is the mean molecular mass of the gas. Solving these two equations gives:

$$l = \frac{RT}{m_A g}$$

which is the equation for the pressure scale height. As the pressure scale height is almost equal to the density scale height of the primary constituent, and because the Knudsen number is the ratio of mean free path and typical density fluctuation scale, this means that the exobase lies in the region where $\mathrm{Kn}(h_{EB}) \simeq 1$.

The fluctuation in the height of the exobase is important because this provides atmospheric drag on satellites, eventually causing them to fall from orbit if no action is taken to maintain the orbit.

Upper Boundary of Earth

In principle, the exosphere covers distances where particles are still gravitationally bound to Earth, i.e. particles still have ballistic orbits that will take them back towards Earth. The upper boundary of the exosphere can be defined as the distance at which the influence of solar radiation pressure on atomic hydrogen exceeds that of Earth's gravitational pull. This happens at half the distance to the Moon (the average distance between Earth and the Moon is 384,400 kilometres (238,900 mi)). The exosphere, observable from space as the geocorona, is seen to extend to at least 10,000 kilometres (6,200 mi) from Earth's surface. The exosphere is a transitional zone between Earth's atmosphere and space.

Moon's Exosphere

On 17 August 2015, based on studies with the Lunar Atmosphere and Dust Environment Explorer (LADEE) spacecraft, NASA scientists reported the detection of neon in the exosphere of the moon.

References

- Miller, Charles (2008). Biological oceanography. 350 Main Street, Malden, MA 02148 USA: Blackwell Publishing Ltd. pp. 60–62. ISBN 978-0-632-05536-4
- Belosludov, V. R.; Subbotin, O. S.; Krupskii, D. S.; Prokuda, O. V.; et al. (2006). "Microscopic model of clathrate compounds". Journal of Physics: Conference Series. 29: 1. Bibcode:2006JPhCS..29....1B. doi:10.1088/1742-6596/29/1/001
- Sparacino-Watkins, Courtney; Stolz, John F.; Basu, Partha (2013-12-16). "Nitrate and periplasmic nitrate reductases". Chem. Soc. Rev. 43 (2): 676–706. doi:10.1039/c3cs60249d. ISSN 1460-4744. PMC 4080430
- Vaclav Smil (2003). The Earth's Biosphere: Evolution, Dynamics, and Change. MIT Press. p. 107. ISBN 978-0-262-69298-4
- "Health | Particulate Matter | Air & Radiation | US EPA". Epa.gov. 17 November 2010. Archived from the original on 2 October 2006. Retrieved 1 February 2015
- Su, Joseph Z.; Kim, Andrew K.; Crampton, George P.; Liu, Zhigang (2001). "Fire Suppression with Inert Gas Agents». Journal of Fire Protection Engineering. 11 (2): 72–87. doi:10.1106/X21V-YQKU-PMKP-XGTP. ISSN 1042-3915
- "Description of Aim-9 Operation". planken.org. Archived from the original on 22 December 2008. Retrieved 1 February 2009.
- Ward, Peter Douglas; Brownlee, Donald (2003). The life and death of planet Earth. Macmillan. pp. 117–128. ISBN 0-8050-7512-7
- Anstie, Francis (1874). "Clinic of the Month: Dr. McKendrick on Ozone". The Practitioner: A Journal of Therapeutics and Public Health. 12 (January-June): 123. Retrieved 13 April 2018
- Pavlov, Alexander A.; et al. (January 2003), "Methane-rich Proterozoic atmosphere?", Geology, 31 (1): 87–90, Bibcode:2003Geo....31...87P, doi:10.1130/0091-7613(2003)031<0087:MRPA>2.0.CO;2
- Gary Polakovic (7 June 2003). "Getting the Cows to Cool It". The Los Angeles Times. Retrieved 4 November 2011

- Streng, A. G. (1961). "Tables of Ozone Properties". Journal of Chemical Engineering Data. 6 (3): 431–436. doi:10.1021/je00103a031
- "Environment Canada - Pollution and Waste - Tracking Pollution in Canada". Ec.gc.ca. 2012-07-05. Archived from the original on 2006-09-24. Retrieved 2012-11-18
- Gale, Joseph (2009). Astrobiology of Earth : the emergence, evolution, and future of life on a planet in turmoil. Oxford: Oxford University Press. ISBN 978-0-19-920580-6

Chapter 3

Köppen Climate Classification

Köppen climate classification is a famous climate classification system first introduced by Wladimir Köppen. According to this system, climate is divided into five major categories or groups, which are tropical, dry, continental, temperate and polar. Each group is represented by a letter of the English alphabet. This chapter introduces the different climates under the Köppen climate classification.

Climate is an average weather condition at a specific place over a long period of time (30-35 years and more), including absolute extremes of temperature, precipitation etc. Factors such as latitude, altitude, prevailing winds, oceancurrents, maritime and continental situations exercise direct influence in determining the climate. The classification of climates of the world is done in order to understand how the different climatic elements form combination in different parts of the world. Although, there are several ways in which the climates of the world can be classified, the lesson is devoted to discuss the Koppen's classification of world climates which is most widely used. In this lesson we will study the measurements, design and combinations of major, main and sub groups of the world climates as developed by Koppen.

Historical Developments in Koppen's Climate Classification

Wladimir Koppen (1846-1940), a Russian born German climatologist and botanist attempted he classification of world climates. Climate classification was first published as 'thermal zones of the earth' in 1884 in Russian language. Scholars E. Volken and S. Bronnimann translated the work in German. With several modifications, Koppen improved his classification later on, in1918 and 1936. The details of climatic classification were published in 'Handbuch der Kilmatologoie'. Later on, German climatologist Rudolf Geiger (1954, 1961) collaborated with Koppen on changes to the scheme of classification. Thus, the classification of world climates is also known as the Koppen-Geiger's climate classification, as the two scholar's co-authored the work, and was widely adopted. Classification is based on the empirical study of the distribution of vegetation across the world.

Vegetation as an Index of Temperature and Precipitation

Koppen assumed that the vegetation is the most representative feature of precipitation and temperature. For example tropical evergreen vegetation represents high precipitation and high temperature. Similarly, alpine vegetation represents high precipitation and low to very low temperature. It is commonly found both in higher latitudes as well as in higher altitudes close to snowline. Deciduous vegetation represents high temperature and moderate precipitation in one season followed by a dry season with moderate to low temperature. Arid climates are marked with moisture deficiencies. While hot deserts have high temperatures and low precipitation, cold deserts have low temperature and low precipitation. The arid climates have xerophyte vegetation.

Vegetation as an index of temperature and precipitation is also used to represent major ecosystems and biomes on the earth. The fact is plotted on a graph to represent temperature along X axis. It is to be noted that annual average temperature temperature is to be plotted in a manner that it begins with highvalues of temperature above 300C and ending with very low temperatures upto less than 100C. Whereas annual average precipitation is plotted along Y axis and it keeps on increasing in a manner that it begins with zero and ends up to more than 400 cm.

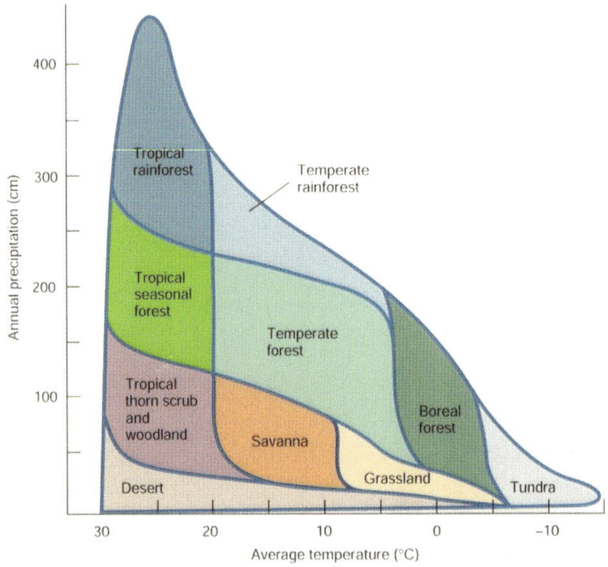

Temperature and Precipitation vs Vegetation

Vegetation response to precipitation and temperature anomalies, particularly during droughts, isof great importance in semi-arid regions, because ecosystem and hydrologic processes depend on vegetation conditions. One of the important indicators of vegetation presence, abundance and vigor is the NDVI (Normalized Difference Vegetation Index). Since NDVI makes use of bio physical interactions where green plant canopies absorb much of the radiation, it seems to be well suited for studies concerned with photosynthetic capacity of vegetation cover. Studies conducted by Prasad, et al. (2008) and Canon Julio et. al.(2011) have indicated that vegetation response to precipitation and temperature anomalies are of great importance.

Delineation of Climatic Regions

For the delineation of climatic regions, Koppen considered the following:

- mean monthly temperature,
- mean annual temperature,
- mean monthly precipitation, and
- mean annual precipitation.

Therefore, Koppen's classification is based on the measurement of temperature and precipitation. Viewing its utility and applicability, FAO published a Global climate map using Koppen classification in 1999.

The Köppen climate classification is one of the most widely used climate classification systems. It was first published by the climatologist Wladimir Köppen (1846-1940) in 1884, with several later modifications by Köppen, notably in 1918 and 1936. Later, the climatologist Rudolf Geiger (1954, 1961) collaborated with Köppen on changes to the classification system, which is thus sometimes called the Köppen–Geiger climate classification system.

The Köppen climate classification divides climates into five main climate groups, with each group being divided based on seasonal precipitation and temperature patterns. The five main groups are *A* (tropical), *B* (dry), *C* (temperate), *D* (continental), and *E* (polar). Each group and subgroup is represented by a letter. All climates are assigned a main group (the first letter). All climates except for those in the *E* group are assigned a seasonal precipitation subgroup (the second letter). For example, *Af* indicates a tropical rainforest climate. The system assigns a temperature subgroup for all groups other than those in the *A* group, indicated by the third letter for climates in *B*, *C*, and *D*, and the second letter for climates in *E*. For example, *Cfb* indicates an oceanic climate with warm summers as indicated by the ending *b*. Climates are classified based on specific criteria unique to each climate type.

Köppen designed the system based on his experience as a botanist, so the main climate groups are based on the different variety of vegetation that grows in climates belonging to each group. In addition to identifying climates, the system can be used to analyze ecosystem conditions and identify the main types of vegetation within climates. Due to its link with the plant life of a region, the system is useful in predicting future changes in plant life within a region.

The Köppen climate classification system has been further modified, within the Trewartha climate classification system in the middle 1960s (revised in 1980). The Trewartha system sought to create a more refined middle latitude climate zone, which was one of the criticisms of the Köppen system (the C climate group was too broad).

Design of Climatic Classification

Koppen's scheme of climatic classification begins with English capital alphabet denoting prevalent vegetation type. It is followed by small English alphabets in second and third place with exception to B and E group of climates. In B and E groups, both letters first and second letters are in capital alphabets. A brief discussion on capital and small letters used in the design of classification is made below.

Capital letters: In all, Koppen has used ten capital alphabets in his scheme of climate classification. An expression of first five capital letters that denote vegetation zones is given below:

A - Mega -thermal,

B - Xerophytes,

C - Meso - thermal,

D - Micro -thermal,

E - Heckisto – thermal.

Next five capital alphabets denote distinct *regional characteristics*. They are as follows:

F - Permanently Frozen

M - Marine

S - Semi Arid/steppe

T - Tundra

W - Waste (desert)

Small letters: Similar to capital alphabets, Koppen has also used ten small alphabets (letters) to denote area specific characteristics that supplement the major climate group. The first six small letters signify seasonal characteristics of precipitation. They are as follows:

Characteristics of seasonal precipitation

(i) Feucht (precipitation round the year) = f

(ii) Hot = h

(iii) Cold = k

(iv) Monsoon = m

(v) Summer dry = s

(vi) Winter dry = w

The level of heat: The next four small letters denote the level of heat:

(i) Warmest month summer temperature > 22° C = a

(ii) Warmest month summer temperature < 22° C = b

(iii) Temperature constantly > 10° C = c

(iv) Average temperature of coldest month < -38° C = d

Besides arrangements of alphabets, Koppen has also used following terms to denote:

POTET =Potential evapotranspiration and PRECIP =Precipitation

Thus, the climatic classification of Koppen is a self- explanatory. It is convenient to place values as per the order of alphabets.

Table: Symbols used in the classification of world climates by Koppen-Geiger

Major Classification & symbols	Main Classification & symbols	Sub Classification and symbols
Group A: Tropical/mega- thermal climates. Vegetation is tropical rain forests.	Af. = Tropical Rainforest Am = Tropical Monson Aw = Tropical Savanna	
Group B: Dry (arid and semi-arid) climates Xerophytes type of vegetation	BW: B =Arid, W = waste h = hot, k – cold BS: B = Arid, S =Semiarid/Steppe	BWh = Hot desert BWk = Cold desert BSh = Semiarid hot BSk =Semiarid cold

Group C: temperate / Mesothermal climates Plants adapted to moderate heat and moisture	Cf= Year round precipitation, a= warmest month temperature $>22^\circ$ C b= warmest month temperature $<22^\circ$ C c= 1 to 3 month's temperature $>10^\circ$ C Cw= winter dry Cs= summer dry	Cfa Cfb Cfc Cwa Cwb Cwc Csa Csb
Group D:Continental / microthermal climates Evergreen deciduous forests and steppes are the natural vegetation.	Df=Humid continental precipitation throughout the year. Dw= Humid continental-winter dry. Dwd= Coldest month temperature$<-38^\circ$ C	Dfa Dfb Dfc Dwa Dwb Dwc Dwd
Group E: Polar Climates/ Hekisto-thermal type of vegetation Mosses, lichens etc. are the natural vegetation.	This type of climate has an average temperature $<10^\circ$C	ET= Mild tundra climate with average temperatures between 0° C and 10° C. ETf= Cold Tundra climate; at least one month with an average temperature $> 0^\circ$ C. EF= Ice caps climate: average temperature in all the months $<0^\circ$ C.

Measurements in Koppen's climatic classification: Koppen's classification of world climates is based on two types of measurements viz. primary and secondary. A brief account of these measurements is given below.

a) Primary measurements: The primary measurements are concerned with broad parameters that distinguish one major climate group with another. These measurements are related to the critical values of temperature and precipitation conditions. The details of critical values have been given against major climate groups in Table.

Table: Primary Measurements

A- Tropical	$T_{cold} \geq 18^\circ$ C
B- Dry	$MAP<10\times P_{threshold}$
C- Meso thermal	$T_{hot} \geq 10^\circ$ C & 0° C $>T_{cold} < 18^\circ$ C
D- Micro- Thermal	$T_{hot} \geq 10^\circ$ C & $T_{cold} \geq 0^\circ$ C
E- Polar	$T_{hot} < 10^\circ$ C
H- High Land	Elevation≥2300 meters

Where: T_{cold} stands for temperature in coldest month, T_{hot} temperature in hottest month, MAP mean annual precipitation and $P_{Threshold}$ for precipitation threshold. $P_{threshold}$ = precipitation threshold, which varies according to the following rules: if 70% of MAP occurs in winter, then $P_{threshold}$ = 2 × MAT, if 70% of MAP occurs in summer, then $P_{threshold}$ = 2 × MAT + 28, otherwise $P_{threshold}$ 2 × MAT + 14

b) Secondary measurements: Similar to primary measurements, Koppen's classification of world climates has also used measurements for secondary or sub group of climates. They have been so designed to distinguish one sub group with another. It also involves some elementary calculations to arrive at the critical values. The details have been given in table.

Table: Secondary measurements

Af	P_{dry}≥60mm	w	P_{wdry}<P_{swet}÷10
Am	Not (Af)P_{dry}≥100mm -MAP÷25	F	Not (Cs Or Cw)
Aw	Not (Af)P_{wdry}<100mm-MAP ÷25	A	T_{hot}≥ 18° C
As	Not (Af)P_{sdry}<100mm -MAP÷25	b	Not (a)T_{mon}10≥ 4
W	MAP<5× $P_{threshold}$	C	Not (a or b or d)
S	MAP≥5× $P_{threshold}$	d	Not (a or b) &T_{cold}< -38° C
h	MAT≥ 18° C	T	T_{hot} > 0° C
K	MAT<10° C	F	T_{hot} ≤0° C
s	P_{sdry}< 40mm & P_{sdry}<P_{wwet}÷3	T	T_{hot} > 0° C (Highland Case)

Where P_{dry} (precipitation in driest month), P_{wdry} (precipitation in winter driest month), P_{sdry} (precipitation in driest summer month) P_{swet} (precipitation in summer wettest month), P_{wwet} (precipitation in wettest month winter) MAT (mean annual temperature), MAP (mean annual Precipitation), T_{hot} (Temperature in hottest month), T_{mon} (monthly temperature), T_{cold} (temperature in coldest month).

Combinations of alphabets as expressions of climates: Based on the arrangements of capital as well as small alphabets, Koppen has developed combinations to explain the world climates. Combinations explaining climate types have been developed in three sets viz. two, three and four alphabet combinations. A brief account of each combination group is presented below.

a) The two alphabet combination: The combination has been used in two distinct ways. The first way is represented when both the alphabets, in the combination, are capital letters. For example 'ET' and 'EF' which denote "Polar Tundra' and 'Polar Frost' climates respectively. The second way of two alphabet combination is that first, in sequence, is a capital letter while second one is a small letter. As usual, the first capital alphabet denotes the major vegetation group and the second small alphabet denotes seasonal characteristics of precipitation. For example: Af (tropical rainforest), As (tropical with summer dry season e.g. in eastern part of Tamil Nadu India) and Aw (tropical climate with winter dry season e.g. large part of peninsular India south of Tropic of Cancer) represent a two alphabet combination.

b) The three alphabet combination: This combination is being used to denote vegetation, precipitation, thermal and locational characteristics of a climate type. It is mostly used to denote 'A' 'C' and 'D' type of climates. For example: 'Amw' (tropical monsoon with short dry winter) 'Cwg' (Meso thermal with dry winter in Gangetic plains of India), and 'Dfc' (micro thermal humid with short summer found in eastern Himalayan region of India).

c) The four alphabet combination: This combinationis commonly used to denote 'B' group (arid and semiarid) of climates only. The first two alphabets are capital letters while remainingtwo are small letters. The first capital alphabet denotes 'B' which is common to all sub climates, while the second capital alphabet could be either 'W' (waste/ desert) or 'S' (semiarid/steppe). The third alphabet in this combination is a small letter and denotes the setting in which arid or semi-arid region lies. It could be either in a hot setting ('h') or in a cold setting ('k'). The fourth alphabet in the combination is again a small letter and denotes seasonal situation of moisture deficiency. The climatic sub group may experience moisture deficiency either a summer season (s) or a winter season (w). For example in case of arid/ desert climates, the subgroup may be either BWhw (hot desert winter dry e.g. Thar desert in India) or BWks (cold desert summer dry) whereas in case of semiarid climates, the sub-group may be either BShw (semi- arid in hot setting- winter dry) or BSks (semi-arid in cold setting-summer dry).

Based on combinational arrangements, Koppen, has divided the world in 5 major, 14 main and 29 sub classes of climates in the world. A detailed scheme of climates of the world as per Koppen - Geiger is given in the table. There could be some additional minor climate groups also.

Table: Classification of world climates based on Koppen-Geiger

Af	BWh	Csa	Cwa	Cfa	Dsa	Dwa	Dfa	ET
Am	BWk	Csb	Cwb	Cfb	Dsb	Dwb	Dfb	ETf
Aw	BSh	-	Cwc	Cfc	Dsc	Dwc	Dfc	Ef
As	BSk	-	-	-	Dsd	Dwd	Dfd	-

Meaning of Symbols

Köppen climate classification scheme symbols description table		
1st	**2nd**	**3rd**
A (Tropical)	f (Rainforest)	
	m (Monsoon)	
	w (Savanna, Wet)	
	s (Savanna, Dry)	
B (Arid)	W (Desert)	
	S (Steppe)	
		h (Hot)
		k (Cold)
		n (With frequent fog)
C (Temperate)	s (Dry summer)	
	w (Dry winter)	
	f (Without dry season)	
		a (Hot summer)
		b (Warm summer)
		c (Cold summer)

D (Cold (continental))	s (Dry summer)	
	w (Dry winter)	
	f (Without dry season)	
		a (Hot summer)
		b (Warm summer)
		c (Cold summer)
		d (Very cold winter)
E (Polar)	T (Tundra)	
	F (Eternal winter (ice cap))	

The Köppen climate classification scheme divides climates into five main climate groups: *A* (tropical), *B* (dry), *C* (temperate), *D* (continental), and *E* (polar). The second letter indicates the seasonal precipitation type, while the third letter indicates the level of heat.

Group A: Tropical (Megathermal) Climates

This type of climate has every month of the year with an average temperature of 18 °C (64.4 °F) or higher, with significant precipitation.

- *Af* = Tropical rainforest climate; average precipitation of at least 60 mm (2.4 in) in every month.
- *Am* = Tropical monsoon climate; driest month (which nearly always occurs at or soon after the "winter" solstice for that side of the equator) with precipitation less than 60 mm (2.4 in), but more than 4% the total annual precipitation.
- *Aw* or *As* = Tropical wet and dry or savanna climate; with the driest month having precipitation less than 60 mm (2.4 in) and less than 4% of the total annual precipitation.

Group B: Dry (Arid and Semiarid) Climates

This type of climate is defined by little precipitation.

Multiply the average annual temperature in Celsius by 20, then add

(a) 280 if 70% or more of the total precipitation is in the spring and summer months (April–September in the Northern Hemisphere, or October–March in the Southern), or

(b) 140 if 30%–70% of the total precipitation is received during the spring and summer, or

(c) 0 if less than 30% of the total precipitation is received during the spring and summer.

If the annual precipitation is less than 50% of this threshold, the classification is BW (arid: desert climate); if it is in the range of 50%–100% of the threshold, the classification is BS (semi-arid: steppe climate).

A third letter can be included to indicate temperature. Originally, h signified low-latitude climate (average annual temperature above 18 °C (64.4 °F)) while k signified middle-latitude climate (average annual temperature below 18 °C), but the more common practice today, especially in the United States, is to use h to mean the coldest month has an average temperature above 0 °C (32 °F), with k denoting that at least one month's averages below 0 °C. The n is used to denote a climate characterized by frequent fog.

- *BWh* = Hot desert climate
- *BWk* = Cold desert climate
- *BWn* = Desert climate with frequent fog
- *BSh* = Hot semi-arid climate
- *BSk* = Cold semi-arid climate
- *BSn* = Semi-arid climate with frequent fog

Group C: Temperate (Mesothermal) Climates

This type of climate has the coldest month averaging between −3 °C (27 °F) and 18 °C (64.4 °F) and at least one month averaging above 10 °C (50 °F). Some modern authors prefer the limit of 0 °C (32 °F) and 18 °C (64.4 °F).

- Cfa = Humid subtropical climate; coldest month averaging above −3 °C (27 °F), at least one month's average temperature above 22 °C (71.6 °F), and at least four months averaging above 10 °C (50 °F). No significant precipitation difference between seasons (neither abovementioned set of conditions fulfilled). No dry months in the summer.
- Cfb = Temperate oceanic climate; coldest month averaging above −3 °C (27 °F), all months with average temperatures below 22 °C (71.6 °F), and at least four months averaging above 10 °C (50 °F). No significant precipitation difference between seasons (neither abovementioned set of conditions fulfilled).
- Cfc = Subpolar oceanic climate; coldest month averaging above −3 °C (27 °F) and 1–3 months averaging above 10 °C (50 °F). No significant precipitation difference between seasons (neither abovementioned set of conditions fulfilled).
- Cwa = Monsoon-influenced humid subtropical climate; coldest month averaging above −3 °C (27 °F), at least one month's average temperature above 22 °C (71.6 °F), and at least four months averaging above 10 °C (50 °F). At least ten times as much rain in the wettest month of summer as in the driest month of winter (alternative definition is 70% or more of average annual precipitation is received in the warmest six months).
- Cwb = Subtropical highland climate or Monsoon-influenced temperate oceanic climate; coldest month averaging above −3 °C (27 °F), all months with average temperatures below 22 °C (71.6 °F), and at least four months averaging above 10 °C (50 °F). At least ten times as much rain in the wettest month of summer as in the driest month of winter (an alternative definition is 70% or more of average annual precipitation received in the warmest six months).
- Cwc = Cold subtropical highland climate or Monsoon-influenced subpolar oceanic climate; coldest month averaging above −3 °C (27 °F) and 1–3 months averaging above 10 °C (50

°F). At least ten times as much rain in the wettest month of summer as in the driest month of winter (alternative definition is 70% or more of average annual precipitation is received in the warmest six months).

- Csa = Hot-summer Mediterranean climate; coldest month averaging above −3 °C (27 °F), at least one month's average temperature above 22 °C (71.6 °F), and at least four months averaging above 10 °C (50 °F). At least three times as much precipitation in the wettest month of winter as in the driest month of summer, and driest month of summer receives less than 30 mm (1.2 in).
- Csb = Warm-summer Mediterranean climate; coldest month averaging above −3 °C (27 °F), all months with average temperatures below 22 °C (71.6 °F), and at least four months averaging above 10 °C (50 °F). At least three times as much precipitation in the wettest month of winter as in the driest month of summer, and driest month of summer receives less than 30 mm (1.2 in).
- Csc = Cool-summer Mediterranean climate; coldest month averaging above −3 °C (27 °F) and 1–3 months averaging above 10 °C (50 °F). At least three times as much precipitation in the wettest month of winter as in the driest month of summer, and driest month of summer receives less than 30 mm (1.2 in).

Group D: Continental (Microthermal) climates

This type of climate has at least one month averaging below −3 °C (27 °F) and at least one month averaging above 10 °C (50 °F). Some modern authors prefer the limit of 0 °C (32 °F).

- Dfa = Hot-summer humid continental climate; coldest month averaging below −3 °C (27 °F), at least one month's average temperature above 22 °C (71.6 °F), and at least four months averaging above 10 °C (50 °F). No significant precipitation difference between seasons (neither abovementioned set of conditions fulfilled).
- Dfb = Warm-summer humid continental climate; coldest month averaging below −3 °C (27 °F), all months with average temperatures below 22 °C (71.6 °F), and at least four months averaging above 10 °C (50 °F). No significant precipitation difference between seasons (neither abovementioned set of conditions fulfilled).
- Dfc = Subarctic climate; coldest month averaging below −3 °C (27 °F) and 1–3 months averaging above 10 °C (50 °F). No significant precipitation difference between seasons (neither abovementioned set of conditions fulfilled).
- Dfd = Extremely cold subarctic climate; coldest month averaging below −38 °C (−36.4 °F) and 1–3 months averaging above 10 °C (50 °F). No significant precipitation difference between seasons (neither abovementioned set of conditions fulfilled).
- Dwa = Monsoon-influenced hot-summer humid continental climate; coldest month averaging below −3 °C (27 °F), at least one month's average temperature above 22 °C (71.6 °F), and at least four months averaging above 10 °C (50 °F). At least ten times as much rain in the wettest month of summer as in the driest month of winter (alternative definition is 70% or more of average annual precipitation is received in the warmest six months).

- Dwb = Monsoon-influenced warm-summer humid continental climate; coldest month averaging below −3 °C (27 °F), all months with average temperatures below 22 °C (71.6 °F), and at least four months averaging above 10 °C (50 °F). At least ten times as much rain in the wettest month of summer as in the driest month of winter (alternative definition is 70% or more of average annual precipitation is received in the warmest six months).
- Dwc = Monsoon-influenced subarctic climate; coldest month averaging below −3 °C (27 °F) and 1–3 months averaging above 10 °C (50 °F). At least ten times as much rain in the wettest month of summer as in the driest month of winter (alternative definition is 70% or more of average annual precipitation is received in the warmest six months).
- Dwd = Monsoon-influenced extremely cold subarctic climate; coldest month averaging below −38 °C (−36.4 °F) and 1–3 months averaging above 10 °C (50 °F). At least ten times as much rain in the wettest month of summer as in the driest month of winter (alternative definition is 70% or more of average annual precipitation is received in the warmest six months).
- Dsa = Mediterranean-influenced hot-summer humid continental climate; coldest month averaging below −3 °C (27 °F), average temperature of the warmest month above 22 °C (71.6 °F) and at least four months averaging above 10 °C (50 °F). At least three times as much precipitation in the wettest month of winter as in the driest month of summer, and driest month of summer receives less than 30 mm (1.2 in).
- Dsb = Mediterranean-influenced warm-summer humid continental climate; coldest month averaging below −3 °C (27 °F), average temperature of the warmest month below 22 °C (71.6 °F) and at least four months averaging above 10 °C (50 °F). At least three times as much precipitation in the wettest month of winter as in the driest month of summer, and driest month of summer receives less than 30 mm (1.2 in).
- Dsc = subarctic climate; coldest month averaging below −3 °C (27 °F) and 1–3 months averaging above 10 °C (50 °F). At least three times as much precipitation in the wettest month of winter as in the driest month of summer, and driest month of summer receives less than 30 mm (1.2 in).
- Dsd = extremely cold, subarctic climate; coldest month averaging below −38 °C (−36.4 °F) and 1–3 months averaging above 10 °C (50 °F). At least three times as much precipitation in the wettest month of winter as in the driest month of summer, and driest month of summer receives less than 30 mm (1.2 in).

Group E: Polar and Alpine (Montane) Climates

This type of climate has every month of the year with an average temperature below 10 °C (50 °F).

- ET = Tundra climate; average temperature of warmest month between 0 °C (32 °F) and 10 °C (50 °F).
- EF = Ice cap climate; eternal winter, with all 12 months of the year with average temperatures below 0 °C (32 °F).

Spatial Pattern of Climatic Classification

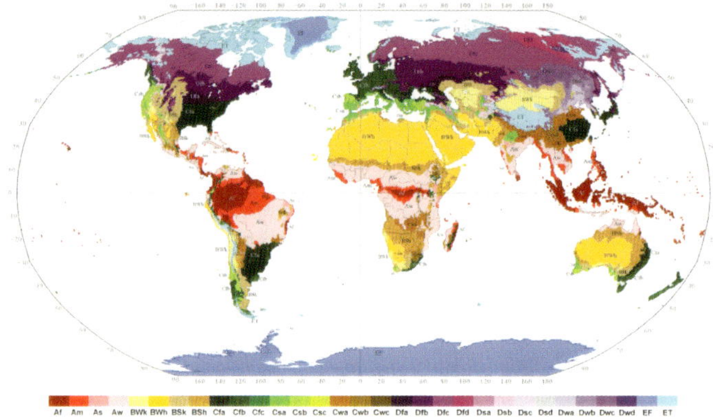

World Koppen's Climate Classification

The spatial pattern of climates can be studied based on the sun's angle of incidence on the earth. On both sides of equator up to tropics A group of climates 132 Understanding Climatology prevail. From equator up to 10° in both the hemispheres sub group of climate Af climate is found. It is closely followed by Am sub group along coastal margins. The two subgroups Aw and As extend upto tropics. While Aw sub group is quite common, As sub group is rare. Arid or B group of hot climates are found along the western margins of the continents. However, along eastern sides of the continents in low latitudes up to 30°, meso thermal or C group of climates are found. These climates are also found along western sides of the continents around coastal margins. D group of climates are found in temperate zones, interior parts of the continents and eastern margins of the continents. E group of climates are usually found in higher latitudes or in higher altitudes. Along arctic and sub- arctic regions, such climates are common. Maritime influences play a decisive and a balancing role in determining a climate and biological situations. Similarly, prevailing winds easterly adds to humidity proportions along eastern margins while westerly winds add to humidity in western margins of the continents. Figure: presents world map of climatic classification as attempted by Koppen – Geiger.

References

- McKnight, Tom L; Hess, Darrel (2000). "Climate Zones and Types". Physical Geography: A Landscape Appreciation. Upper Saddle River, NJ: Prentice Hall. ISBN 0-13-020263-0
- Resh, V.H.; L.A. Bêche; J.E. Lawrence; R.D. Mazor; E.P. McElravy; A.H. Purcell; S.M. Carlson (2013). "Long-term Population and Community Patterns of Benthic Macroinvertebrates and Fishes in Northern California Mediterranean-climate Streams". Journal of the North American Benthological Society. 719: 93–118. doi:10.1007/s10750-012-1373-9. Retrieved 10 November 2013
- Cereceda, P.; Larrain, H.; osses, P.; Farias, M.; Egaña, I. (2008). "The climate of the coast and fog zone in the Tarapacá Region, Atacama Desert, Chile". Atmospheric Research. 87 (3-4): 301–311. Bibcode:2008AtmRe..87..301C. doi:10.1016/j.atmosres.2007.11.011. Retrieved 21 January 2018
- McKnight, Tom L; Hess, Darrel (2000). "Climate Zones and Types". Physical Geography: A Landscape Appreciation. Upper Saddle River, NJ: Prentice Hall. ISBN 0-13-020263-0
- Peel, M. C.; Finlayson, B. L.; McMahon, T. A. (March 1, 2007). "Updated world map of the Koppen-Geiger climate classification" (PDF). hydrol-earth-syst-sci.net. University of Melbourne: Hydrology and Earth System Sciences. pp. 1633–1644. Retrieved May 8, 2017

- McKnight, Tom L; Hess, Darrel (2000). "Climate Zones and Types". Physical Geography: A Landscape Appreciation. Upper Saddle River, NJ: Prentice Hall. ISBN 0-13-020263-0
- Lawrence, J.E.; K.B. Lunde; R.D. Mazor; L.A. Bêche; E.P. McElravy; V.H. Resh (2010). "Long-Term Macroinvertebrate Responses to Climate Change: Implications for Biological Assessment in Mediterranean-Climate Streams". Journal of the North American Benthological Society. 29(4): 1424–1440. doi:10.1899/09-178.1. Archived from the original on 13 July 2015. Retrieved 10 November 2013
- Longton, R.E. (1988). Biology of Polar Bryophytes and Lichen. Studies in Polar Research. Cambridge University Press. p. 20. ISBN 0-521-25015-3
- "Average Conditions Accra, Ghana". BBC Weather. May 2011. Archived from the original on 21 October 2006. Retrieved 9 September2009
- Peel, M. C.; B. L. Finlayson; T. A. McMahon (2007). "Updated world map of the Köppen-Geiger climate classification» (PDF). Hydrology and Earth System Sciences. 11: 1633–1644. Bibcode:2007HESS...11.1633P. doi:10.5194/hess-11-1633-2007
- "Estadistica Climatologica Tomo III (pg 512–537)" (PDF) (in Spanish). Dirección General de Aeronáutica Civil. March 2001. Retrieved 15 January 2013
- Lugo, A. E. (1999). "The Holdridge life zones of the conterminous United States in relation to ecosystem mapping" (PDF). Journal of Biogeography. 26: 1025–1038. doi:10.1046/j.1365-2699.1999.00329.x. Retrieved 27 May 2015
- Vijendra K. Boken; Arthur P. Cracknell; Ronald L. Heathcote (2005-03-24). Monitoring and Predicting Agricultural Drought : A Global Study: A Global Study. Oxford University Press. p. 349. ISBN 0198036787. Retrieved 2015-02-18

Chapter 4

Climate Change and Global Warming

Global warming refers to the rise in earth's average temperature and the effects it has on the environment. Climate change is the change in weather patterns over an extended time period. This chapter presents in detail the phenomenon of climate change, global warming and greenhouse gases. It also explores the physical impacts of climate change, its effect on ecosystems, plant biodiversity, etc.

Climate Change

In simpler terms, climate change may be expressed as any substantial change in the Earth's climate that lasts for an extended period of time. Normally, minimum thirtyyears' time period are being considered for analysing or predicting any climatic condition of a place. Two major elements of climate are temperature and precipitation.According to the Inter-Governmental Panel on Climate Change (IPCC), climate change refers to:

"any change in climate over time, whether due to natural variability or as a result of human activity".

The United Nations Framework Convention on Climate Change (UNFCC) in Article 1defined it as:

"climate change refers to a change in climate which is attributed directly or indirectly to human activity that alters the composition of the global atmosphere and which is in addition to natural climate variability observed over comparable time period".

Scientists actively work to understand past and future climate by using observations and theoretical models. A climate record—extending deep into the Earth's past—has been assembled, and continues to be built up, based on geological evidence from borehole temperature profiles, cores removed from deep accumulations of ice, floral and faunal records, glacial and periglacial processes, stable-isotope and other analyses of sediment layers, and records of past sea levels. More recent data are provided by the instrumental record. General circulation models, based on the physical sciences, are often used in theoretical approaches to match past climate data, make future projections, and link causes and effects in climate change.

Factors that can shape climate are called climate forcings or "forcing mechanisms". These can be either "internal" or "external". Internal forcing mechanisms are natural processes within the climate system itself (e.g., the thermohaline circulation). External forcing mechanisms can be either anthropogenic caused by humans - (e.g. increased emissions of greenhouse gases and dust) or natural (e.g., changes in solar output, the earth's orbit, volcano eruptions).

Physical evidence to observe climate change includes a range of parameters. Global records of surface temperature are available beginning from the mid-late 19th century. For earlier periods, most of the evidence is indirect—climatic changes are inferred from changes in proxies, indicators that reflect climate, such as vegetation, ice cores, dendrochronology, sea level change, and glacial geology. Other physical evidence includes arctic sea ice decline, cloud cover and precipitation, vegetation, animals and historical and archaeological evidence.

Terminology

The most general definition of climate change is a change in the statistical properties (principally its mean and spread) of the climate system when considered over long periods of time, regardless of cause. Accordingly, fluctuations over periods shorter than a few decades, such as El Niño, do not represent climate change.

The term "climate change" is often used to refer specifically to anthropogenic climate change (also known as global warming). Anthropogenic climate change is caused by human activity, as opposed to changes in climate that may have resulted as part of Earth's natural processes. In this sense, especially in the context of environmental policy, the term climate change has become synonymous with anthropogenic global warming. Within scientific journals, global warming refers to surface temperature increases while climate change includes global warming and everything else that increasing greenhouse gas levels affect.

A related term, "climatic change", was proposed by the World Meteorological Organization (WMO) in 1966 to encompass all forms of climatic variability on time-scales longer than 10 years, but regardless of cause. During the 1970s, the term climate change replaced climatic change to focus on anthropogenic causes, as it became clear that human activities had a potential to drastically alter the climate. Climate change was incorporated in the title of the Intergovernmental Panel on Climate Change (IPCC) and the UN Framework Convention on Climate Change (UNFCCC). Climate change is now used as both a technical description of the process, as well as a noun used to describe the problem.

Sources of Climate Change

Do you know climate change had happened many a times on the Earth's history since its origin? Then, you must be thinking that why we are making so much hue and cry about this. This is because climate change in today's context is anthropogenic or in simpler term created by human activities. Evidences suggest that earlier climate change occurred due to natural processes. Therefore, we will discuss briefly about various natural sources responsible for climate change during different geological periods of earth's history. We will also discuss in details climate change due to anthropogenic activities that have been taking place in recent times.

The sources or the causes for climate change could be grouped into two:

- Natural and
- Anthropogenic

Natural Sources of Climate Change

Some of the factors mentioned below are responsible for long term climate change whereas others are responsible for short term climate change. Major natural sources of climate change aresolar variability, changes in Earth's orbit and tilt, plate tectonics and biological evolution.

Let us discuss briefly above mentioned factors one by one.

Solar Variability: This happens due to variation in solar energy output. You might have read that the energy output of the sun varies slightly over time. This has been measured by the dataobtained through satellite. The variation in the total output has been observed both throughout the day as well as during the solar cycle. These variations can directly affect Earth's climate. But the exact role that solar variability plays is still a controversial and not unanimously accepted by the scientific community.

Changes in Earth's Orbit and Tilt: This was another major natural source responsible for climate change. You might have read that earth generally changes its orbit over a 100,000-year cycle.It normally shifts from almost a circular to elliptical orbit. This has led to changes in the tilt of Earth's axis between 21.8 and 24.4 degrees. As a result of which there has been change in the planet's solar energy budget.In other words, there used to be a change in the amount of solar radiation that different areas of the Earth receive. The northern hemisphere is now closer to the sun in winter and farther away in summer. As a result of which it receives five percent less sunlight in summer than 12,000 years ago. More tilt leads to greater seasonal variations. This is known as Milankovitch cycle. Milankovitch was a Serbian geophysicist and astronomer who theorized this above mentioned phenomena.

Plate Tectonics: As a geography student, you might have read about continental drift and plate tectonic theory in your course on Physical Geography or specifically in geomorphology. As described by both the theories, millions of years ago most of the Earth's land mass was a single continent known as Pangea and surrounded by one huge water bodies known as panthalas. Over time, portions of it drifted apart, gradually forming the continents and ocean basins existing today. The redistribution of land mass and ocean area has had a major effect on global climate.

Biological Evolution: This natural process took place mainly during the evolution of photosynthetic organism. Do you know that several billion years ago Earth's pre-life atmosphere was mainly covered by carbon dioxide and methane? The gradual shift in the atmospheric concentrations fromcarbon dioxide and methane to nitrogen and oxygen took place because photosynthetic organisms consumed carbon dioxide and generated oxygen.

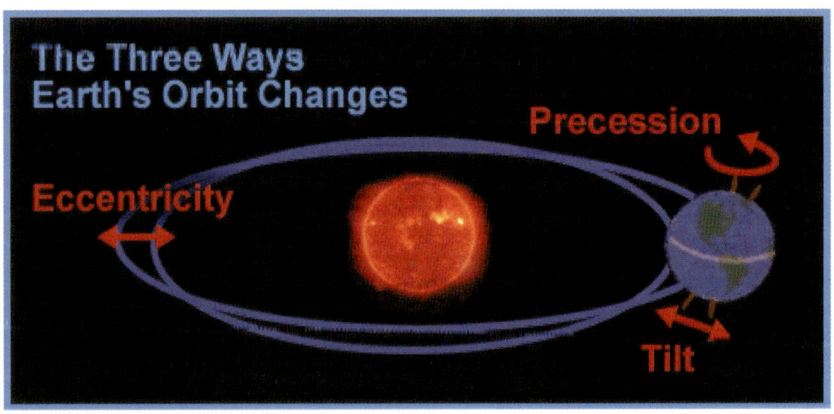

Climate Change due to Changes in Earth's Orbit and Tilt

Causes

On the broadest scale, the rate at which energy is received from the Sun and the rate at which it is lost to space determine the equilibrium temperature and climate of Earth. This energy is distributed around the globe by winds, ocean currents, and other mechanisms to affect the climates of different regions.

Factors that can shape climate are called climate forcings or "forcing mechanisms". These include processes such as variations in solar radiation, variations in the Earth's orbit, variations in the albedo or reflectivity of the continents, atmosphere, and oceans, mountain-building and continental drift and changes in greenhouse gas concentrations. There are a variety of climate change feedbacks that can either amplify or diminish the initial forcing. Some parts of the climate system, such as the oceans and ice caps, respond more slowly in reaction to climate forcings, while others respond more quickly. There are also key threshold factors which when exceeded can produce rapid change.

Forcing mechanisms can be either "internal" or "external". Internal forcing mechanisms are natural processes within the climate system itself (e.g., the thermohaline circulation). External forcing mechanisms can be either anthropogenic (e.g. increased emissions of greenhouse gases and dust) or natural (e.g., changes in solar output, the earth's orbit, volcano eruptions).

Whether the initial forcing mechanism is internal or external, the response of the climate system might be fast (e.g., a sudden cooling due to airborne volcanic ash reflecting sunlight), slow (e.g. thermal expansion of warming ocean water), or a combination (e.g., sudden loss of albedo in the Arctic Ocean as sea ice melts, followed by more gradual thermal expansion of the water). Therefore, the climate system can respond abruptly, but the full response to forcing mechanisms might not be fully developed for centuries or even longer.

Internal Forcing Mechanisms

Scientists generally define the five components of earth's climate system to include atmosphere, hydrosphere, cryosphere, lithosphere (restricted to the surface soils, rocks, and sediments), and biosphere. Natural changes in the climate system ("internal forcings") result in internal "climate variability". Examples include the type and distribution of species, and changes in ocean-atmosphere circulations.

Ocean-atmosphere Variability

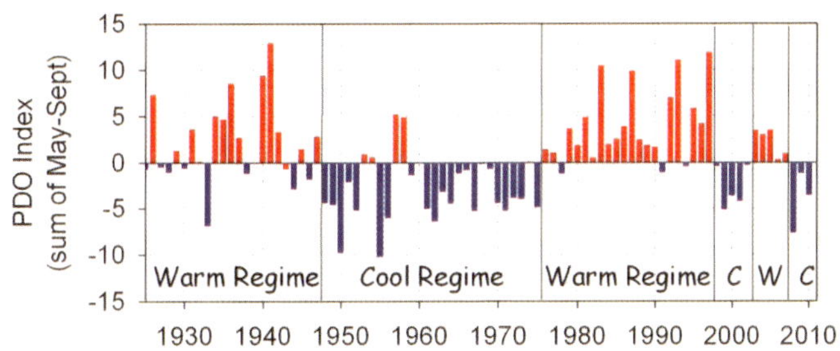

Pacific decadal oscillation 1925 to 2010

The ocean and atmosphere can work together to spontaneously generate internal climate variability that can persist for years to decades at a time. Examples of this type of variability include the El Niño-Southern Oscillation, the Pacific decadal oscillation, and the Atlantic Multidecadal Oscillation. These variations can affect global average surface temperature by redistributing heat between the deep ocean and the atmosphere and/or by altering the cloud/water vapor/sea ice distribution which can affect the total energy budget of the earth.

The oceanic aspects of these circulations can generate variability on centennial timescales due to the ocean having hundreds of times more mass than in the atmosphere, and thus very high thermal inertia. For example, alterations to ocean processes such as thermohaline circulation play a key role in redistributing heat in the world's oceans. Due to the long timescales of this circulation, ocean temperature at depth is still adjusting to effects of the Little Ice Age which occurred between the 1600 and 1800s.

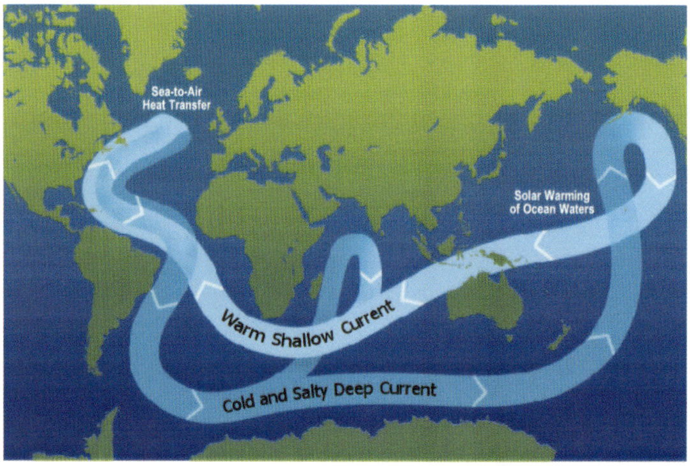

A schematic of modern thermohaline circulation. Tens of millions of years ago, continental-plate movement formed a land-free gap around Antarctica, allowing the formation of the ACC, which keeps warm waters away from Antarctica

Life

Life affects climate through its role in the carbon and water cycles and through such mechanisms as albedo, evapotranspiration, cloud formation, and weathering. Examples of how life may have affected past climate include:

- Glaciation 2.3 billion years ago triggered by the evolution of oxygenic photosynthesis, which depleted the atmosphere of the greenhouse gas carbon dioxide and introduced free oxygen.
- Another glaciation 300 million years ago ushered in by long-term burial of decomposition-resistant detritus of vascular land-plants (creating a carbon sink and forming coal)
- Termination of the Paleocene-Eocene Thermal Maximum 55 million years ago by flourishing marine phytoplankton
- Reversal of global warming 49 million years ago by 800,000 years of arctic azolla blooms
- Global cooling over the past 40 million years driven by the expansion of grass-grazer ecosystems

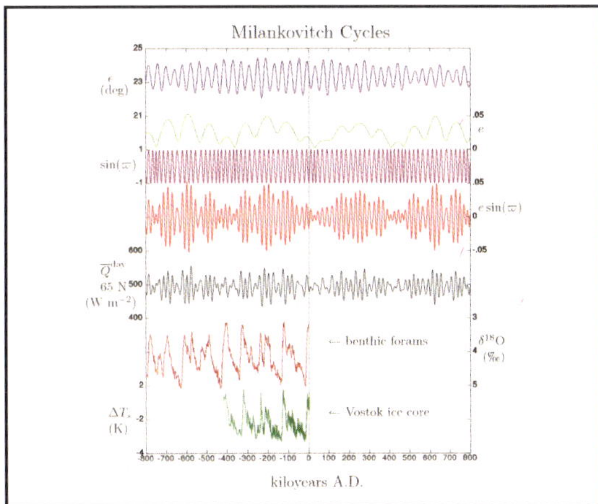

Milankovitch cycles from 800,000 years ago in the past to 800,000 years in the future

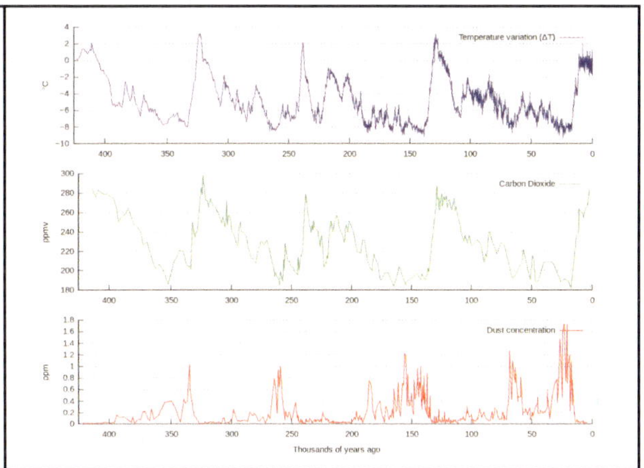

Variations in CO_2, temperature and dust from the Vostok ice core over the last 450,000 years

External Forcing Mechanisms

Human Influences

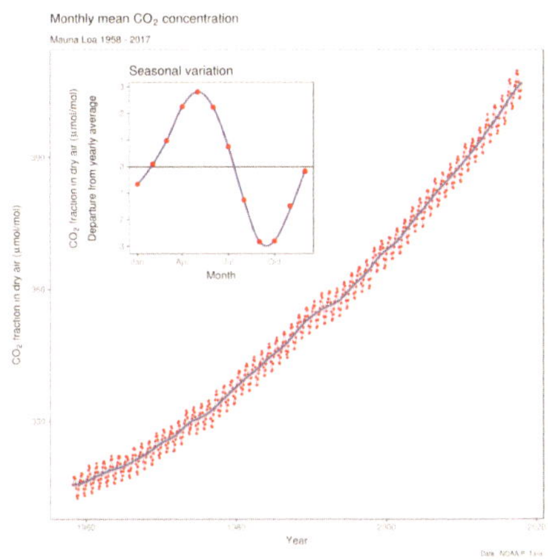

Increase in atmospheric CO_2 levels

In the context of climate variation, anthropogenic factors are human activities which affect the climate. The scientific consensus on climate change is "that climate is changing and that these changes are in large part caused by human activities," and it "is largely irreversible".

"..., there is a strong, credible body of evidence, based on multiple lines of research, documenting that climate is changing and that these changes are in large part caused by human activities. While much remains to be learned, the core phenomenon, scientific questions, and hypotheses have been examined thoroughly and have stood firm in the face of serious scientific debate and careful evaluation of alternative explanations."

— *United States National Research Council, Advancing the Science of Climate Change*

Of most concern in these anthropogenic factors is the increase in CO_2 levels. This is due to emissions from fossil fuel combustion, followed by aerosols (particulate matter in the atmosphere), and the CO_2 released by cement manufacture. Other factors, including land use, ozone depletion, animal husbandry (ruminant animals such as cattle produce methane, as do termites), and deforestation, are also of concern in the roles they play – both separately and in conjunction with other factors – in affecting climate, microclimate, and measures of climate variables.

Orbital Variations

Slight variations in Earth's motion lead to changes in the seasonal distribution of sunlight reaching the Earth's surface and how it is distributed across the globe. There is very little change to the area-averaged annually averaged sunshine; but there can be strong changes in the geographical and seasonal distribution. The three types of kinematic change are variations in Earth's eccentricity, changes in the tilt angle of Earth's axis of rotation, and precession of Earth's axis. Combined together, these produce Milankovitch cycles which affect climate and are notable for their correlation to glacial and interglacial periods, their correlation with the advance and retreat of the Sahara, and for their appearance in the stratigraphic record.

The IPCC notes that Milankovitch cycles drove the ice age cycles, CO_2 followed temperature change "with a lag of some hundreds of years", and that as a feedback amplified temperature change. The depths of the ocean have a lag time in changing temperature (thermal inertia on such scale). Upon seawater temperature change, the solubility of CO_2 in the oceans changed, as well as other factors affecting air-sea CO_2 exchange.

Solar Output

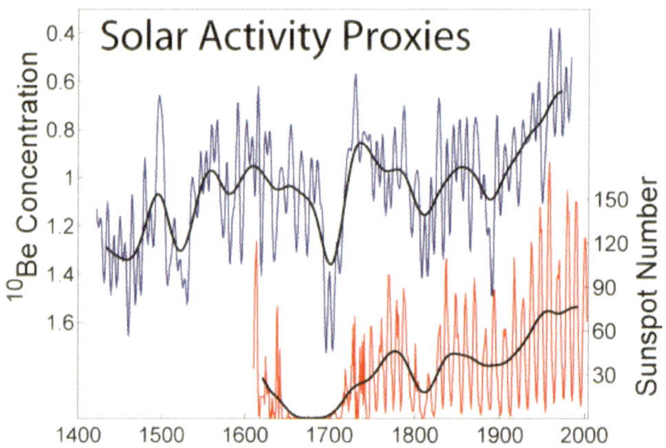

Variations in solar activity during the last several centuries based on observations of sunspots and beryllium isotopes. The period of extraordinarily few sunspots in the late 17th century was the Maunder minimum

The Sun is the predominant source of energy input to the Earth. Other sources include geothermal energy from the Earth's core, tidal energy from the Moon and heat from the decay of radioactive compounds. Both long- and short-term variations in solar intensity are known to affect global climate.

Three to four billion years ago, the Sun emitted only 75% as much power as it does today. If the atmospheric composition had been the same as today, liquid water should not have existed on Earth. However, there is evidence for the presence of water on the early Earth, in the Hadean and Archean eons, leading to what is known as the faint young Sun paradox. Hypothesized solutions to this paradox include a vastly different atmosphere, with much higher concentrations of greenhouse gases than currently exist. Over the following approximately 4 billion years, the energy output of the Sun increased and atmospheric composition changed. The Great Oxygenation Event – oxygenation of the atmosphere around 2.4 billion years ago – was the most notable alteration. Over the next five billion years from the present, the Sun's ultimate death as it becomes a red giant and then a white dwarf will have large effects on climate, with the red giant phase possibly ending any life on Earth that survives until that time.

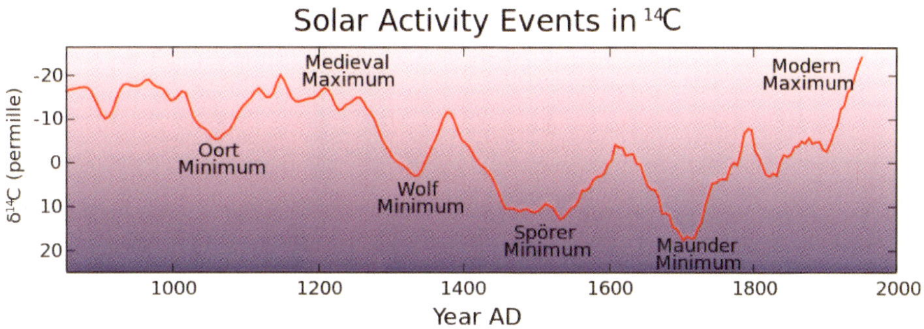

Solar activity events recorded in radiocarbon

Solar output varies on shorter time scales, including the 11-year solar cycle and longer-term modulations. Solar intensity variations, possibly as a result of the Wolf, Spörer, and the Maunder Minima, are considered to have been influential in triggering the Little Ice Age. This event extended from 1550 to 1850 A.D. and was marked by relative cooling and greater glacier extent than the centuries before and afterward. Solar variation may also have affected some of the warming observed from 1900 to 1950. The cyclical nature of the Sun's energy output is not yet fully understood; it differs from the very slow change that is happening within the Sun as it ages and evolves.

Some studies point toward solar radiation increases from cyclical sunspot activity affecting global warming, and climate may be influenced by the sum of all effects (solar variation, anthropogenic radiative forcings, etc.).

A 2010 study suggests "that the effects of solar variability on temperature throughout the atmosphere may be contrary to current expectations."

In 2011, CERN announced the initial results from its CLOUD experiment in the *Nature* journal. The results indicate that ionisation from cosmic rays significantly enhances aerosol formation in the presence of sulfuric acid and water, but in the lower atmosphere where ammonia is also required, this is insufficient to account for aerosol formation and additional trace vapours must be involved. The next step is to find more about these trace vapours, including whether they are of natural or human origin.

Volcanism

The eruptions considered to be large enough to affect the Earth's climate on a scale of more than 1 year are the ones that inject over 100,000 tons of SO_2 into the stratosphere. This is due to the optical properties of SO_2 and sulfate aerosols, which strongly absorb or scatter solar radiation, creating a global layer of sulfuric acid haze. On average, such eruptions occur several times per century, and cause cooling (by partially blocking the transmission of solar radiation to the Earth's surface) for a period of a few years.

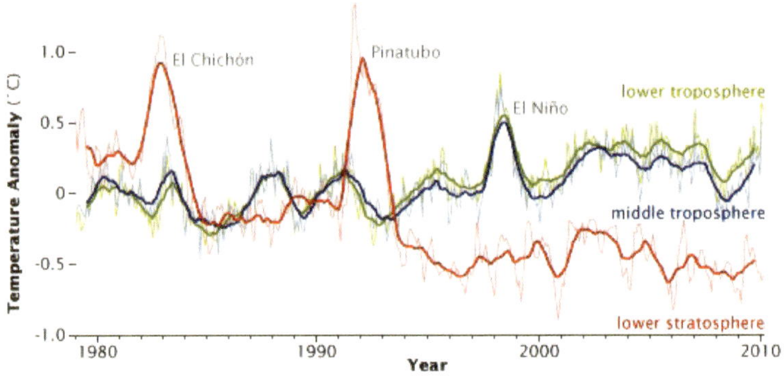

In atmospheric temperature from 1979 to 2010, determined by MSU NASA satellites, effects appear from aerosols released by major volcanic eruptions (El Chichón and Pinatubo). El Niño is a separate event, from ocean variability

The eruption of Mount Pinatubo in 1991, the second largest terrestrial eruption of the 20th century, affected the climate substantially, subsequently global temperatures decreased by about 0.5 °C (0.9 °F) for up to three years. Thus, the cooling over large parts of the Earth reduced surface temperatures in 1991–93, the equivalent to a reduction in net radiation of 4 watts per square meter. The Mount Tambora eruption in 1815 caused the Year Without a Summer. Much larger eruptions, known as large igneous provinces, occur only a few times every fifty – one hundred million years – through flood basalt, and caused in Earth past global warming and mass extinctions.

Small eruptions, with injections of less than 0.1 Mt of sulfur dioxide into the stratosphere, affect the atmosphere only subtly, as temperature changes are comparable with natural variability. However, because smaller eruptions occur at a much higher frequency, they too significantly affect Earth's atmosphere.

Seismic monitoring maps current and future trends in volcanic activities, and tries to develop early warning systems. In climate modelling the aim is to study the physical mechanisms and feedbacks of volcanic forcing.

Volcanoes are also part of the extended carbon cycle. Over very long (geological) time periods, they release carbon dioxide from the Earth's crust and mantle, counteracting the uptake by sedimentary rocks and other geological carbon dioxide sinks. The US Geological Survey estimates are that volcanic emissions are at a much lower level than the effects of current human activities, which generate 100–300 times the amount of carbon dioxide emitted by volcanoes. A review of published studies indicates that annual volcanic emissions of carbon dioxide, including amounts released from mid-ocean ridges, volcanic arcs, and hot spot volcanoes, are only the equivalent of 3 to 5 days of human-caused output. The annual amount put out by human activities may be greater than the amount released by supererruptions, the most recent of which was the Toba eruption in Indonesia 74,000 years ago.

Although volcanoes are technically part of the lithosphere, which itself is part of the climate system, the IPCC explicitly defines volcanism as an external forcing agent.

Plate Tectonics

Over the course of millions of years, the motion of tectonic plates reconfigures global land and ocean areas and generates topography. This can affect both global and local patterns of climate and atmosphere-ocean circulation.

The position of the continents determines the geometry of the oceans and therefore influences patterns of ocean circulation. The locations of the seas are important in controlling the transfer of heat and moisture across the globe, and therefore, in determining global climate. A recent example of tectonic control on ocean circulation is the formation of the Isthmus of Panama about 5 million years ago, which shut off direct mixing between the Atlantic and Pacific Oceans. This strongly affected the ocean dynamics of what is now the Gulf Stream and may have led to Northern Hemisphere ice cover. During the Carboniferous period, about 300 to 360 million years ago, plate tectonics may have triggered large-scale storage of carbon and increased glaciation. Geologic evidence points to a "megamonsoonal" circulation pattern during the time of the supercontinent Pangaea, and climate modeling suggests that the existence of the supercontinent was conducive to the establishment of monsoons.

The size of continents is also important. Because of the stabilizing effect of the oceans on temperature, yearly temperature variations are generally lower in coastal areas than they are inland. A larger supercontinent will therefore have more area in which climate is strongly seasonal than will several smaller continents or islands.

Other Mechanisms

The Earth receives an influx of ionized particles known as cosmic rays from a variety of external sources, including the Sun. A hypothesis holds that an increase in the cosmic ray flux would increase the ionization in the atmosphere, leading to greater cloud cover. This, in turn, would tend to cool the surface. The non-solar cosmic ray flux may vary as a result of a nearby supernova event, the solar system passing through a dense interstellar cloud, or the oscillatory movement of the Sun's position with respect to the galactic plane. The latter can increase the flux of high-energy cosmic rays coming from the Virgo cluster.

Evidence exists that the Chicxulub impact some 66 million years ago had severely affected the Earth's climate. Large quantities of sulfate aerosols were kicked up into the atmosphere, decreasing global temperatures by up to 26 °C and producing sub-freezing temperatures for a period of 3–16 years. The recovery time for this event took more than 30 years.

Sources of Evidences about Climate Change

When we discuss about the science of climate change we generally look for evidences and causes responsible for climate change. As a result of which the scientific community including geographers look for three key elements. These three key elements are "the study of past climates, the measurement of current climate change and the modelling and projections of future climate scenario."

Scientists all over the world have been systematically compiling and analysing different climate related parameters to ascertain about the changing climate. They have also made an attempt to ascertain that the present climate change is human induced. The IPCC, since its inception in 1988 has been working as the international scientific organization coordinating global climate change research, climate forecasts, and policy formulation by engagingin collaboration of scientists and policy experts from many disciplines across the globe. Therefore, this assertion is based on certain facts rather than any speculation or politically motivated. Some of the sources of evidences are as follows:

Evidences of Climate Change

Climate scientists have been constantly engaged in gathering information on key parameters of climate that indicates that some of these parameters have been unequivocally reflectwarming of climate. These evidences are gathered from various sources for last hundred years as well as in extensive details for the last two to three decades. Various sources used for gathering such information are innumerable weather stations across the globe, satellites, ships, aircrafts, weather balloons and buoys. These evidences are collated and trend analysis has been undertaken by numerous scientists engaged at national and international level.

At the international level, these researches are compiled and global trends have been reported by the Intergovernmental Panel on Climate Change (IPCC) in the form of Assessment Reports. IPCC since its origin in 1988, has already been published these assessment reports. So far, five reports have been published in chronological order. The latest Fifth Assessment Reports have been published between September 2013 and April 2014.

Increasing Temperatures over Land and Ocean Surfaces

If we analyse temperature data in terms of hemisphere since 1880, it has been observed that the years with the warmest land-surface temperatures were 2005 and 2010 in the Northern Hemisphere. In the Southern Hemisphere, 2009 was the warmest. It has also been observed that the period from 2000 to 2010 was the warmest decade since 1880. The data from long-term climate reconstructions of temperature point to the present time as the warmest in the last 120,000 years. These reconstructions also suggest that the increase in temperature during the twentieth century is extremely likely (within a confidence interval of greater than 95%–100 %) the largest to occur in any century over the past 1000 years. Record-setting summer daytime temperatures are being recorded in many countries.

According to National Oceanic Administration of America (NOAA) oceantemperatures have also been rising. According to NOAA sea surface temperatures increased at an average rate of 0.07 C° (0.13 F°) per year from 1901 to 2012 as oceans absorbed atmospheric heat. This rise is reflected in measurements of upper-ocean heat content, which includes the upper 700 m (2296 ft.) of ocean.

The instrumental temperature record from surface stations was supplemented by radiosonde balloons, extensive atmospheric monitoring by the mid-20th century, and, from the 1970s on, with global satellite data as well. Taking the record as a whole, most of the 20th century had been unprecedentedly warm, while the 19th and 17th centuries were quite cool.

The $^{18}O/^{16}O$ ratio in calcite and ice core samples used to deduce ocean temperature in the distant past is an example of a temperature proxy method, as are other climate metrics noted in subsequent categories.

Melting of Glacial and Sea Ice

You might have read in the newspapers, magazines or heard from TV, Radio etc. that glacier in Greenland and Antarctica have been receding due to fast rate of melting.Is this a fact? If yes, what are the reasons responsible for such situation? As discussed in the previous paragraph that there has been increase in surface temperature both at land and sea. This rise in temperature has a direct relationship with the melting of glacial and sea ice.

As temperatures rise in Earth's atmosphere, glaciers are losing ice mass, shrinking in size. This process is known as glacial retreat. Earth's two largest ice sheets, in Greenland and Antarctica, are also losing icemass. This is evident from the satellite records. Analysis of satellite data of July 2012 revealed that 97% of the ice sheet's surface was melting. This was the greatest extent in the 30-year record of satellite measurements. Another analysis related to summer melt of the Greenland ice sheet has also revealed that it has increased 30% from 1979 to 2006. On the basis of these evidences, scientists now estimate that between one- and two-thirds of Arctic permafrost will thaw over the next 200 years. As you know, these permafrost reserves took thousands of years to form.

Similar kind of observation has also been noticed in relation to sea ice. Maximum sea ice cover in the world is found in the Arctic Sea. The extent of Arctic Sea ice varies over the course of a year. Every summer, some amount of sea ice thaws whereas in winter, the ice refreezes. It has been observed with the help of satellite data that the minimum extent of summer sea-ice occurs in September whereas maximum extent of winter sea-ice occurs in the months of February or early March. However, analysis of satellite datarevealed that this has declined since 1979. September sea ice is declining at a rate of 11% per decade incomparison to the 1979– 2000 average and reached its lowest extent in the modern record in 2012. The accelerating decline of summer sea ice, in association with record losses of sea ice in 2007 and 2012, suggests that summer sea ice may disappear sooner than predicted by most models. Some scientists have also estimated an ice-free summer Arctic Ocean within the next few decades.

Glaciers are considered among the most sensitive indicators of climate change. Their size is determined by a mass balance between snow input and melt output. As temperatures warm, glaciers retreat unless snow precipitation increases to make up for the additional melt; the converse is also true.

Glaciers grow and shrink due both to natural variability and external forcings. Variability in temperature, precipitation, and englacial and subglacial hydrology can strongly determine the evolution of a glacier in a particular season. Therefore, one must average over a decadal or longer time-scale and/or over many individual glaciers to smooth out the local short-term variability and obtain a glacier history that is related to climate.

A world glacier inventory has been compiled since the 1970s, initially based mainly on aerial photographs and maps but now relying more on satellites. This compilation tracks more than 100,000 glaciers covering a total area of approximately 240,000 km^2, and preliminary estimates indicate that the remaining ice cover is around 445,000 km^2. The World Glacier Monitoring Service collects data annually on glacier retreat and glacier mass balance. From this data, glaciers worldwide have been found to be shrinking significantly, with strong glacier retreats in the 1940s, stable or growing conditions during the 1920s and 1970s, and again retreating from the mid-1980s to the present.

The most significant climate processes since the middle to late Pliocene (approximately 3 million years ago) are the glacial and interglacial cycles. The present interglacial period (the Holocene) has lasted about 11,700 years. Shaped by orbital variations, responses such as the rise and fall of continental ice sheets and significant sea-level changes helped create the climate. Other changes, including Heinrich events, Dansgaard–Oeschger events and the Younger Dryas, however, illustrate how glacial variations may also influence climate without the orbital forcing.

Glaciers leave behind moraines that contain a wealth of material—including organic matter, quartz, and potassium that may be dated—recording the periods in which a glacier advanced and retreated. Similarly, by tephrochronological techniques, the lack of glacier cover can be identified by the presence of soil or volcanic tephra horizons whose date of deposit may also be ascertained.

Data from NASA's Grace satellites show that the land ice sheets in both Antarctica and Greenland have been losing mass since 2002. Both ice sheets have seen an acceleration of ice mass loss since 2009.

The decline in Arctic sea ice, both in extent and thickness, over the last several decades is further evidence for rapid climate change. Sea ice is frozen seawater that floats on the ocean surface. It covers millions of square kilometers in the polar regions, varying with the seasons. In the Arctic, some sea ice remains year after year, whereas almost all Southern Ocean or Antarctic sea ice melts away and reforms annually. Satellite observations show that Arctic sea ice is now declining at a rate of 13.2 percent per decade, relative to the 1981 to 2010 average. The 2007 Arctic summer sea ice retreat was unprecedented. Decades of shrinking and thinning in a warm climate has put the Arctic sea ice in a precarious position, it is now vulnerable to atmospheric anomalies. "Both extent and volume anomaly fluctuate little from January to July and then decrease steeply in August and September". This decrease is because of lessened ice production as a result of the unusually high SAT. During the Arctic summer, a slower rate of sea ice production is the same as a faster rate of sea ice melting.

Rising Sea-Level

This is another important indicator of climate change. It has been observed that sea level is rising more quickly than the prediction simulated by most of the climate models. During the last century, sea level rose 17–21 cm (6.7–8.3 in.). However, this rise is not uniform across the globe. A greater rise has been observed in some areas like Atlantic coast of U.S. than at any time during the past 2000 years. How did the scientists arrive at such conclusion? Do you have any idea about the tools used for assessing the rise in sea level? Generally we use tidal gauges and satellites to generate such kind of data. Tidal gauge records from 1901 to 2010 show that sea level rose at a rate of 1.7 mm (0.07 in.) per year. Satellite data for the period 1993 - 2013 show that sea level rose 3.16 mm (0.12 in.) per year. This rise is primary due to two major factors that are presently contributing to sea-level rise. About two-thirds of the rise comes from the melting of glaciers and ice sheets whereas the rest one third comes from the thermal expansion of seawater due to absorption of atmospheric heat and expand in volume.

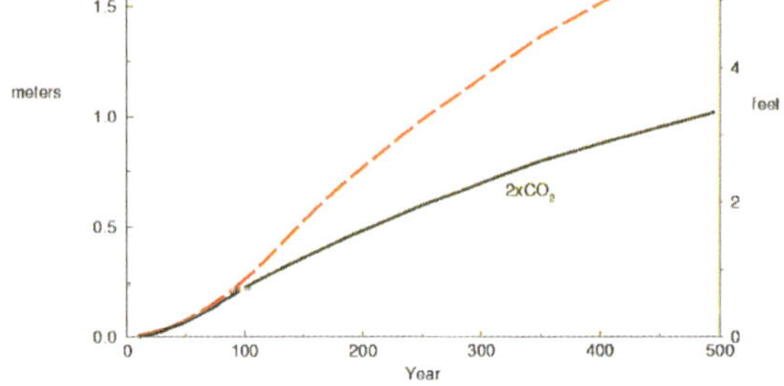

The estimated change in sea level caused by carbon dioxide emissions

Global sea level change for much of the last century has generally been estimated using tide gauge measurements collated over long periods of time to give a long-term average. More recently, altimeter measurements — in combination with accurately determined satellite orbits — have provided an improved measurement of global sea level change. To measure sea levels prior to instrumental measurements, scientists have dated coral reefs that grow near the surface of the ocean, coastal sediments, marine terraces, ooids in limestones, and nearshore archaeological remains. The predominant dating methods used are uranium series and radiocarbon, with cosmogenic radionuclides being sometimes used to date terraces that have experienced relative sea level fall. In the early Pliocene, global temperatures were 1–2°C warmer than the present temperature, yet sea level was 15–25 meters higher than today.

According to recent studies, global-mean sea level rose by 195 mm during the period from 1870 to 2004. Since 2004, satellite-based records indicate that there has been a further 43 mm of global-mean sea levels rise, as of July 2017.

Increasing Humidity and Extreme Events

It has been observed that global average specific humidity has increased by about 0.1 g of water vapour per kilogram of air per decade since 1973. This change is consistent with rising air temperatures, since warm air has a greater capacity to absorb water vapour. A greater amount of water vapour in the atmosphere affects weather in a number of ways and can lead to "extreme" events involving temperature, precipitation, and storm intensity. According to the World Meteorological Organization, the decade from 2001 to 2010 showed evidence of a worldwide increase in extreme events, notably heat waves, increased precipitation, and floods. However, to establish a strong linkage between trends related to extreme weather and climate change requires data for a longer time frame than what is now available.

These above mentioned change have differential impacts in terms of geographical locations, sectors of economy and socio-economic groups. Therefore, some of the major impacts/consequences of climate change are mentioned for two levels. One is at macro scale i.e. at global level and the other at micro-level i.e. at national level.

Cloud Cover and Precipitation

Past precipitation can be estimated in the modern era with the global network of precipitation gauges. Surface coverage over oceans and remote areas is relatively sparse, but, reducing reliance on interpolation, satellite clouds and precipitation data has been available since the 1970s. Quantification of climatological variation of precipitation in prior centuries and epochs is less complete but approximated using proxies such as marine sediments, ice cores, cave stalagmites, and tree rings. In July 2016 scientists published evidence of increased cloud cover over polar regions, as predicted by climate models.

Climatological temperatures substantially affect cloud cover and precipitation. For instance, during the Last Glacial Maximum of 18,000 years ago, thermal-driven evaporation from the oceans onto continental landmasses was low, causing large areas of extreme desert, including polar deserts (cold but with low rates of cloud cover and precipitation). In contrast, the world's climate was cloudier and wetter than today near the start of the warm Atlantic Period of 8000 years ago.

Estimated global land precipitation increased by approximately 2% over the course of the 20th century, though the calculated trend varies if different time endpoints are chosen, complicated by ENSO and other oscillations, including greater global land cloud cover precipitation in the 1950s and 1970s than the later 1980s and 1990s despite the positive trend over the century overall. Similar slight overall increase in global river runoff and in average soil moisture has been perceived.

Vegetation

A change in the type, distribution and coverage of vegetation may occur given a change in the climate. Some changes in climate may result in increased precipitation and warmth, resulting in improved plant growth and the subsequent sequestration of airborne CO_2. A gradual increase in warmth in a region will lead to earlier flowering and fruiting times, driving a change in the timing of life cycles of dependent organisms. Conversely, cold will cause plant bio-cycles to lag. Larger, faster or more radical changes, however, may result in vegetation stress, rapid plant loss and desertification in certain circumstances. An example of this occurred during the Carboniferous Rainforest Collapse (CRC), an extinction event 300 million years ago. At this time vast rainforests covered the equatorial region of Europe and America. Climate change devastated these tropical rainforests, abruptly fragmenting the habitat into isolated 'islands' and causing the extinction of many plant and animal species.

Dendroclimatology is the analysis of tree ring growth patterns to determine past climate variations. Wide and thick rings indicate a fertile, well-watered growing period, while thin, narrow rings indicate a period of lower rainfall and less-than-ideal growing conditions.

Forest Genetic Resources

Even though this is a field with many uncertainties, it is expected that over the next 50 years climate changes will have an effect on the diversity of forest genetic resources and thereby on the distribution of forest tree species and the composition of forests. Diversity of forest genetic resources enables the potential for a species (or a population) to adapt to climatic changes and related future challenges such as temperature changes, drought, pests, diseases and forest fire. However, species are not naturally capable to adapt in the pace of which the climate is changing and the increasing temperatures will most likely facilitate the spread of pests and diseases, creating an additional threat to forest trees and their populations. To inhibit these problems human interventions, such as transfer of forest reproductive material, may be needed.

Pollen Analysis

Palynology is the study of contemporary and fossil palynomorphs, including pollen. Palynology is used to infer the geographical distribution of plant species, which vary under different climate conditions. Different groups of plants have pollen with distinctive shapes and surface textures, and since the outer surface of pollen is composed of a very resilient material, they resist decay. Changes in the type of pollen found in different layers of sediment in lakes, bogs, or river deltas indicate changes in plant communities. These changes are often a sign of a changing climate. As an example, palynological studies have been used to track changing vegetation patterns throughout the Quaternary glaciations and especially since the last glacial maximum.

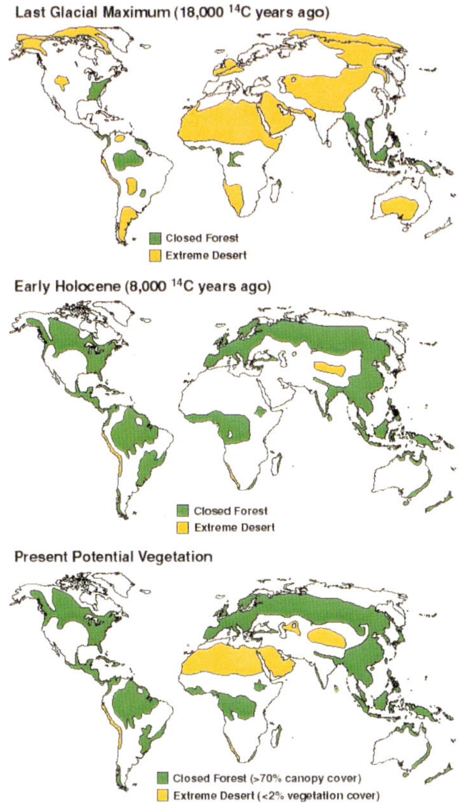

Top: Arid ice age climate
Middle: Atlantic Period, warm and wet
Bottom: Potential vegetation in climate now if not for human effects like agriculture

Animals

Remains of beetles are common in freshwater and land sediments. Different species of beetles tend to be found under different climatic conditions. Given the extensive lineage of beetles whose genetic makeup has not altered significantly over the millennia, knowledge of the present climatic range of the different species, and the age of the sediments in which remains are found, past climatic conditions may be inferred. The studies of the impact in vertebrates are few mainly from developing countries, where there are the fewest studies; between 1970 and 2012, vertebrates declined by 58 percent, with freshwater, marine, and terrestrial populations declining by 81, 36, and 35 percent, respectively.

Similarly, the historical abundance of various fish species has been found to have a substantial relationship with observed climatic conditions. Changes in the primary productivity of autotrophs in the oceans can affect marine food webs.

Historical and Archaeological Evidence

Climate change in the recent past may be detected by corresponding changes in settlement and agricultural patterns. Archaeological evidence, oral history and historical documents can offer insights into past changes in the climate. Climate change effects have been linked to the rise and also the collapse of various civilizations.

Global Warming

Global warming, also referred to as climate change, is the observed century-scale rise in the average temperature of the Earth's climate system and its related effects. Multiple lines of scientific evidence show that the climate system is warming. Many of the observed changes since the 1950s are unprecedented in the instrumental temperature record, which extends back to the mid-19th century, and in paleoclimate proxy records covering thousands of years.

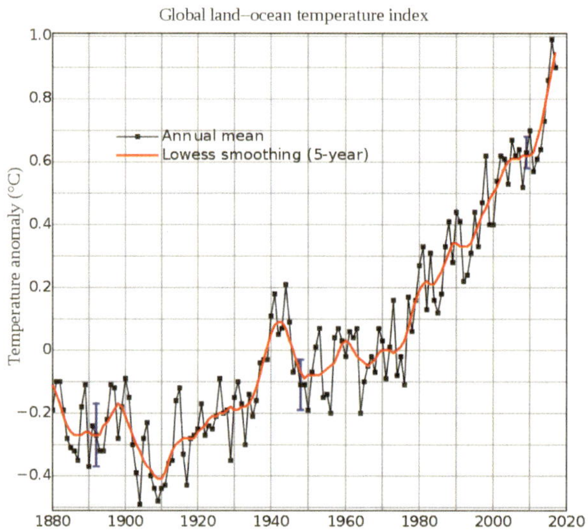

Global mean surface-temperature change from 1880 to 2017, relative to the 1951–1980 mean. The black line is the global annual mean, and the red line is the five-year local regression line. The blue uncertainty bars show a 95% confidence interval

In 2013, the Intergovernmental Panel on Climate Change (IPCC) Fifth Assessment Report concluded that "It is *extremely likely* that human influence has been the dominant cause of the observed warming since the mid-20th century." The largest human influence has been the emission of greenhouse gases such as carbon dioxide, methane and nitrous oxide. Climate model projections summarized in the report indicated that during the 21st century, the global surface temperature is likely to rise a further 0.3 to 1.7 °C (0.5 to 3.1 °F) in the lowest emissions scenario, and 2.6 to 4.8 °C (4.7 to 8.6 °F) in the highest emissions scenario. These findings have been recognized by the national science academies of the major industrialized nations and are not disputed by any scientific body of national or international standing.

Future climate change and associated impacts will differ from region to region. Anticipated effects include increasing global temperatures, rising sea levels, changing precipitation, and expansion of deserts in the subtropics. Warming is expected to be greater over land than over the oceans and greatest in the Arctic, with the continuing retreat of glaciers, permafrost and sea ice. Other likely changes include more frequent extreme weather events such as heat waves, droughts, heavy rainfall with floods and heavy snowfall; ocean acidification; and species extinctions due to shifting temperature regimes. Effects significant to humans include the threat to food security from decreasing crop yields and the abandonment of populated areas due to rising sea levels. Because the climate system has a large "inertia" and greenhouse gases will remain in the atmosphere for a long time, many of these effects will persist for not only decades or centuries, but for tens of thousands of years to come.

Possible societal responses to global warming include mitigation by emissions reduction, adaptation to its effects, building systems resilient to its effects, and possible future climate engineering. Most countries are parties to the United Nations Framework Convention on Climate Change (UNFCCC), whose ultimate objective is to prevent dangerous anthropogenic climate change. Parties to the UNFCCC have agreed that deep cuts in emissions are required and that global warming should be limited to well below 2.0 °C (3.6 °F) compared to pre-industrial levels, with efforts made to limit warming to 1.5 °C (2.7 °F).

Public reactions to global warming and concern about its effects are also increasing. A global 2015 Pew Research Center report showed that a median of 54% of all respondents asked consider it "a very serious problem". Significant regional differences exist, with Americans and Chinese (whose economies are responsible for the greatest annual CO_2 emissions) among the least concerned.

Observed Temperature Changes

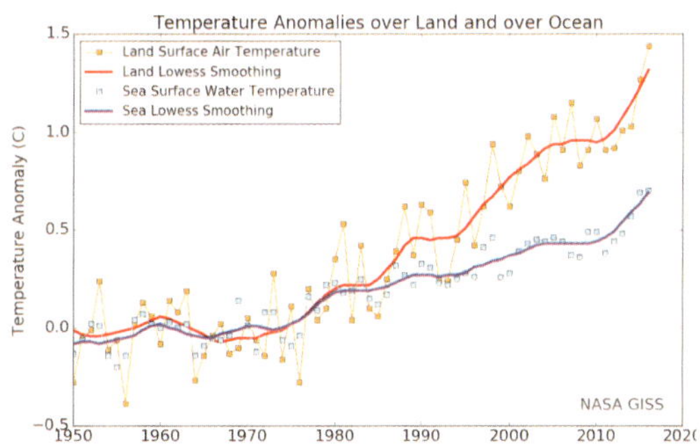

Annual (thin lines) and five-year lowess smooth (thick lines) for the temperature anomalies averaged over the Earth's land area (red line) and sea surface temperature anomalies (blue line) averaged over the part of the ocean that is free of ice at all times (open ocean)

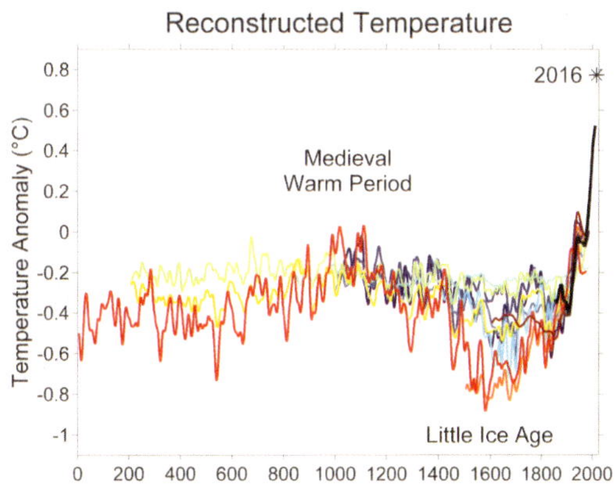

Two millennia of mean surface temperatures according to different reconstructions from climate proxies, each smoothed on a decadal scale, with the instrumental temperature record overlaid in black

In the period from 1880 to 2012, the global average (land and ocean) surface temperature has increased by 0.85 [0.65 to 1.06] °C, multiple independently produced datasets confirm. In the period from 1906 to 2005, Earth's average surface temperature rose by 0.74±0.18 °C. The rate of warming almost doubled in the last half of that period (0.13±0.03 °C per decade, against 0.07±0.02 °C per decade). Although the popular press often reports the increase of the average near-surface atmospheric temperature as *the* measure of global warming, most of the additional energy stored in the climate system since 1970 has accumulated in the oceans. The rest has melted ice and warmed the continents and the atmosphere.

Since 1979, the average temperature of the lower troposphere has increased between 0.12 and 0.135 °C (0.216 and 0.243 °F) per decade, satellite temperature measurements confirm. Climate proxies show the temperature to have been relatively stable over the one or two thousand years before 1850, with regionally varying fluctuations such as the Medieval Warm Period and the Little Ice Age.

The warming evident in the instrumental temperature record is consistent with a wide range of observations, as documented by many independent scientific groups. Examples include sea level rise, widespread melting of snow and land ice, increased heat content of the oceans, increased humidity, and the earlier timing of spring events, e.g., the flowering of plants. The probability that these changes could have occurred by chance is virtually zero.

Regional Trends and Short-term Fluctuations

Global warming refers to global averages. Because it is not a uniform phenomenon, effects can vary by region. Since 1979, global average land temperatures have increased about twice as fast as global average ocean temperatures (0.25 °C per decade against 0.13 °C per decade). Ocean temperatures increase more slowly than land temperatures because of the larger effective heat capacity of the oceans and because oceans lose more heat by evaporation. Since the beginning of industrialisation in the eighteenth century, the temperature difference between the hemispheres has increased due to feedbacks from melting of sea ice and snow in the North, and because there is more land in the northern hemisphere. In the past one hundred years, average arctic temperatures have been increasing at almost twice the rate of the rest of the world. At least one region – the southeastern part of the United States – has experienced cooler than normal temperatures.

Although more greenhouse gases are emitted in the Northern than in the Southern Hemisphere, this fact does not contribute to the difference in warming because the major greenhouse gases persist long enough to diffuse within as well as between the hemispheres.

There are different ways in which a climate can be forced to change, but because the climate system has large thermal inertia it can take centuries – or even longer – for the climate to fully adjust. One climate commitment study concluded that if greenhouse gases were stabilized at year 2000 levels, surface temperatures would still increase by about one-half degree Celsius, and another found that if they were stabilized at 2005 levels, surface warming could exceed a whole degree Celsius. Some of this surface warming will be driven by past natural forcings which are still seeking equilibrium in the climate system. One study using a highly simplified climate model indicates these past natural forcings may account for as much as 64% of the committed 2050 surface warming and their influence will fade with time compared to the human contribution.

Global temperature is subject to short-term fluctuations that overlay long-term trends and can temporarily mask them. The relative stability in surface temperature from 2002 to 2009, which has since been dubbed the global warming hiatus by the media and some scientists, is an example of such an episode. 2015 updates to account for differing methods of measuring ocean surface temperature measurements show a positive trend over the recent decade.

Warmest Years vs. Overall Trend

Sixteen of the seventeen warmest years on record have occurred since 2000. While record-breaking years attract considerable public interest, individual years are less significant than the overall trend. Some climatologists have criticized the attention that the popular press gives to "warmest year" statistics. In particular, ocean oscillations such as the El Niño Southern Oscillation (ENSO) can cause temperatures of a given year to be abnormally warm or cold for reasons unrelated to the overall trend of climate change. Gavin Schmidt stated: "the long-term trends or the expected sequence of records are far more important than whether any single year is a record or not."

Initial Causes of Temperature Changes (External Forcings)

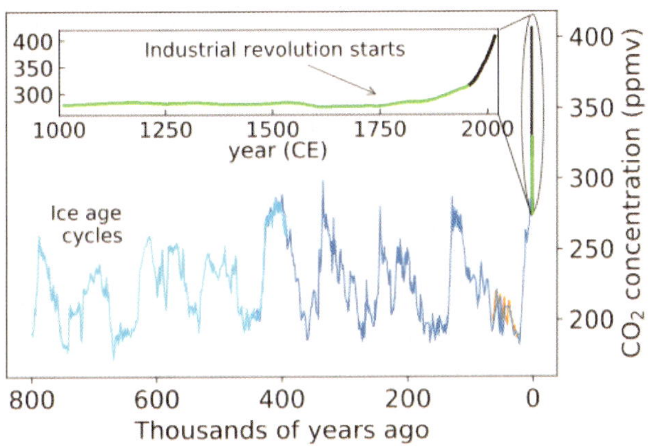

CO_2 concentrations over the last 800,000 years

By itself, the climate system may generate random changes in global temperatures for years to decades at a time, but long-term changes emanate only from so-called *external forcings*. These forcings are "external" to the climate system, but not necessarily external to Earth. Examples of external forcings include changes in the composition of the atmosphere (e.g., increased concentrations of greenhouse gases), solar luminosity, volcanic eruptions, and variations in Earth's orbit around the Sun.

Greenhouse Gases

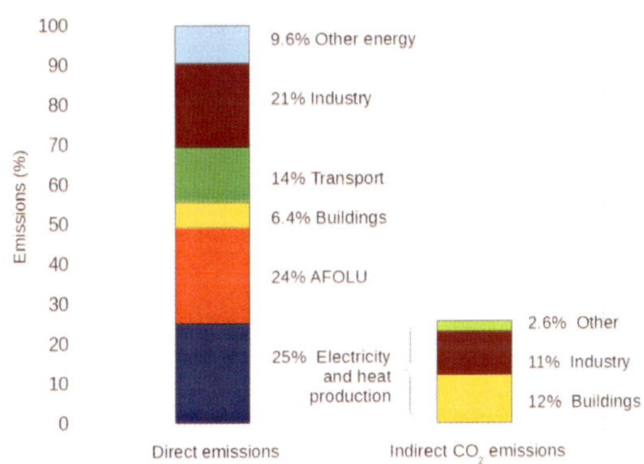

Annual world greenhouse gas emissions, in 2010, by sector.
Percentage share of global cumulative energy-related CO_2 emissions between 1751 and 2012 across different regions

The greenhouse effect is the process by which absorption and emission of infrared radiation by gases in a planet's atmosphere warm its lower atmosphere and surface. It was proposed by Joseph Fourier in 1824, discovered in 1860 by John Tyndall, was first investigated quantitatively by Svante Arrhenius in 1896, and its scientific description was developed in the 1930s through 1960s by Guy Stewart Callendar.

On Earth, an atmosphere containing naturally occurring amounts of greenhouse gases causes air temperature near the surface to be warmer by about 33 °C (59 °F) than it would be in their absence. Without the Earth's atmosphere, the Earth's average temperature would be well below the freezing temperature of water. The major greenhouse gases are water vapour, which causes about 36–70% of the greenhouse effect; carbon dioxide (CO_2), which causes 9–26%; methane (CH_4), which causes 4–9%; and ozone (O_3), which causes 3–7%. Clouds also affect the radiation balance through cloud forcings similar to greenhouse gases.

Human activity since the Industrial Revolution has increased the amount of greenhouse gases in the atmosphere, leading to increased radiative forcing from CO_2, methane, tropospheric ozone, CFCs and nitrous oxide. According to work published in 2007, the concentrations of CO_2 and methane had increased by 36% and 148% respectively since 1750. These levels are much higher than at any time during the last 800,000 years, the period for which reliable data has been extracted from ice cores. Less direct geological evidence indicates that CO_2 values higher than this were last seen about 20 million years ago.

Fossil fuel burning has produced about three-quarters of the increase in CO_2 from human activity over the past 20 years. The rest of this increase is caused mostly by changes in land-use, particularly deforestation. Another significant non-fuel source of anthropogenic CO_2 emissions is the calcination of limestone for clinker production, a chemical process which releases CO_2. Estimates of global CO_2 emissions in 2011 from fossil fuel combustion, including cement production and gas flaring, was 34.8 billion tonnes (9.5 ± 0.5 PgC), an increase of 54% above emissions in 1990. Coal burning was responsible for 43% of the total emissions, oil 34%, gas 18%, cement 4.9% and gas flaring 0.7%.

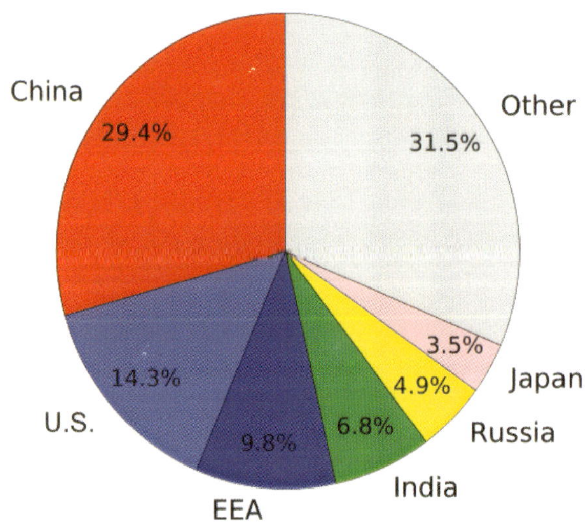

Global carbon dioxide emissions by country

In May 2013, it was reported that readings for CO_2 taken at the world's primary benchmark site in Mauna Loa surpassed 400 ppm. According to professor Brian Hoskins, this is likely the first time CO_2 levels have been this high for about 4.5 million years. Monthly global CO_2 concentrations exceeded 400 ppm in March 2015, probably for the first time in several million years. On 12 November 2015, NASA scientists reported that human-made carbon dioxide continues to increase above levels not seen in hundreds of thousands of years; currently, about half of the carbon dioxide released from the burning of fossil fuels is not absorbed by vegetation and the oceans and remains in the atmosphere.

Over the last three decades of the twentieth century, gross domestic product per capita and population growth were the main drivers of increases in greenhouse gas emissions. CO_2 emissions are continuing to rise due to the burning of fossil fuels and land-use change. Emissions can be attributed to different regions. Attributions of emissions due to land-use change are subject to considerable uncertainty.

Emissions scenarios, estimates of changes in future emission levels of greenhouse gases, have been projected that depend upon uncertain economic, sociological, technological, and natural developments. In most scenarios, emissions continue to rise over the century, while in a few, emissions are reduced. Fossil fuel reserves are abundant, and will not limit carbon emissions in the 21st century. Emission scenarios, combined with modelling of the carbon cycle, have been used to produce estimates of how atmospheric concentrations of greenhouse gases might change in the future. Using the six IPCC SRES "marker" scenarios, models suggest that by the year 2100, the atmospheric concentration of CO_2 could range between 541 and 970 ppm. This is 90–250% above the concentration in the year 1750.

The popular media and the public often confuse global warming with ozone depletion, i.e., the destruction of stratospheric ozone (e.g., the ozone layer) by chlorofluorocarbons. Although there are a few areas of linkage, the relationship between the two is not strong. Reduced stratospheric ozone has had a slight cooling influence on surface temperatures, while increased tropospheric ozone has had a somewhat larger warming effect.

Aerosols and Soot

Global dimming, a gradual reduction in the amount of global direct irradiance at the Earth's surface, was observed from 1961 until at least 1990. Solid and liquid particles known as *aerosols*, produced by volcanoes and human-made pollutants, are thought to be the main cause of this dimming. They exert a cooling effect by increasing the reflection of incoming sunlight. The effects of the products of fossil fuel combustion – CO_2 and aerosols – have partially offset one another in recent decades, so that net warming has been due to the increase in non-CO_2 greenhouse gases such as methane. Radiative forcing due to aerosols is temporally limited due to the processes that remove aerosols from the atmosphere. Removal by clouds and precipitation gives tropospheric aerosols an atmospheric lifetime of only about a week, while stratospheric aerosols can remain for a few years. Carbon dioxide has a lifetime of a century or more, and as such, changes in aerosols will only delay climate changes due to carbon dioxide. Black carbon is second only to carbon dioxide for its contribution to global warming (contribution being estimated at 17 to 20%, whereas carbon dioxide contributes 40 to 45% to global warming).

Ship tracks can be seen as lines in these clouds over the Atlantic Ocean on the East Coast of the United States. Atmospheric particles from these and other sources could have a large effect on climate through the aerosol indirect effect

In addition to their direct effect by scattering and absorbing solar radiation, aerosols have indirect effects on the Earth's radiation budget. Sulfate aerosols act as cloud condensation nuclei and thus lead to clouds that have more and smaller cloud droplets. These clouds reflect solar radiation more efficiently than clouds with fewer and larger droplets, a phenomenon known as the Twomey effect. This effect also causes droplets to be of more uniform size, which reduces growth of raindrops and makes the cloud more reflective to incoming sunlight, known as the Albrecht effect. Indirect effects are most noticeable in marine stratiform clouds, and have very little radiative effect on convective clouds. Indirect effects of aerosols represent the largest uncertainty in radiative forcing.

Soot may either cool or warm Earth's climate system, depending on whether it is airborne or deposited. Atmospheric soot directly absorbs solar radiation, which heats the atmosphere and cools the surface. In isolated areas with high soot production, such as rural India, as much as 50% of surface warming due to greenhouse gases may be masked by atmospheric brown clouds. When deposited, especially on glaciers or on ice in arctic regions, the lower surface albedo can also directly heat the surface. The influences of atmospheric particles, including black carbon, are most pronounced in the tropics and sub-tropics, particularly in Asia, while the effects of greenhouse gases are dominant in the extratropics and southern hemisphere.

Solar Activity

Since 1978, solar irradiance has been measured by satellites. These measurements indicate that the Sun's radiative output has not increased since then, so the warming that occurred in the past 40 years cannot be attributed to an increase in solar energy reaching the Earth.

Climate models have been used to examine the role of the Sun in recent climate change. Models are unable to reproduce the rapid warming observed in recent decades when only taking into account variations in solar output and volcanic activity. Models are, however, able to simulate the observed 20th century changes in temperature when they include all of the most important external forcings, consisting of both human influences and natural forcings.

Another line of evidence for the Sun's non-attributability is the differing temperature changes at different levels in the Earth's atmosphere. According to basic physical principles, the greenhouse effect produces warming of the lower atmosphere (the troposphere), but cooling of the upper atmosphere (the stratosphere). If solar variations were responsible for the observed warming, warming of both the troposphere and the stratosphere would be expected.

Variations in Earth's Orbit

The tilt of the Earth's axis and the shape of its orbit around the Sun vary slowly over tens of thousands of years. This changes climate by changing the seasonal and latitudinal distribution of incoming solar energy at Earth's surface. During the last few thousand years, this phenomenon contributed to a slow cooling trend at high latitudes of the Northern Hemisphere during summer, a trend that was reversed by greenhouse-gas-induced warming during the 20th century. Orbital cycles favorable for glaciation are not expected within the next 50,000 years.

Climate Change Feedback

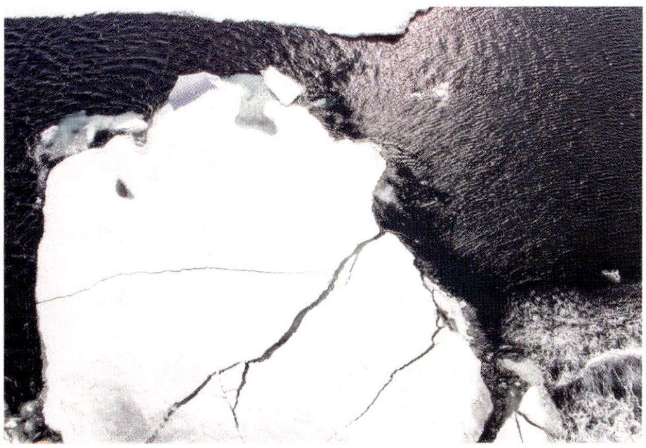

The dark ocean surface reflects only 6 percent of incoming solar radiation, instead sea ice reflects 50 to 70 percent

The climate system includes a range of *feedbacks*, which alter the response of the system to changes in external forcings. Positive feedbacks increase the response of the climate system to an initial forcing, while negative feedbacks reduce it.

There are a range of feedbacks in the climate system, including water vapour, changes in ice-albedo (snow and ice cover affect how much the Earth's surface absorbs or reflects incoming sunlight), clouds, and changes in the Earth's carbon cycle (e.g., the release of carbon from soil). The main negative feedback is the energy the Earth's surface radiates into space as infrared radiation. According to the Stefan-Boltzmann law, if the absolute temperature (as measured in kelvins) doubles, radiated energy increases by a factor of 16 (2 to the 4th power).

Feedbacks are an important factor in determining the sensitivity of the climate system to increased atmospheric greenhouse gas concentrations. Other factors being equal, a higher *climate sensitivity* means that more warming will occur for a given increase in greenhouse gas forcing. Uncertainty over the effect of feedbacks is a major reason why different climate models project different magnitudes of warming for a given forcing scenario. More research is needed to understand the role of clouds and carbon cycle feedbacks in climate projections.

The IPCC projections previously mentioned span the "likely" range (greater than 66% probability, based on expert judgement) for the selected emissions scenarios. However, the IPCC's projections do not reflect the full range of uncertainty. The lower end of the "likely" range appears to be better constrained than the upper end.

An observation based study on future climate change, on the soil carbon feedback, conducted since 1991 in Harvard, suggests release of about 190 petagrams of soil carbon, the equivalent of the past two decades of greenhouse gas emissions from fossil fuel burning, until 2100 from the top 1-meter of Earth's soils, due to changes in microbial communities under elevated temperatures. Climate models do not account for this possible feedback mechanism.

Climate Models

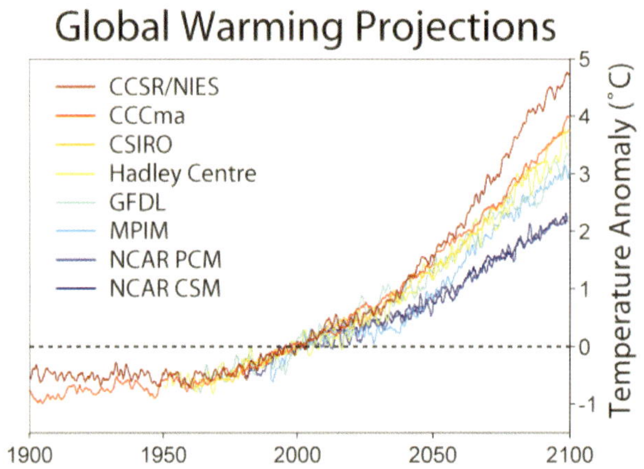

Calculations of global warming prepared in or before 2001 from a range of climate models under the SRES A2 emissions scenario, which assumes no action is taken to reduce emissions and regionally divided economic development

A climate model is a representation of the physical, chemical and biological processes that affect the climate system. Such models are based on scientific disciplines such as fluid dynamics and thermodynamics as well as physical processes such as radiative transfer. The models may be used to predict a range of variables such as local air movement, temperature, clouds, and other atmospheric properties; ocean temperature, salt content, and circulation; ice cover on land and sea; the transfer of heat and moisture from soil and vegetation to the atmosphere; and chemical and biological processes, among others.

Although researchers attempt to include as many processes as possible, simplifications of the actual climate system are inevitable because of the constraints of available computer power and limitations in knowledge of the climate system. Results from models can also vary due to different greenhouse gas inputs and the model's climate sensitivity. For example, the uncertainty in IPCC's 2007 projections is caused by the use of multiple models with differing sensitivity to greenhouse gas concentrations, the use of differing estimates of humanity's future greenhouse gas emissions, any additional emissions from climate feedbacks that were not included in the models IPCC used to prepare its report, i.e., greenhouse gas releases from permafrost.

The models do not assume the climate will warm due to increasing levels of greenhouse gases.

Instead the models predict how greenhouse gases will interact with radiative transfer and other physical processes. Warming or cooling is thus a result, not an assumption, of the models.

Clouds and their effects are especially difficult to predict. Improving the models' representation of clouds is therefore an important topic in current research. Another prominent research topic is expanding and improving representations of the carbon cycle.

Models are also used to help investigate the causes of recent climate change by comparing the observed changes to those that the models project from various natural and human causes. Although these models do not unambiguously attribute the warming that occurred from approximately 1910 to 1945 to either natural variation or human effects, they do indicate that the warming since 1970 is dominated by anthropogenic greenhouse gas emissions.

The physical realism of models is tested by examining their ability to simulate contemporary or past climates. Climate models produce a good match to observations of global temperature changes over the last century, but do not simulate all aspects of climate. Not all effects of global warming are accurately predicted by the climate models used by the IPCC. Observed Arctic shrinkage has been faster than that predicted. Precipitation increased proportionally to atmospheric humidity, and hence significantly faster than global climate models predict. Since 1990, sea level has also risen considerably faster than models predicted it would.

Effects

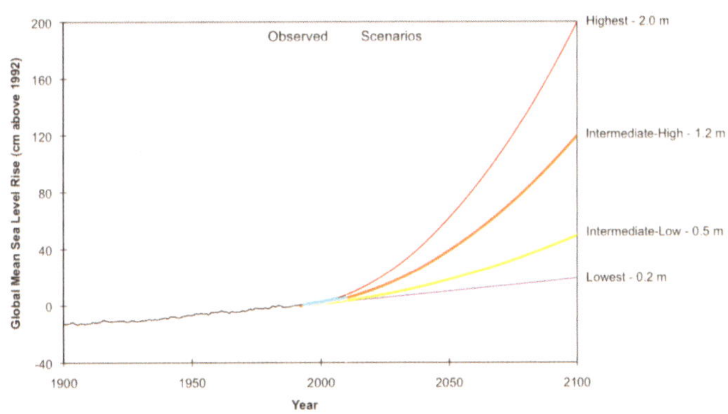

Projections of global mean sea level rise by Parris and others. Probabilities have not been assigned to these projections. Therefore, none of these projections should be interpreted as a "best estimate" of future sea level rise

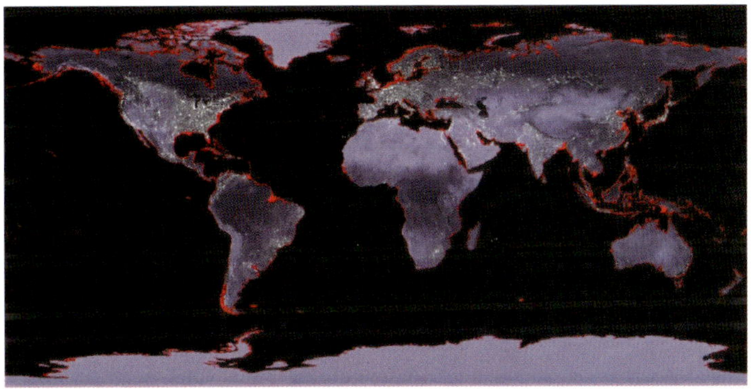

Map of the Earth with a six-meter sea level rise represented in red

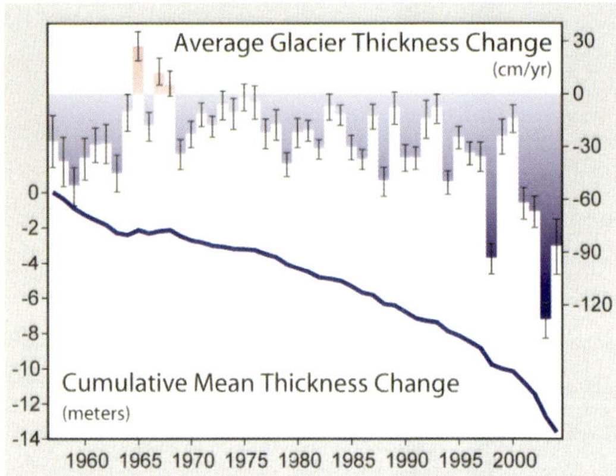

Sparse records indicate that glaciers have been retreating since the early 1800s. In the 1950s measurements began that allow the monitoring of glacial mass balance, reported to the World Glacier Monitoring Service (WGMS) and the National Snow and Ice Data Center (NSIDC)

As the climate change melts sea ice, the U.S. Geological Survey projects that two-thirds of polar bears will disappear by 2050

Environmental and biological

The environmental effects of global warming are broad and far reaching. They include the following diverse effects:

- Arctic sea ice decline, sea level rise, retreat of glaciers: Global warming has led to decades of shrinking and thinning in a warm climate that has put the Arctic sea ice in a precarious position, it is now vulnerable to atmospheric anomalies. Projections of declines in Arctic sea ice vary. Recent projections suggest that Arctic summers could be ice-free (defined as ice extent less than 1 million square km) as early as 2025–2030. The sea level rise since 1993 has been estimated to have been on average 2.6 mm and 2.9 mm per year ± 0.4 mm. Additionally, sea level rise has accelerated from 1995 to 2015. Over the 21st century, the IPCC projects for a high emissions scenario, that global mean sea level could rise by 52–98 cm.

- Extreme weather, extreme events, tropical cyclones: Data analysis of extreme events from 1960 until 2010 suggests that droughts and heat waves appear simultaneously with increased frequency. Extremely wet or dry events within the monsoon period have increased since 1980. Projections suggest a probable increase in the frequency and severity of some extreme weather events, such as heat waves.

- Ecosystem changes, changes in ocean properties: In terrestrial ecosystems, the earlier timing of spring events, as well as poleward and upward shifts in plant and animal ranges, have been linked with high confidence to recent warming. It is expected that most ecosystems will be affected by higher atmospheric CO_2 levels, combined with higher global temperatures. Overall, it is expected that climate change will result in the extinction of many species and reduced diversity of ecosystems. The physical effect of global warming on oceans include an increase in acidity, and a reduction of oxygen levels (ocean deoxygenation). Increases in atmospheric CO_2 concentrations have led to an increase in dissolved CO_2 and thus ocean acidity, measured by lower pH values. Ocean acidification could threaten coral reefs, fisheries, protected species, and other natural resources of value to society.

- Long-term effects of global warming, runaway climate change: On the timescale of centuries to millennia, the magnitude of global warming will be determined primarily by anthropogenic CO_2 emissions. This is due to carbon dioxide's very long lifetime in the atmosphere. Long-term effects also include a response from the Earth's crust, due to ice melting and deglaciation, in a process called post-glacial rebound, when land masses are no longer depressed by the weight of ice. This could lead to landslides and increased seismic and volcanic activities. Tsunamis could be generated by submarine landslides caused by warmer ocean water thawing ocean-floor permafrost or releasing gas hydrates.
- Abrupt climate change, cold blob (North Atlantic): Climate change could result in global, large-scale changes in natural and social systems. Examples include ocean acidification caused by increased atmospheric concentrations of carbon dioxide, and the long-term melting of ice sheets, which contributes to sea level rise. Some large-scale changes could occur abruptly, i.e., over a short time period, and might also be irreversible. Examples of abrupt climate change are the rapid release of methane and carbon dioxide from permafrost, which would lead to amplified global warming. Another example is the possibility for the Atlantic Meridional Overturning Circulation to slow- or shutdown. This could trigger cooling in the North Atlantic, Europe, and North America. It would particularly affect areas such as the British Isles, France and the Nordic countries, which are warmed by the North Atlantic drift.

Social Systems

The effects of climate change on human systems, mostly due to warming or shifts in precipitation patterns, or both, have been detected worldwide. The future social impacts of climate change will be uneven across the world. Many risks are expected to increase with higher magnitudes of global warming. All regions are at risk of experiencing negative impacts. Low-latitude, less developed areas face the greatest risk. A study from 2015 concluded that economic growth (gross domestic product) of poorer countries is much more impaired with projected future climate warming, than previously thought. In small islands and mega deltas, inundation as a result of sea level rise is expected to threaten vital infrastructure and human settlements. This could lead to issues of homelessness in countries with low-lying areas such as Bangladesh, as well as statelessness for populations in countries such as the Maldives and Tuvalu.

Examples of impacts of global warming on humans include:

- A meta-analysis concluded in 2014 that each degree of temperature rise will increase violence by up to 20%, which includes fist fights, violent crimes, civil unrest or wars.

- Estimates in 2015 based on the IPCC A1B emission scenario from additional greenhouse gases released from permafrost, found associated impact damages to the economy to be US$43 trillion.
- Crop production will probably be negatively affected in low latitude countries, while effects at northern latitudes may be positive or negative. Global warming of around 4.6 °C relative to pre-industrial levels could pose a large risk to global and regional food security. Production of wheat and maize globally has been impacted by climate change. While crop production has increased in some mid-latitude regions such as the UK and Northeast China, economic losses due to extreme weather events have increased globally.
- Generally impacts on public health will be more negative than positive. Impacts include: the effects of extreme weather, leading to injury and loss of life; and indirect effects, such as undernutrition brought on by crop failures. There has been a shift from cold- to heat-related mortality in some regions as a result of warming.
- Livelihoods of indigenous peoples of the Arctic have been altered by climate change, and there is emerging evidence of climate change impacts on livelihoods of indigenous peoples in other regions. Regional impacts of climate change are now observable at more locations than before, on all continents and across ocean regions.

Regional

The Arctic, Africa, small islands and Asian megadeltas are regions that are likely to be especially affected by future climate change. Africa is one of the most vulnerable continents to climate variability and change because of multiple existing stresses and low adaptive capacity. Existing stresses include poverty, political conflicts, and ecosystem degradation. By 2050, between 350 million and 600 million people are projected to experience increased water stress due to climate change. Climate variability and change is projected to severely compromise agricultural production, including access to food, across Africa.

Responses

Mitigation

Mitigation of climate change are actions to reduce greenhouse gas emissions, or enhance the capacity of carbon sinks to absorb greenhouse gases from the atmosphere. There is a large potential for future reductions in emissions by a combination of activities, including energy conservation and increased energy efficiency; the use of low-carbon energy technologies, such as renewable energy, nuclear energy, and carbon capture and storage; and enhancing carbon sinks through, for example, reforestation and preventing deforestation. A 2015 report by Citibank concluded that transitioning to a low carbon economy would yield positive return on investments.

Near- and long-term trends in the global energy system are inconsistent with limiting global warming at below 1.5 or 2 °C, relative to pre-industrial levels. Pledges made as part of the Cancún agreements are broadly consistent with having a likely chance (66 to 100% probability) of limiting global warming (in the 21st century) at below 3 °C, relative to pre-industrial levels.

In limiting warming at below 2 °C, more stringent emission reductions in the near-term would allow for less rapid reductions after 2030. Many integrated models are unable to meet the 2 °C target if pessimistic assumptions are made about the availability of mitigation technologies.

Adaptation

Other policy responses include adaptation to climate change. Adaptation to climate change may be planned, either in reaction to or anticipation of climate change, or spontaneous, i.e., without government intervention. Planned adaptation is already occurring on a limited basis. The barriers, limits, and costs of future adaptation are not fully understood.

A concept related to adaptation is *adaptive capacity*, which is the ability of a system (human, natural or managed) to adjust to climate change (including climate variability and extremes) to moderate potential damages, to take advantage of opportunities, or to cope with consequences. Unmitigated climate change (i.e., future climate change without efforts to limit greenhouse gas emissions) would, in the long term, be likely to exceed the capacity of natural, managed and human systems to adapt.

Environmental organizations and public figures have emphasized changes in the climate and the risks they entail, while promoting adaptation to changes in infrastructural needs and emissions reductions.

Climate Engineering

Climate engineering (sometimes called *geoengineering* or *climate intervention*) is the deliberate modification of the climate. It has been investigated as a possible response to global warming, e.g. by NASA and the Royal Society. Techniques under research fall generally into the categories solar radiation management and carbon dioxide removal, although various other schemes have been suggested. A study from 2014 investigated the most common climate engineering methods and concluded they are either ineffective or have potentially severe side effects and cannot be stopped without causing rapid climate change.

Society and Culture

Political Discussion

Most countries in the world are parties to the United Nations Framework Convention on Climate Change (UNFCCC). The ultimate objective of the Convention is to prevent dangerous human interference of the climate system. As stated in the Convention, this requires that greenhouse gas concentrations are stabilized in the atmosphere at a level where ecosystems can adapt naturally to climate change, food production is not threatened, and economic development can proceed in a sustainable fashion. The Framework Convention was agreed on in 1992, but global emissions have risen since then.

During negotiations, the G77 (a lobbying group in the United Nations representing 133 developing countries) pushed for a mandate requiring developed countries to "take the lead" in reducing their emissions. This was justified on the basis that the developed countries' emissions had contributed most to the cumulation of greenhouse gases in the atmosphere, per-capita emissions (i.e., emissions per head of population) were still relatively low in developing countries, and the emissions of developing countries would grow to meet their development needs.

This mandate was sustained in the Kyoto Protocol to the Framework Convention, which entered into legal effect in 2005. In ratifying the Kyoto Protocol, most developed countries accepted legally binding commitments to limit their emissions. These first-round commitments expired in 2012. United States President George W. Bush rejected the treaty on the basis that "it exempts 80% of the world, including major population centres such as China and India, from compliance, and would cause serious harm to the US economy."

At the 15th UNFCCC Conference of the Parties, held in 2009 at Copenhagen, several UNFCCC Parties produced the Copenhagen Accord. Parties associated with the Accord (140 countries, as of November 2010) aim to limit the future increase in global mean temperature to below 2 °C. The 16th Conference of the Parties (COP16) was held at Cancún in 2010. It produced an agreement, not a binding treaty, that the Parties should take urgent action to reduce greenhouse gas emissions to meet a goal of limiting global warming to 2 °C above pre-industrial temperatures. It also recognized the need to consider strengthening the goal to a global average rise of 1.5 °C.

Scientific Discussion

The discussion continues in scientific articles that are peer-reviewed and assessed by scientists who work in the relevant fields and participate in the Intergovernmental Panel on Climate Change. The scientific consensus as of 2013 stated in the IPCC Fifth Assessment Report is that it "is extremely likely that human influence has been the dominant cause of the observed warming since the mid-20th century". A 2008 report by the U.S. National Academy of Sciences stated that most scientists by then agreed that observed warming in recent decades was primarily caused by human activities increasing the amount of greenhouse gases in the atmosphere. In 2005 the Royal Society stated that while the overwhelming majority of scientists were in agreement on the main points, some individuals and organizations opposed to the consensus on urgent action needed to reduce greenhouse gas emissions had tried to undermine the science and work of the IPCC. National science academies have called on world leaders for policies to cut global emissions.

In the scientific literature, there is a strong consensus that global surface temperatures have increased in recent decades and that the trend is caused mainly by human-induced emissions of greenhouse gases. No scientific body of national or international standing disagrees with this view. In November 2017, a second warning to humanity signed by 15,364 scientists from 184 countries stated that "the current trajectory of potentially catastrophic climate change due to rising greenhouse gases from burning fossil fuels, deforestation, and agricultural production — particularly from farming ruminants for meat consumption" is "especially troubling". A July 2017 study published in *Environmental Research Letters* asserts that the most significant action individuals could make to mitigate their own carbon footprint is to have fewer children, followed by living vehicle free, forgoing air travel and adopting a plant-based diet.

Public Opinion and Disputes

The global warming controversy refers to a variety of disputes, substantially more pronounced in the popular media than in the scientific literature, regarding the nature, causes, and consequences of global warming. The disputed issues include the causes of increased global average air temperature, especially since the mid-20th century, whether this warming trend is unprecedented or within normal climatic variations, whether humankind has contributed significantly to it, and whether the increase is completely or partially an artefact of poor measurements. Additional disputes concern estimates of climate sensitivity, predictions of additional warming, and what the consequences of global warming will be.

Global warming was the cover story in *Ms.* magazine

In the United States from about 1990 onwards, American conservative think tanks had begun challenging the legitimacy of global warming as a social problem. They challenged the scientific evidence, argued that global warming would have benefits, and asserted that proposed solutions would do more harm than good. Some people dispute aspects of climate change science. Organizations such as the libertarian Competitive Enterprise Institute, conservative commentators, and some companies such as ExxonMobil have challenged IPCC climate change scenarios, funded scientists who disagree with the scientific consensus, and provided their own projections of the economic cost of stricter controls. On the other hand, some fossil fuel companies have scaled back their efforts in recent years, or even called for policies to reduce global warming. Global oil companies have begun to acknowledge climate change exists and is caused by human activities and the burning of fossil fuels.

The global warming problem came to international public attention in the late 1980s. Polling groups began to track opinions on the subject, at first mainly in the United States. The longest consistent polling, by Gallup in the US, found relatively small deviations of 10% or so from 1998 to 2015 in opinion on the seriousness of global warming, but with increasing polarization between those concerned and those unconcerned.

By 2010, with 111 countries surveyed, Gallup determined that there had been a substantial decrease since 2007–2008 in the number of Americans and Europeans who viewed global warming as a serious threat. In the US, just a little over half the population (53%) viewed it as a serious concern for either themselves or their families; this was 10 points below the 2008 poll (63%).Latin America had the biggest rise in concern: 73% said global warming was a serious threat to their families. This global poll also found that people were more likely to attribute global warming to human activities than to natural causes, except in the US where nearly half (47%) of the population attributed global warming to natural causes.

A March–May 2013 survey by Pew Research Center for the People & the Press polled 39 countries about global threats. According to 54% of those questioned, global warming featured top of the perceived global threats.

Greenhouse Gases

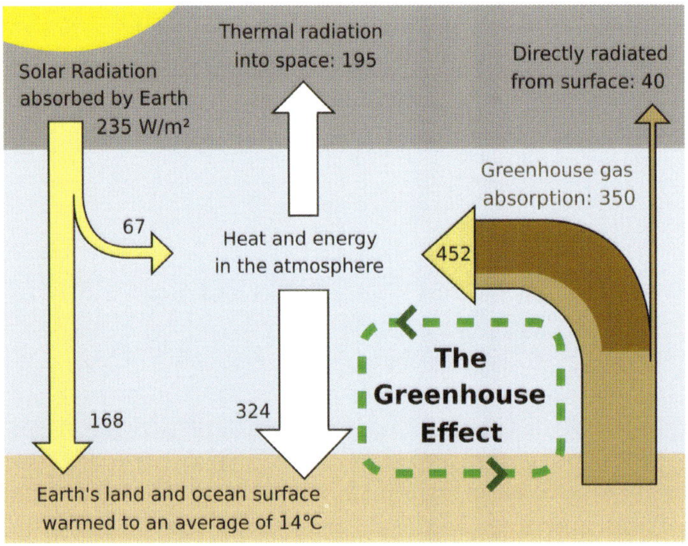

Greenhouse effect schematic showing energy flows between space, the atmosphere, and Earth's surface. Energy influx and emittance are expressed in watts per square meter (W/m^2)

A greenhouse gas is a gas in an atmosphere that absorbs and emits radiant energy within the thermal infrared range. This process is the fundamental cause of the greenhouse effect. The primary greenhouse gases in Earth's atmosphere are water vapor, carbon dioxide, methane, nitrous oxide, and ozone. Without greenhouse gases, the average temperature of Earth's surface would be about −18 °C (0 °F), rather than the present average of 15 °C (59 °F). In the Solar System, the atmospheres of Venus, Mars and Titan also contain gases that cause a greenhouse effect.

Human activities since the beginning of the Industrial Revolution (around 1750) have produced a 40% increase in the atmospheric concentration of carbon dioxide (CO_2), from 280 ppm in 1750 to 406 ppm in early 2017. This increase has occurred despite the uptake of more than half of the emissions by various natural "sinks" involved in the carbon cycle. The vast majority of anthropogenic carbon dioxide emissions (i.e., emissions produced by human activities) come from combustion of fossil fuels, principally coal, oil, and natural gas, with comparatively modest additional contributions coming from deforestation, changes in land use, soil erosion, and agriculture.

It has been estimated that if greenhouse gas emissions continue at their present rate, Earth's surface temperature could exceed historical values as early as 2047, with potentially harmful effects on ecosystems, biodiversity and the livelihoods of people worldwide. Recent estimates also suggest that at current emission rates the Earth could pass a threshold of 2 °C global warming, which the United Nations' IPCC designated as the upper limit to avoid "dangerous" global warming, by 2036.

Gases in Earth's Atmosphere

Greenhouse Gases

Greenhouse gases are those that absorb and emit infrared radiation in the wavelength range emitted by Earth. In order, the most abundant greenhouse gases in Earth's atmosphere are:

- Water vapor (H_2O)
- Carbon dioxide (CO_2)
- Methane (CH_4)
- Nitrous oxide (N_2O)
- Ozone (O_3)
- Chlorofluorocarbons (CFCs)
- Hydrofluorocarbons (incl. HCFCs and HFCs)

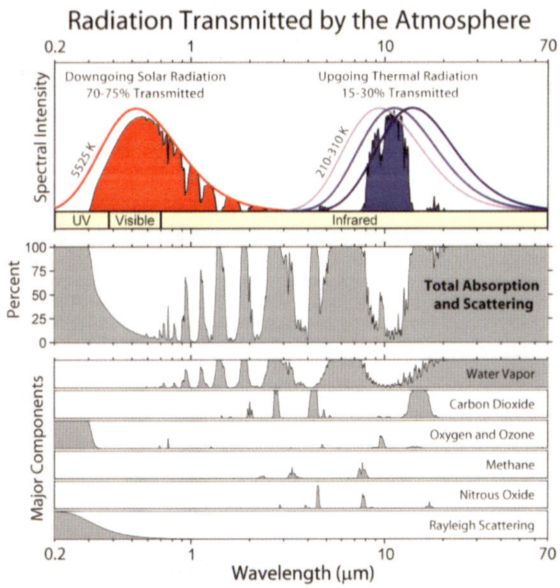

Atmospheric absorption and scattering at different wavelengths of electromagnetic waves. The largest absorption band of carbon dioxide is not far from the maximum in the thermal emission from ground, and it partly closes the window of transparency of water; hence its major effect

Atmospheric concentrations of greenhouse gases are determined by the balance between sources (emissions of the gas from human activities and natural systems) and sinks (the removal of the gas from the atmosphere by conversion to a different chemical compound). The proportion of an emission remaining in the atmosphere after a specified time is the "airborne fraction" (AF). The *annual airborne fraction* is the ratio of the atmospheric increase in a given year to that year's total emissions. As of 2006 the annual airborne fraction for CO_2 was about 0.45. The annual airborne fraction increased at a rate of 0.25 ± 0.21% per year over the period 1959–2006.

Non-greenhouse Gases

The major atmospheric constituents, nitrogen (N_2), oxygen (O_2), and argon (Ar), are not greenhouse gases because molecules containing two atoms of the same element such as N_2 and O_2 and monatomic molecules such as argon (Ar) have no net change in the distribution of their electrical charges when they vibrate. Hence they are almost totally unaffected by infrared radiation. Although molecules containing two atoms of different elements such as carbon monoxide (CO) or hydrogen chloride (HCl) absorb infrared radiation, these molecules are short-lived in the atmosphere owing to their reactivity and solubility. Therefore, they do not contribute significantly to the greenhouse effect and often are omitted when discussing greenhouse gases.

Indirect Radiative Effects

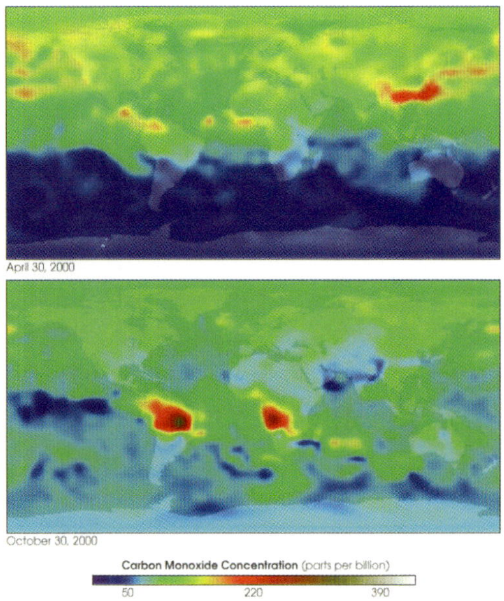

The false colors in this image represent concentrations of carbon monoxide in the lower atmosphere, ranging from about 390 parts per billion (dark brown pixels), to 220 parts per billion (red pixels), to 50 parts per billion (blue pixels)

Some gases have indirect radiative effects (whether or not they are greenhouse gases themselves). This happens in two main ways. One way is that when they break down in the atmosphere they produce another greenhouse gas. For example, methane and carbon monoxide (CO) are oxidized to give carbon dioxide (and methane oxidation also produces water vapor, as discussed later). Oxidation of CO to CO_2 directly produces an unambiguous increase in radiative forcing although the reason is subtle. The peak of the thermal IR emission from Earth's surface is very close to a strong vibrational absorption band of CO_2 (15 microns, or 667 cm^{-1}). On the other hand, the single CO vibrational band only absorbs IR at much shorter wavelengths (4.7 microns, or 2145 cm^{-1}), where the emission of radiant energy from Earth's surface is at least a factor of ten lower. Oxidation of methane to CO_2, which requires reactions with the OH radical, produces an instantaneous reduction in radiative absorption and emission since CO_2 is a weaker greenhouse gas than methane, although CO_2 has a longer lifetime. As described below this is not the whole story, since the oxidations of CO and CH_4 are intertwined by both consuming OH radicals. In any case, the calculation of the total radiative effect needs to include both the direct and indirect forcing.

A second type of indirect effect happens when chemical reactions in the atmosphere involving these gases change the concentrations of greenhouse gases. For example, the destruction of non-methane volatile organic compounds (NMVOCs) in the atmosphere can produce ozone. The size of the indirect effect can depend strongly on where and when the gas is emitted.

Methane has a number of indirect effects in addition to forming CO_2. First, the main chemical that reacts with methane in the atmosphere is the hydroxyl radical (OH), thus more methane means that the concentration of OH goes down. Effectively, methane increases its own atmospheric lifetime and therefore its overall radiative effect. The second effect is that the oxidation of methane can produce ozone. Third, as well as making CO_2, the oxidation of methane produces water; this is a major source of water vapor in the stratosphere, which is otherwise very dry. CO and NMVOCs also produce CO_2 when they are oxidized. They remove OH from the atmosphere, and this leads to higher concentrations of methane. The surprising effect of this is that the global warming potential of CO is three times that of CO_2. The same process that converts NMVOCs to carbon dioxide can also lead to the formation of tropospheric ozone. Halocarbons have an indirect effect because they destroy stratospheric ozone. Finally, hydrogen can lead to ozone production and CH_4 increases as well as producing water vapor in the stratosphere.

Contribution of Clouds to Earth's Greenhouse Effect

The major non-gas contributor to Earth's greenhouse effect, clouds, also absorb and emit infrared radiation and thus have an effect on radiative properties of the greenhouse gases. Clouds are water droplets or ice crystals suspended in the atmosphere.

Do you know, from where these GHGs are emanating? Practically, these emissions are coming from almost all the sectors of our economy. Let us discuss four major GHGs and their sources of origin due to anthropogenic activities. The four major GHGs are Carbon Dioxide (CO_2), Methane (CH_4), Nitrous Oxide (N_2O) and Fluorinated gases (mostly HFCs). Let us now discuss about major sources of emission of these four principal greenhouse gases particularly arising out of human activities.

Carbon Dioxide (CO_2): The major sources of CO_2 are mainly originated from burning coal, oil, and gas (about 75%). About 20% of the total CO_2 emissions come from deforestation and decomposition of peat lands, crop residues, and organic materials in agricultural soils. Smaller amounts are produced from turning oil and gas into plastics and other compounds that eventually are decomposed into CO_2 again (about 3%) as well as from manufacture of cement through decomposition of one of the main ingredients, limestone (about 3%).

Methane (CH_4): It comes from a variety of sources.The largest source is livestock, particularly cattle and sheep (25%). This is followed by leaks from extraction, processing, and distribution of natural gas (15%). Other important sources are rice cultivation (12%), associated gas from coal production (10%), and decomposition of organic waste in waste water treatment (9%) and landfills (7%).

Nitrous Oxide (N_2O): It mainly comes from fertilized grasslands and croplands, where nitrogen fertilizers are decomposed in the soil (35%).This is followed by animal waste (26%). Surface water polluted with nitrogen accounts for about 15%. Small amounts come from chemical factories, such as those for nylon production (5%) and waste water treatment (2%). A small quantity of N_2O (about 1% of the total) comes from Cars with catalytic converters.

Fluorinated Gases (mostly HFCs): These gases are mainly emitted from air conditioners in cars and refrigerators, as well as from the production of industrial chemicals. SF6 is mainly used as an insulator in electrical equipment.

Anthropogenic Greenhouse Gases

Since about 1750 human activity has increased the concentration of carbon dioxide and other greenhouse gases. Measured atmospheric concentrations of carbon dioxide are currently 100 ppm higher than pre-industrial levels. Natural sources of carbon dioxide are more than 20 times greater than sources due to human activity, but over periods longer than a few years natural sources are closely balanced by natural sinks, mainly photosynthesis of carbon compounds by plants and marine plankton. As a result of this balance, the atmospheric mole fraction of carbon dioxide remained between 260 and 280 parts per million for the 10,000 years between the end of the last glacial maximum and the start of the industrial era.

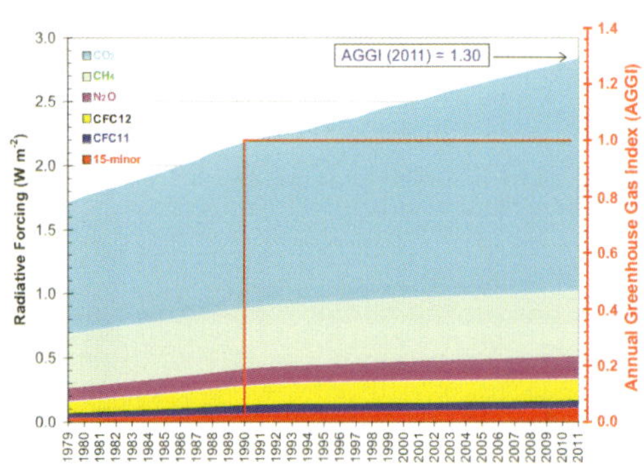

This graph shows changes in the annual greenhouse gas index (AGGI) between 1979 and 2011. The AGGI measures the levels of greenhouse gases in the atmosphere based on their ability to cause changes in Earth's climate

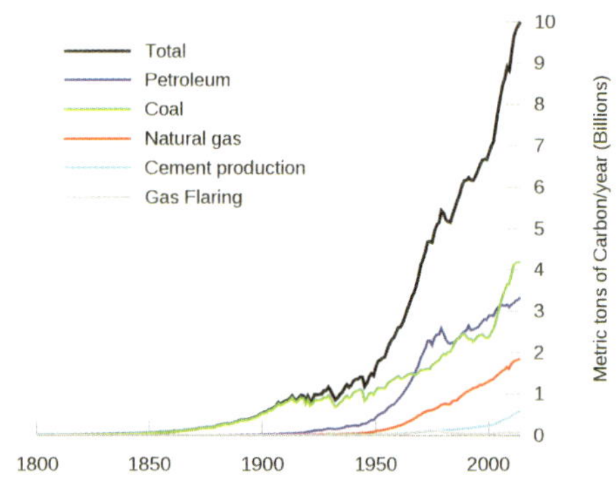

Modern global CO_2 emissions from the burning of fossil fuels

It is likely that anthropogenic (i.e., human-induced) warming, such as that due to elevated greenhouse gas levels, has had a discernible influence on many physical and biological systems. Future warming is projected to have a range of impacts, including sea level rise, increased frequencies and severities of some extreme weather events, loss of biodiversity, and regional changes in agricultural productivity.

The main sources of greenhouse gases due to human activity are:

- Burning of fossil fuels and deforestation leading to higher carbon dioxide concentrations in the air. Land use change (mainly deforestation in the tropics) account for up to one third of total anthropogenic CO_2 emissions.
- Livestock enteric fermentation and manure management, paddy rice farming, land use and wetland changes, man-made lakes, pipeline losses, and covered vented landfill emissions leading to higher methane atmospheric concentrations. Many of the newer style fully

vented septic systems that enhance and target the fermentation process also are sources of atmospheric methane.

- Use of chlorofluorocarbons (CFCs) in refrigeration systems, and use of CFCs and halons in fire suppression systems and manufacturing processes.
- Agricultural activities, including the use of fertilizers, that lead to higher nitrous oxide (N_2O) concentrations.

The seven sources of CO_2 from fossil fuel combustion are (with percentage contributions for 2000–2004):

Seven main fossil fuel combustion sources	Contribution (%)
Liquid fuels (e.g., gasoline, fuel oil)	36%
Solid fuels (e.g., coal)	35%
Gaseous fuels (e.g., natural gas)	20%
Cement production	3 %
Flaring gas industrially and at wells	< 1%
Non-fuel hydrocarbons	< 1%
"International bunker fuels" of transport not included in national inventories	4 %

Carbon dioxide, methane, nitrous oxide (N_2O) and three groups of fluorinated gases (sulfur hexafluoride (SF_6), hydrofluorocarbons (HFCs), and perfluorocarbons (PFCs)) are the major anthropogenic greenhouse gases, and are regulated under the Kyoto Protocol international treaty, which came into force in 2005. Emissions limitations specified in the Kyoto Protocol expired in 2012. The Cancún agreement, agreed in 2010, includes voluntary pledges made by 76 countries to control emissions. At the time of the agreement, these 76 countries were collectively responsible for 85% of annual global emissions.

Although CFCs are greenhouse gases, they are regulated by the Montreal Protocol, which was motivated by CFCs' contribution to ozone depletion rather than by their contribution to global warming. Note that ozone depletion has only a minor role in greenhouse warming though the two processes often are confused in the media. On 15 October 2016, negotiators from over 170 nations meeting at the summit of the United Nations Environment Programme reached a legally binding accord to phase out hydrofluorocarbons (HFCs) in an amendment to the Montreal Protocol.

Sectors

Tourism

According to UNEP global tourism is closely linked to climate change. Tourism is a significant contributor to the increasing concentrations of greenhouse gases in the atmosphere. Tourism accounts for about 50% of traffic movements. Rapidly expanding air traffic contributes about 2.5% of the production of CO_2. The number of international travelers is expected to increase from 594 million in 1996 to 1.6 billion by 2020, adding greatly to the problem unless steps are taken to reduce emissions.

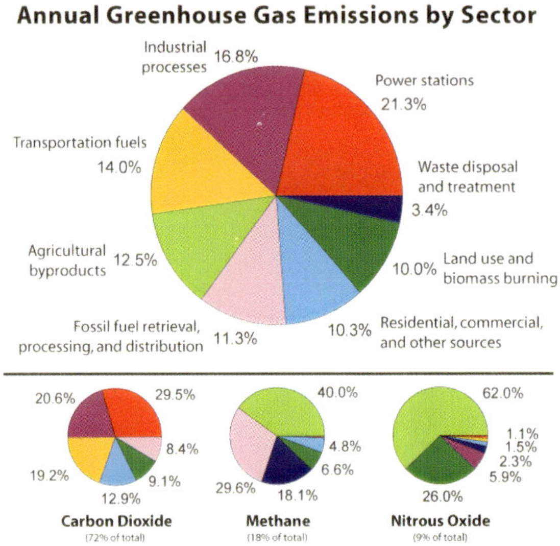

This figure shows the relative fraction of anthropogenic greenhouse gases coming from each of eight categories of sources, as estimated by the Emission Database for Global Atmospheric Research version 3.2, fast track 2000 project

These values are intended to provide a snapshot of global annual greenhouse gas emissions in the year 2000. The top panel shows the sum over all anthropogenic greenhouse gases, weighted by their global warming potential over the next 100 years. This consists of 72% carbon dioxide, 18% methane, 8% nitrous oxide and 1% other gases. Lower panels show the comparable information for each of these three primary greenhouse gases, with the same coloring of sectors as used in the top chart. Segments with less than 1% fraction are not labeled.

Road Haulage

The road haulage industry plays a part in production of CO_2, contributing around 20% of the UK's total carbon emissions a year, with only the energy industry having a larger impact at around 39%. Average carbon emissions within the haulage industry are falling—in the thirty-year period from 1977 to 2007, the carbon emissions associated with a 200-mile journey fell by 21 percent; NOx emissions are also down 87 percent, whereas journey times have fallen by around a third. Due to their size, HGVs often receive criticism regarding their CO_2 emissions; however, rapid development in engine technology and fuel management is having a largely positive effect.

Global Warming Potential of Anthropogenic Greenhouse Gases

Do you know what is the Global Warming Potential (GWP) of anthropogenic GHGs and how do we calculate it? GWP was developed for comparing the global warming impacts of different gases. In other words, it is a measure to calculate how much energy the emissions of 1 ton of a gas will absorb over a given period of time, relative to the emissions of 1 ton of carbon dioxide (CO_2). Therefore, calculation of GWP takes into account both the radiative properties of a particular greenhouse gas molecule and its lifetime existence in the atmosphere, once emitted. The larger the GWP value, the more is the warming capacity of that particular given gas incomparison to CO_2 over that particular time period. The time period usually used for GWPs is 100 years. This is the atmospheric life time of CO_2.

GWP of CO_2 is 1 because it is the gas being used as the reference. On the other handmethane is considerably more short-lived in the atmosphere than CO_2, persisting for decades rather than centuries. This measurement is helpful for formulation of mitigation policies.If we need to avoid a dangerous short-term climate tipping point, we might focus more effort on reducing methane because it is a particularly potent, if short lived, greenhouse gas. On the other hand, if our goal is to stabilize long-term greenhouse gas concentrations, we would be better served by focusing purely on CO_2 emissions.

Table: Atmospheric Lifetime and GWP of Major Greenhouse Gases

Sr. No.	Green House Gases	Atmospheric Lifetime in Years	Global Warming Potential
1	Carbon dioxide (CO_2)	Approx. 100	1
2	Methane (CH_4)	12	23
3	Nitrous oxide (N_2O)	120	310
4	Hydrofluorocarbons (HFCs)	1.5 to 264	140 to 11,700
5	Perfluorocarbons (PFCs)	3,200 to 50,000	6,500 to 9,000
6	Sulphur hexafluoride (SF_6)	3, 200	23900

To have a better understanding about the global warming and climate change due to GHG emission by anthropogenic activities, we have to understand the processes of radiative forcing.

Radiative Forcing

Theoretically, we can say that Earth's energy balance is zero. What does this mean? This means that the amount of energy arriving at Earth's surface is equal to the amount of energy eventually radiated back to space. However, Earth's climate has cycled through periods where this balance is not achieved and Earth systems are either gaining or losing heat. The term radiative forcing is also known as climate forcing. In other words, any change in the radiative balance caused by changes in atmospheric composition. Therefore, it describes the amount by which some perturbation causes Earth's energy balance to deviate from zero. A positive forcing indicates a warming condition whereas a negative forcing indicates cooling.

Scientists at National Oceanic and Atmospheric Administration (NOAA) have measured the radiative forcing caused by major long-lived greenhouse gases for the period 1979-2015. This is quantified in watts of energy affecting Earth's energy budget. Figure, shows Annual Greenhouse Gas Index, as measured by NOAA. According to the figure the Annual Greenhouse Gas Index value reached 1.32 in 2012. Do you know what does this indicator convey? This indicator converts the total radiative forcing for each gas into an index by using the ratio of the Radiative Forcing (RF) for a particular year compared with the RF in 1990 (the baseline year). The graph shows that RF has increased steadily for all gases, with the proportion attributed to CO_2 increasing the most.

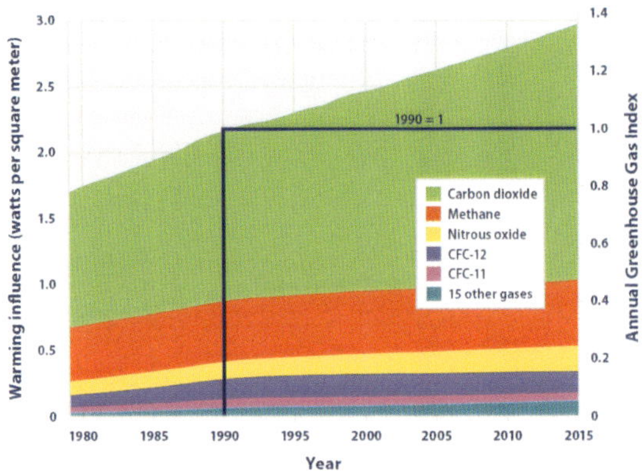

Radiative Forcing Caused by Major Long-Lived GHGs 1979-2015

According to the Report given by Working Group I of the Fifth Assessment Report of the IPCC which deals with the Physical Science Basis show that there is a difference in the scientific understanding of different RF effects. It has been observed that emissions can result either in an increase or a decrease in RF at the global scale. Over the past 260 years (1750-2010), emissions of CO_2, CH_4, N_2O, F-gases, black carbon, CO, and NMVOC all resulted in an increase in RF. On the other hand, emissions of SO_2, organic carbon and mineral dust all contributed to a decrease in RF. Emissions of halocarbons had both a positive and negative impact on RF. Also, the emissions of NOX and NH_3 have had both a positive and negative RF effect, but with a negative net impact on RF. Figure 2 further highlights that interactions between aerosols and clouds resulted in a negative RF, but that the contribution of individual emitted compounds within mixes of aerosols is unknown.

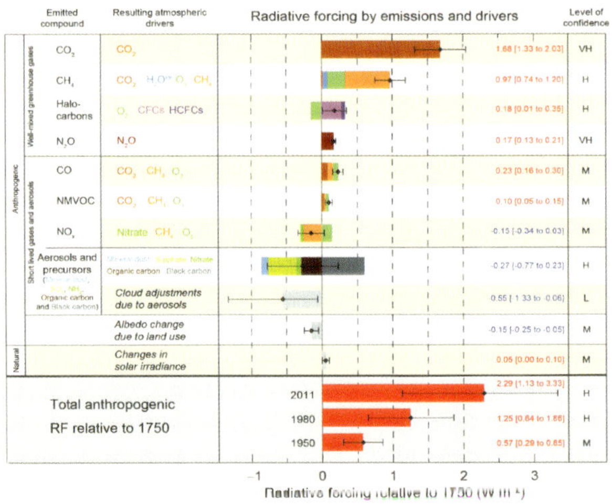

Radiative Forcing Estimates in 2011

Anthropogenic Greenhouse Gas Emission and Climate Change

The increasing concentration of atmospheric CO_2 and other greenhouse gases due to anthropogenic activities is the major cause for global climate change has now been established. According to an analysis undertaken by Mann and Kump has revealed that out of all the major greenhouse gases, percentage share of CO_2 emission is highest. Therefore, it plays a significant role in human induced climate change. If we analyse contribution of major GHGs, it has been observed that CO_2

accounted for more than three fourth of the emission (about 77%). The other GHGs which plays a significant role are methane (about 14%), nitrous oxide (about 8%), and the chlorofluorocarbons constitute the remaining 1%. The same has also been reported by IPCC Fourth Assessment Report (AR4).

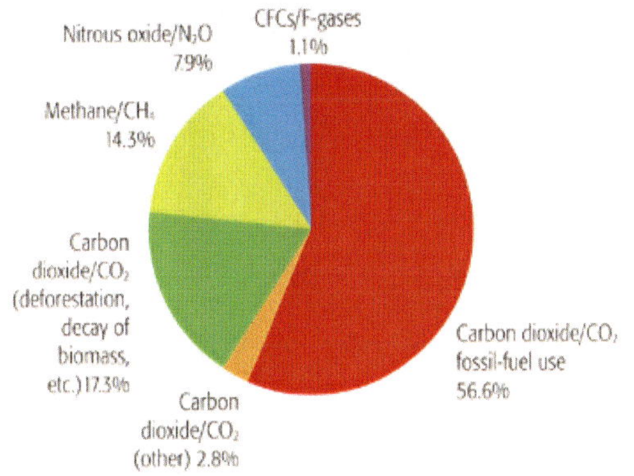

Greenhouse Gas Emissions by Gas

According to IPCC Fifth Assessment Report titled 'Climate Change 2014 Synthesis Report,cumulative anthropogenic CO_2 emissions to the atmosphere between 1750 and 2011 were 2040 ± 310 $GtCO_2$ (Gt=Gigatonne, one gigatonne is equal to one billion tonnes). Out of the total cumulative anthropogenic CO_2 emissions, about 40% have remained in the atmosphere (880 ± 35 $GtCO_2$). The remaining was removed from the atmosphere and stored in plants and soils on land and in the ocean. These vegetation and soil on land and in oceans are major source of carbon sinks. If we analyse CO_2 emission over the time period, it has been observed that about half of the anthropogenic CO_2 emissions between 1750 and 2011 have occurred in the last 40 years. The trend analysis has also revealed that total anthropogenic GHG emissions have continued to increase over 1970 to 2010 with larger absolute increases between 2000 and 2010. This has happened despite of a growing number of climate change mitigation policies implemented across the world.

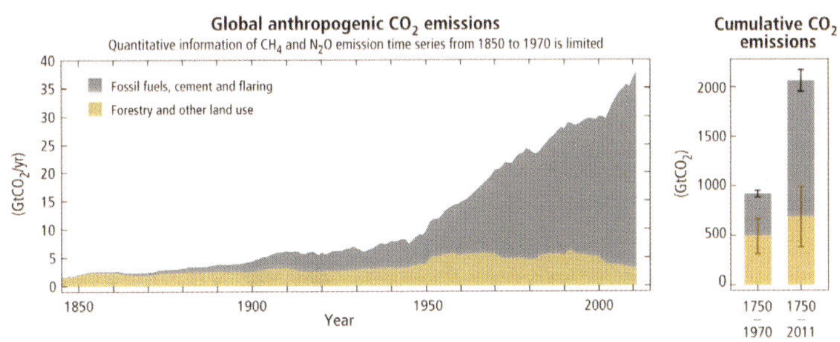

Trends of Global Anthropogenic CO_2 Emissions: 1750-2011

Anthropogenic GHG emissions in 2010 have reached 49 ± 4.5 GtCO2-eq/yr. Emissions of CO_2 from fossil fuel combustion and industrial processes contributed about 78% of the total GHG emissions increase from 1970 to 2010, with a similar percentage contribution for the increase during the period2000 to 2010. As mentioned earlier, at globallevel, economic and population growth continued to be the most important drivers of increases in CO_2 emissions from fossil fuel combustion. The contribution of population growth between 2000 and 2010 remained roughly identical to the previous three decades, while the contribution of economic growth has risen sharply. Increased use of coal has reversed the long-standing trend of gradual decarbonisation of the world's energy supply. Do you know what decarbonisation is? Decarbonisation refers to reducingthe carbon intensity of energy. Therefore, major strategies for reducing the greenhouse gas emissions are to control the release of greenhouse gases from fossil fuel combustion, land-use change and the burning of vegetation. If we succeed in doing so,then this would lead to decrease in the projected rate and magnitude of warming. In other words, future climate change would be determined by historic, current and future emissions of these greenhouse gases.

Effects of Climate Change

Remedial Measures to Overcome the Effects of Climate Change

In climate change discourse, there are two approaches to address human induced climate change. These are mitigation and adaptation. Mitigation has the long history in the climate policy, whereas the adaptation has recently gained importance.

The Concept of Mitigation and Adaptation: The concept 'mitigation' in general means the reduction of the atmospheric GHGs, and hence, we can avoid the likelihood of the occurrence of the climatic variability and extreme events. IPCC defines mitigation as*"an anthropogenic intervention to reduce the sources or enhance the sinks of greenhouse gases."* On the other hand, the notion 'adaptation' in general refers to the individual, communities, and societies to adjust their activities, life courses and location to take an opportunity, to get advantage from the fluxes of the social-ecological systems. The climate change literature views it as *"the adjustment in human and natural systems to actual or expected climatic stimuli, which can reduce the negative impacts and take advantage of the positive"* (UNFCC 1992).

The Need for Mitigation and Adaptation: As mentioned above, climate change has severe non-linear impacts on the wellbeing of the human society. Many developing nations have already experienced weather related extreme events in terms of floods, droughts, heat waves and tropical cyclones that are more frequent or intense than previous experiences. In general, it affects different sectors, such as fresh water resources and their management, food and fibre and forest products, coastal system and low lying areas, and health etc. The resulting impacts will have significant consequences on the environment, production systems and livelihood from future climate variability and change. Importantly, the developing nations are facing more burdens as compared to the developed nations. Meanwhile, Stern has estimated "if we don't act, the overall damage cost will be equivalent to at least 5 percent of GDP now and forever, and if wider range of risks and impacts is taken into account, the estimates of damage could rise to 20 percent of the GDP or more".

Physical Impacts of Climate Change

This section is about the physical impacts of climate change. For some of these physical impacts, their effect on social and economic systems are also described.

In their usage, "climate change" refers to a change in the state of the climate that can be identified by changes in the mean and/or variability of its properties, and that persists for extended periods, typically decades or longer. The climate change referred to may be due to natural causes and/or the result of human activity

Global Warming

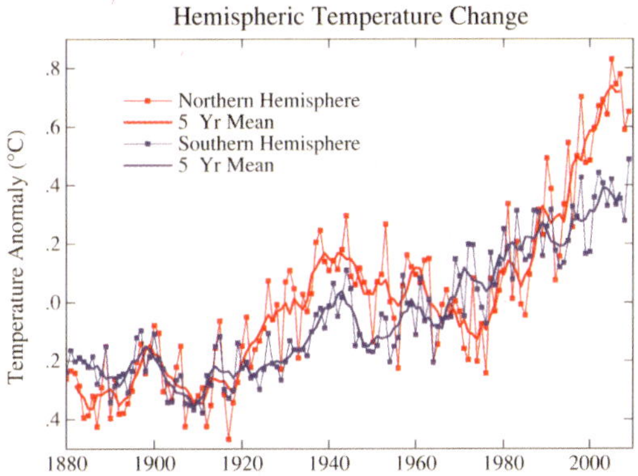

Global mean surface temperature difference from the average for 1880–2009

Global surface temperatures have increased about 0.74 °C (plus or minus 0.18 °C) since the late-19th century, and the linear trend for the past 50 years of 0.13 °C (plus or minus 0.03 °C) per decade is nearly twice that for the past 100 years. The warming has not been globally uniform. Some areas have, in fact, cooled slightly over the last century. The recent warmth has been greatest over North America and Eurasia between 40 and 70°N. Lastly, seven of the eight warmest years on record have occurred since 2001 and the 10 warmest years have all occurred since 1995.

Consistency of Evidence for Warming

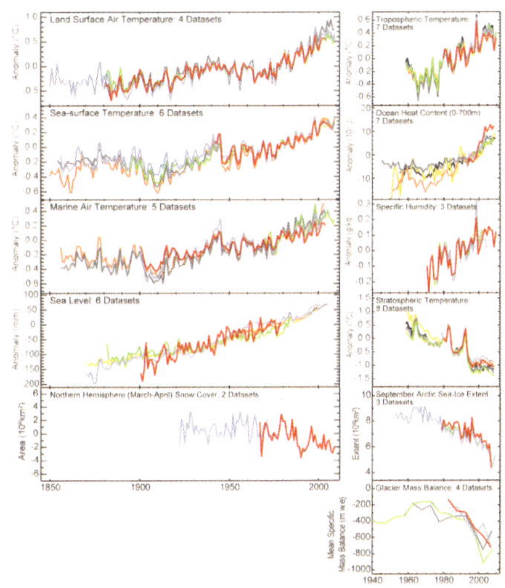

11 key indicators of global warming

Thousands of land and ocean temperature measurements are recorded each day around the globe. This includes measurements from climate reference stations, weather stations, ships, buoys and autonomous gliders in the oceans. These surface measurements are also supplemented with satellite measurements. These measurements are processed, examined for random and systematic errors, and then finally combined to produce a time series of global average temperature change. A number of agencies around the world have produced datasets of global-scale changes in surface temperature using different techniques to process the data and remove measurement errors that could lead to false interpretations of temperature trends. The warming trend that is apparent in all of the independent methods of calculating global temperature change is also confirmed by other independent observations, such as the melting of mountain glaciers on every continent, reductions in the extent of snow cover, earlier blooming of plants in spring, a shorter ice season on lakes and rivers, ocean heat content, reduced Arctic sea ice, and rising sea levels.

Global Average Temperature

Global average temperature is one of the most-cited indicators of global climate change, and shows an increase of approximately 1.4 °F since the early 20th Century. The global surface temperature is based on air temperature data over land and sea-surface temperatures observed from ships, buoys and satellites. There is a clear long-term global warming trend, while each individual year does not always show a temperature increase relative to the previous year, and some years show greater changes than others. These year-to-year fluctuations in temperature are due to natural processes, such as the effects of El Niños, La Niñas, and the eruption of large volcanoes. Notably, the 20 warmest years have all occurred since 1981, and the 10 warmest have all occurred in the past 12 years.

Diurnal Temperature

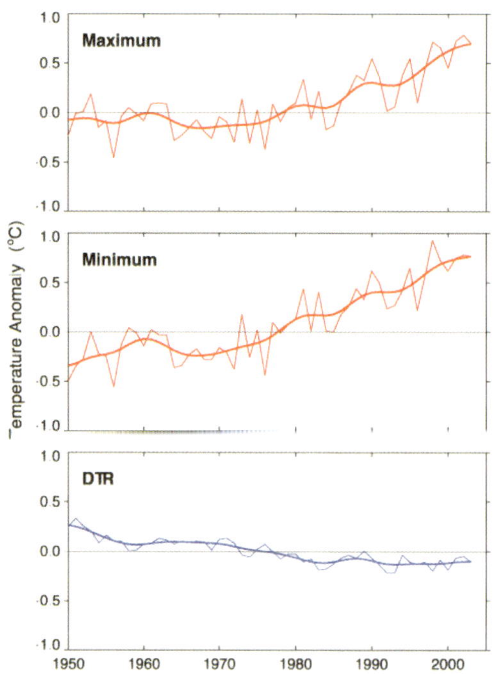

Annual anomalies of maximum and minimum temperatures and diurnal temperature range (DTR) (°C) relative to the 1961 to 1990 mean, averaged for the 71 percent of global land areas where data are available for 1950 to 2004. The smooth curves show decadal variations

There has been a general, but not global, tendency toward reduced diurnal temperature range (DTR: the difference between daily high or maximum and daily low or minimum temperatures) over about 70% of the global land mass since the middle of the 20th century. However, for the period 1979–2005 the DTR shows no trend since the trend in both maximum and minimum temperatures for the same period are virtually identical; both showing a strong warming signal. A variety of factors likely contribute to this change in DTR, particularly on a regional and local basis, including changes in cloud cover, atmospheric water vapor, land use and urban effects.

Indirect Indicators of Warming

Indirect indicators of warming such as borehole temperatures, snow cover, and glacier recession data, are in substantial agreement with the more direct indicators of recent warmth. Evidence such as changes in glacial mass balance (the amount of snow and ice contained in a glacier) is useful since it not only provides qualitative support for existing meteorological data, but glaciers often exist in places too remote to support meteorological stations. The records of glacial advance and retreat often extend back further than weather station records, and glaciers are usually at much higher altitudes than weather stations, allowing scientists more insight into temperature changes higher in the atmosphere.

Effects on Weather

Increasing temperature is likely to lead to increasing precipitation but the effects on storms are less clear. Extratropical storms partly depend on the temperature gradient, which is predicted to weaken in the northern hemisphere as the polar region warms more than the rest of the hemisphere. It is possible that the Polar and Ferrel cells in one or both hemispheres will weaken and eventually disappear, which would cause the Hadley cell to cover the whole planet. This would greatly decrease the temperature gradient between the arctic and the tropics, and cause the earth to flip to a hothouse state.

Precipitation

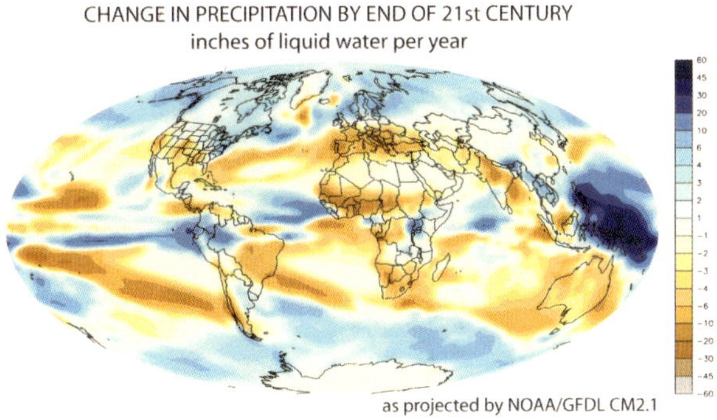

Projected change in annual average precipitation by the end of the 21st century, based on a medium emissions scenario (SRES A1B)

Historically (i.e., over the 20th century), subtropical land regions have been mostly semi-arid, while most subpolar regions have had an excess of precipitation over evaporation. Future global warming is expected to be accompanied by a reduction in rainfall in the subtropics and an increase in precipitation in subpolar latitudes and some equatorial regions. In other words, regions which are dry at present will generally become even drier, while regions that are currently wet will generally become even wetter. This projection does not apply to every locale, and in some cases can be modified by local conditions. Drying is projected to be strongest near the poleward margins of the subtropics (for example, South Africa, southern Australia, the Mediterranean, and the south-western U.S.), a pattern that can be described as a poleward expansion of these semi-arid zones.

This large-scale pattern of change is a robust feature present in nearly all of the simulations conducted by the world's climate modeling groups for the 4th Assessment of the Intergovernmental Panel on Climate Change (IPCC), and is also evident in observed 20th century precipitation trends.

Changes in regional climate are expected to include greater warming over land, with most warming at high northern latitudes, and least warming over the Southern Ocean and parts of the North Atlantic Ocean.

Future changes in precipitation are expected to follow existing trends, with reduced precipitation over subtropical land areas, and increased precipitation at subpolar latitudes and some equatorial regions.

A 2015 study published in *Nature Climate Change*, states:

> About 18% of the moderate daily precipitation extremes over land are attributable to the observed temperature increase since pre-industrial times, which in turn primarily results from human influence. For 2 °C of warming the fraction of precipitation extremes attributable to human influence rises to about 40%. Likewise, today about 75% of the moderate daily hot extremes over land are attributable to warming. It is the most rare and extreme events for which the largest fraction is anthropogenic, and that contribution increases non-linearly with further warming.

Extreme Events

Fire

Fire is a major agent for conversion of biomass and soil organic matter to CO_2. There is a large potential for future alteration in the terrestrial carbon balance through altered fire regimes. With high confidence projected that:

- An increase in global mean temperature of about 0 to 2 °C by 2100 relative to the 1990–2000 period would result in increased fire frequency and intensity in many areas.
- An increase in the region of 2 °C or above would lead to increased frequency and intensity of fires.

Sensitivity to fires in areas that were already vulnerable has been steadily increasing. In high altitude temperate areas, increased temperature is causing snow pack to melt sooner and in greater quantities. The number of days that of higher stream flow caused by snowmelt in the Mississippi, Missouri, and Ohio rivers has been increasing in recent years. The substantial amount of snow that remains atop mountains year around is also disappearing. This leads to the surrounding densely forested areas becoming more dry and staying dry for longer periods of time. In the 1970s, the length of a fire season, which is the period of the year fires are most likely to occur, was about five months. Today, the period is usually seven months, extending into the springtime mud season. In addition, many areas are experiencing higher than normal droughts. Between 2011 and 2014, California experienced the driest period in its recorded history and more than 100 million trees died in the drought, creating areas of dead, dry wood. The decrease in rainfall is also going to increase the risk of wildfire by allowing the fire access to drier fuels. Dry foliage is more susceptible to a wildfire trigger. Wildfire specialists use foliar moisture content to determine how susceptible an area is to a wildfire. In the United States, 2015 was the most destructive year on record for wildfires, with a total of 10,125,149 total acres destroyed by fires. 2017 was the second worst year on record with 10,026,086 acres destroyed. The Thomas Fire occurred in 2017 and was the largest fire in California's history.

The increasing frequency of wildfires as a result of climate change will also lead to an increase in the amount of CO_2 in the atmosphere. This will, in turn, increase the temperature and the frequency of hot days, which will further increase fire danger. It was forecasted that double levels of CO_2, would bring a greater risk of wildfires to Australia, especially the Australian outback. All of the eight sites tested projected an increase in fire danger as a result of CO_2 level increase and all but one projected a longer fire season. The largest population center said to be affected is Alice Springs, a city deep in the Outback.

Extreme Weather

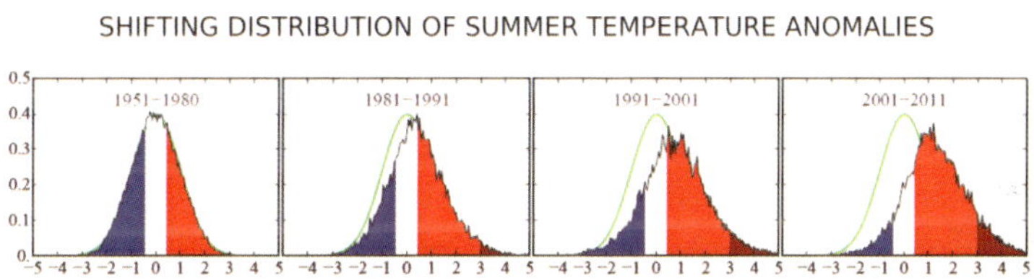

Frequency of occurrence (vertical axis) of local June–July–August temperature anomalies (relative to 1951–1980 mean) for Northern Hemisphere land in units of local standard deviation (horizontal axis)

According to Hansen *et al.* (2012), the distribution of anomalies has shifted to the right as a consequence of global warming, meaning that unusually hot summers have become more common. This is analogous to the rolling of a die: cool summers now cover only half of one side of a six-sided die, white covers one side, red covers four sides, and an extremely hot (red-brown) anomaly covers half of one side.

IPCC projected that in the future, over most land areas, the frequency of warm spells or heat waves would very likely increase. Other likely changes are listed below:

- Increased areas will be affected by drought
- There will be increased intense tropical cyclone activity
- There will be increased incidences of extreme high sea level (excluding tsunamis)

Storm strength leading to extreme weather is increasing, such as the power dissipation index of hurricane intensity. Kerry Emanuel writes that hurricane power dissipation is highly correlated with temperature, reflecting global warming. However, a further study by Emanuel using current model output concluded that the increase in power dissipation in recent decades cannot be completely attributed to global warming. Hurricane modeling has produced similar results, finding that hurricanes, simulated under warmer, high-CO_2 conditions, are more intense, however, hurricane frequency will be reduced. Worldwide, the proportion of hurricanes reaching categories 4 or 5 – with wind speeds above 56 metres per second – has risen from 20% in the 1970s to 35% in the 1990s. Precipitation hitting the US from hurricanes has increased by 7% over the 20th century. The extent to which this is due to global warming as opposed to the Atlantic Multidecadal Oscillation is unclear. Some studies have found that the increase in sea surface temperature may be offset by an increase in wind shear, leading to little or no change in hurricane activity. Hoyos *et al.* (2006) have linked the increasing trend in number of category 4 and 5 hurricanes for the period 1970–2004 directly to the trend in sea surface temperatures.

Thomas Knutson and Robert E. Tuleya of NOAA stated in 2004 that warming induced by greenhouse gas may lead to increasing occurrence of highly destructive category-5 storms. In 2008, Knutson *et al.* found that Atlantic hurricane and tropical storm frequencies could reduce under future greenhouse-gas-induced warming. Vecchi and Soden find that wind shear, the increase of which acts to inhibit tropical cyclones, also changes in model-projections of global warming. There are projected increases of wind shear in the tropical Atlantic and East Pacific associated with the deceleration of the Walker circulation, as well as decreases of wind shear in the western and central Pacific. The study does not make claims about the net effect on Atlantic and East Pacific hurricanes of the warming and moistening atmospheres, and the model-projected increases in Atlantic wind shear.

The World Meteorological Organization explains that "though there is evidence both for and against the existence of a detectable anthropogenic signal in the tropical cyclone climate record to date, no firm conclusion can be made on this point." They also clarified that "no individual tropical cyclone can be directly attributed to climate change."

A substantially higher risk of extreme weather does not necessarily mean a noticeably greater risk of slightly-above-average weather. However, the evidence is clear that severe weather and moderate rainfall are also increasing. Increases in temperature are expected to produce more intense convection over land and a higher frequency of the most severe storms.

Using the Palmer Drought Severity Index, a 2010 study by the National Center for Atmospheric Research projects increasingly dry conditions across much of the globe in the next 30 years, possibly reaching a scale in some regions by the end of the century that has rarely, if ever, been observed in modern times.

Coumou *et al.* (2013) estimated that global warming had increased the probability of local record-breaking monthly temperatures worldwide by a factor of 5. This was compared to a baseline climate in which no global warming had occurred. Using a medium global warming scenario, they project that by 2040, the number of monthly heat records globally could be more than 12 times greater than that of a scenario with no long-term warming.

Increased Evaporation

Over the course of the 20th century, evaporation rates have reduced worldwide; this is thought by many to be explained by global dimming. As the climate grows warmer and the causes of global dimming are reduced, evaporation will increase due to warmer oceans. Because the world is a closed system this will cause heavier rainfall, with more erosion. This erosion, in turn, can in vulnerable tropical areas (especially in Africa) lead to desertification. On the other hand, in other areas, increased rainfall lead to growth of forests in dry desert areas.

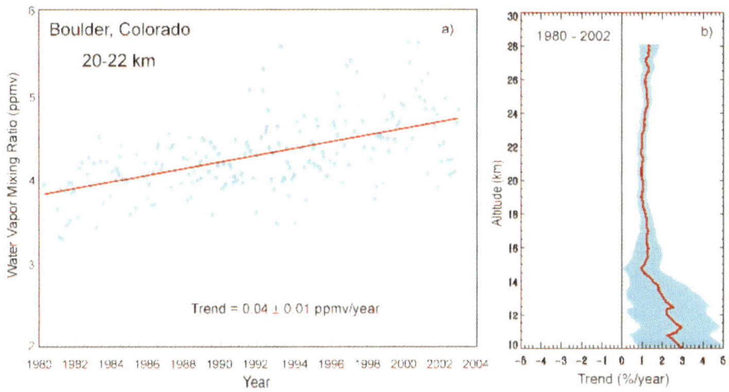

Increasing water vapor at Boulder, Colorado

Scientists have found evidence that increased evaporation could result in more extreme weather as global warming progresses. The IPCC Third Annual Report says: "Global average water vapor concentration and precipitation are projected to increase during the 21st century. By the second half of the 21st century, it is likely that precipitation will have increased over northern mid- to high latitudes and Antarctica in winter. At low latitudes there are both regional increases and decreases over land areas. Larger year-to-year variations in precipitation are very likely over most areas where an increase in mean precipitation is projected."

Increased Freshwater Flow

Research based on satellite observations, published in October 2010, shows an increase in the flow of freshwater into the world's oceans, partly from melting ice and partly from increased precipitation driven by an increase in global ocean evaporation. The increase in global freshwater flow, based on data from 1994 to 2006, was about 18%. Much of the increase is in areas which already experience high rainfall. One effect, as perhaps experienced in the 2010 Pakistan floods, is to overwhelm flood control infrastructure.

Regional Climate Change

General Effects

In a literature assessment, Hegerl *et al.* (2007) assessed evidence for attributing observed climate change. They concluded that since the middle of the 20th century, it was likely that human influences had significantly contributed to surface temperature increases in every continent except Antarctica. The magazine *Scientific American* reported on December 23, 2008, that the 10 places most affected by climate change were Darfur, the Gulf Coast, Italy, northern Europe, the Great Barrier Reef, island nations, Washington, D.C., the Northwest Passage, the Alps, and Uganda.

Northern Hemisphere

In the northern hemisphere, the southern part of the Arctic region (home to 4,000,000 people) has experienced a temperature rise of 1 °C to 3 °C (1.8 °F to 5.4 °F) over the last 50 years. Canada, Alaska and Russia are experiencing initial melting of permafrost. This may disrupt ecosystems and by increasing bacterial activity in the soil lead to these areas becoming carbon sources instead of carbon sinks. A study (published in *Science*) of changes to eastern Siberia's permafrost suggests that it is gradually disappearing in the southern regions, leading to the loss of nearly 11% of Siberia's nearly 11,000 lakes since 1971. At the same time, western Siberia is at the initial stage where melting permafrost is creating new lakes, which will eventually start disappearing as in the east. Furthermore, permafrost melting will eventually cause methane release from melting permafrost peat bogs.

Polar Regions

Anisimov *et al.* (2007) assessed the literature on impacts of climate change in Polar regions. Model projections showed that Arctic terrestrial ecosystems and the active layer (the top layer of soil or rock in permafrost that is subjected to seasonal freezing and thawing) would be a small sink for carbon (i.e., net uptake of carbon) over this century. These projections were viewed as being uncertain. It was judged that increased emissions of carbon from thawing of permafrost could occur. This would lead to an amplification of warming.

Atmosphere

An enhanced greenhouse effect is expected to cause cooling in higher parts of the atmosphere. Cooling of the lower stratosphere (about 49,000-79,500 ft.) since 1979 is shown by both satellite Microwave sounding unit and radiosonde data, but is larger in the radiosonde data likely due to uncorrected errors in the radiosonde data.

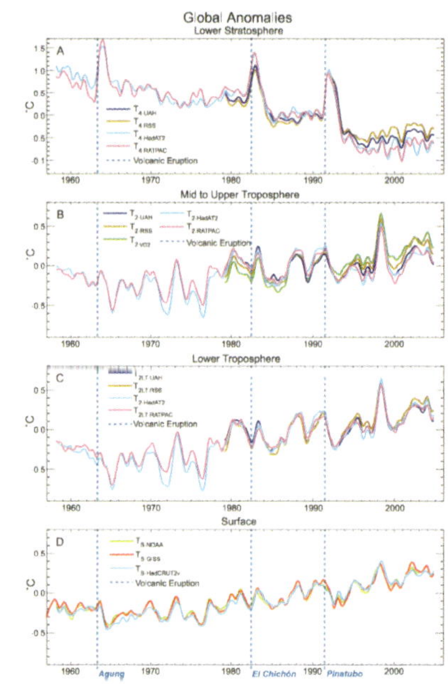

Temperature trends in the lower stratosphere, mid to upper troposphere, lower troposphere, and surface, 1957-2005

A contraction of the thermosphere has been observed as a possible result in part due to increased carbon dioxide concentrations, the strongest cooling and contraction occurring in that layer during solar minimum. The most recent contraction in 2008–2009 was the largest such since at least 1967.

Recent evidence suggests that warming of the tropical oceans since a "tipping point" in 2000 may have acted as a negative feedback, reducing the observed warming during the 2000s (decade). As warming and evaporation above the Pacific Ocean, temperatures in the lower stratosphere near the tropopause declined due to both greenhouse gases and ozone-depleting substances, reducing water vapor levels and removing its warming effect, with vapor concentrations below 2.2 ppmv as measured by the HALOE instrument on the Upper Atmosphere Research Satellite, in the lower stratosphere of the tropics between 5°N - 5°S first being observed since 2001, although a reversal in this pattern is also likely. The water vapor in the stratosphere arrives through tall thunderstorms, while 15% of this vapor is delivered by tropical cyclones, and through chemical breakdown of methane into water vapor and carbon dioxide, both of which are greenhouse gases. The vapor is frozen out of the stratosphere as more of the water is subjected to temperatures that freeze it out of the stratosphere. Water vapor concentrations in the lower stratosphere have declined by 10% (0.4 ppmv) since 2000, reducing warming during the decade by 25%. A rapid cooling of 4 °C to 6 °C also occurred in the lower stratosphere in the mid-1990s, while the rate of ocean warming increased. During the 1990s, increased stratospheric water vapor led to a 30% increase in warming. After 2000, the sea surface temperatures of the tropical Western Pacific, where a warm pool of water exists and where temperatures are heavily influenced by ENSO, between 10°N - 10°S and 139° - 171° longitude became anti-correlated with temperatures at the tropopause in the same latitudes between 171° - 200° longitude, both measured since the early 1980s; although the correlation had been previously positive, since 2000 the SST anomalies increased while tropopause temperatures decreased. A sharp increase in average SSTs within the Western Pacific warm pool by more than 0.25 °C in 2000, which has since stabilized, occurred as the "cold point" temperature of the study area at the tropopause experienced a significant reduction. This resulted in less water vapor from tropical thunderstorms entering the stratosphere. However, prior to 2000, increases in average Western Pacific SSTs had resulted in increases in tropopause cold point temperatures.

Geophysical Systems

Biogeochemical Cycles

Climate change can have an effect on the carbon cycle in an interactive "feedback" process . A feedback exists where an initial process triggers changes in a second process that in turn influences the initial process. A positive feedback intensifies the original process, and a negative feedback reduces it (IPCC). Models suggest that the interaction of the climate system and the carbon cycle is one where the feedback effect is positive.

Using the A2 SRES emissions scenario, Schneider *et al.* found that this effect led to additional warming by 2100, relative to the 1990–2000 period, of 0.1 to 1.5 °C. This estimate was made with high confidence. The climate projections made in the IPCC Fourth Assessment Report of 1.1 to 6.4 °C account for this feedback effect. On the other hand, with medium confidence, Schneider *et al.* (2007) commented that additional releases of GHGs were possible from permafrost, peat lands, wetlands, and large stores of marine hydrates at high latitudes.

Gas Hydrates

Gas hydrates are ice-like deposits containing a mixture of water and gas, the most common gas of which is methane. Gas hydrates are stable under high pressures and at relatively low temperatures and are found underneath the oceans and permafrost regions. Future warming at intermediate depths in the world's oceans, as predicted by climate models, will tend to destabilize gas hydrates resulting in the release of large quantities of methane. On the other hand, projected rapid sea level rise in the coming centuries associated with global warming will tend to stabilize marine gas hydrate deposits.

Carbon Cycle

Models have been used to assess the effect that climate change will have on the carbon cycle. In the Coupled Climate-Carbon Cycle Model Intercomparison Project, eleven climate models were used. Observed emissions were used in the models and future emission projections were based on the IPCC SRES A2 emissions scenario.

Unanimous agreement was found among the models that future climate change will reduce the efficiency of the land and ocean carbon cycle to absorb human-induced CO_2. As a result, a larger fraction of human-induced CO_2 will stay airborne if climate change controls the carbon cycle. By the end of the 21st century, this additional CO_2 in the atmosphere varied between 20 and 220 ppm for the two extreme models, with most models lying between 50 and 100 ppm. This additional CO_2 led to a projected increase in warming of between 0.1 and 1.5 °C.

Cryosphere

Northern Hemisphere average annual snow cover has declined in recent decades. This pattern is consistent with warmer global temperatures. Some of the largest declines have been observed in the spring and summer months.

Sea Ice

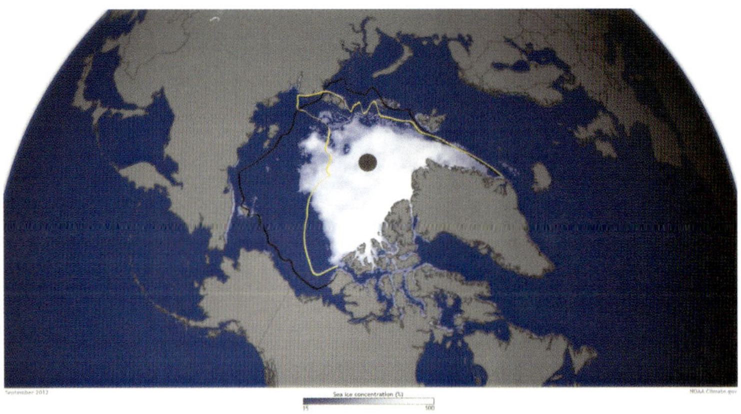

Record low Arctic sea ice extent for September 2012

As the climate warms, snow cover and sea ice extent decrease. Large-scale measurements of sea-ice have only been possible since the satellite era, but through looking at a number of different satellite estimates, it has been determined that September Arctic sea ice has decreased between

1973 and 2007 at a rate of about -10% +/- 0.3% per decade. Sea ice extent for September for 2012 was by far the lowest on record at 3.29 million square kilometers, eclipsing the previous record low sea ice extent of 2007 by 18%. The age of the sea ice is also an important feature of the state of the sea ice cover, and for the month of March 2012, older ice (4 years and older) has decreased from 26% of the ice cover in 1988 to 7% in 2012. Sea ice in the Antarctic has shown very little trend over the same period, or even a slight increase since 1979. Though extending the Antarctic sea-ice record back in time is more difficult due to the lack of direct observations in this part of the world.

In a literature assessment, Meehl *et al.* found that model projections for the 21st century showed a reduction of sea ice in both the Arctic and Antarctic. The range of model responses was large. Projected reductions were accelerated in the Arctic. Using the high-emission A2 SRES scenario, some models projected that summer sea ice cover in the Arctic would disappear entirely by the latter part of the 21st century.

Glacier Retreat and Disappearance

Warming temperatures lead to the melting of glaciers and ice sheets. IPCC found that, on average, mountain glaciers and snow cover had decreased in both the northern and southern hemispheres. This widespread decrease in glaciers and ice caps has contributed to observed sea level rise.

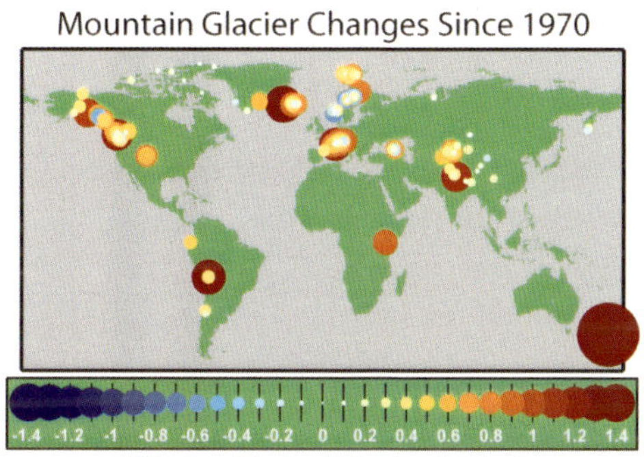

A map of the change in thickness of mountain glaciers since 1970. Thinning in orange and red, thickening in blue

As stated above, the total volume of glaciers on Earth is declining sharply. Glaciers have been retreating worldwide for at least the last century; the rate of retreat has increased in the past decade. Only a few glaciers are actually advancing (in locations that were well below freezing, and where increased precipitation has outpaced melting). The progressive disappearance of glaciers has implications not only for a rising global sea level, but also for water supplies in certain regions of Asia and South America.

With very high or high confidence, IPCC made a number of projections related to future changes in glaciers:

- Mountainous areas in Europe will face glacier retreat
- In Latin America, changes in precipitation patterns and the disappearance of glaciers will significantly affect water availability for human consumption, agriculture, and energy production
- In Polar regions, there will be reductions in glacier extent and the thickness of glaciers.

In historic times, glaciers grew during a cool period from about 1550 to 1850 known as the Little Ice Age. Subsequently, until about 1940, glaciers around the world retreated as the climate warmed. Glacier retreat declined and reversed in many cases from 1950 to 1980 as a slight global cooling occurred. Since 1980, glacier retreat has become increasingly rapid and ubiquitous, and has threatened the existence of many of the glaciers of the world. This process has increased markedly since 1995.

Excluding the ice caps and ice sheets of the Arctic and Antarctic, the total surface area of glaciers worldwide has decreased by 50% since the end of the 19th century. Currently glacier retreat rates and mass balance losses have been increasing in the Andes, Alps, Pyrenees, Himalayas, Rocky Mountains and North Cascades.

The loss of glaciers not only directly causes landslides, flash floods and glacial lake overflow, but also increases annual variation in water flows in rivers. Glacier runoff declines in the summer as glaciers decrease in size, this decline is already observable in several regions. Glaciers retain water on mountains in high precipitation years, since the snow cover accumulating on glaciers protects the ice from melting. In warmer and drier years, glaciers offset the lower precipitation amounts with a higher meltwater input.

Some world regions, such as the French Alps, already show signs of an increase in landslide frequency.

Of particular importance are the Hindu Kush and Himalayan glacial melts that comprise the principal dry-season water source of many of the major rivers of the Central, South, East and Southeast Asian mainland. Increased melting would cause greater flow for several decades, after which "some areas of the most populated regions on Earth are likely to 'run out of water'" as source glaciers are depleted. The Tibetan Plateau contains the world's third-largest store of ice. Temperatures there are rising four times faster than in the rest of China, and glacial retreat is at a high speed compared to elsewhere in the world.

According to a Reuters report, the Himalayan glaciers that are the sources of Asia's biggest rivers—Ganges, Indus, Brahmaputra, Yangtze, Mekong, Salween and Yellow—could diminish as temperatures rise. Approximately 2.4 billion people live in the drainage basin of the Himalayan rivers. India, China, Pakistan, Bangladesh, Nepal and Myanmar could experience floods followed by droughts in coming decades. The Indus, Ganges and Brahmaputra river basins support 700 million people in Asia. In India alone, the Ganges provides water for drinking and farming for more than 500 million people. It has to be acknowledged, however, that increased seasonal runoff of Himalayan glaciers led to increased agricultural production in northern India throughout the 20th century. Research studies suggest that climate change will have a marked affect on meltwater in the Indus Basin.

The recession of mountain glaciers, notably in Western North America, Franz-Josef Land, Asia, the Alps, the Pyrenees, Indonesia and Africa, and tropical and sub-tropical regions of South America, has been used to provide qualitative support to the rise in global temperatures since the late 19th century. Many glaciers are being lost to melting further raising concerns about future local water resources in these glaciated areas. In Western North America the 47 North Cascade glaciers observed all are retreating.

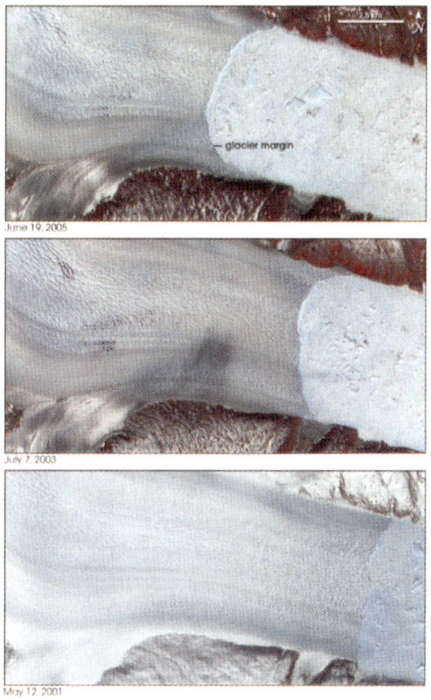

Retreat of the Helheim Glacier, Greenland

Despite their proximity and importance to human populations, the mountain and valley glaciers of temperate latitudes amount to a small fraction of glacial ice on the earth. About 99% is in the great ice sheets of polar and subpolar Antarctica and Greenland. These continuous continental-scale ice sheets, 3 kilometres (1.9 mi) or more in thickness, cap the polar and subpolar land masses. Like rivers flowing from an enormous lake, numerous outlet glaciers transport ice from the margins of the ice sheet to the ocean.

Glacier retreat has been observed in these outlet glaciers, resulting in an increase of the ice flow rate. In Greenland the period since the year 2000 has brought retreat to several very large glaciers that had long been stable. Three glaciers that have been researched, Helheim, Jakobshavn Isbræ and Kangerdlugssuaq Glaciers, jointly drain more than 16% of the Greenland Ice Sheet. Satellite images and aerial photographs from the 1950s and 1970s show that the front of the glacier had remained in the same place for decades. But in 2001 it began retreating rapidly, retreating 7.2 km (4.5 mi) between 2001 and 2005. It has also accelerated from 20 m (66 ft)/day to 32 m (105 ft)/day. Jakobshavn Isbræ in western Greenland had been moving at speeds of over 24 m (79 ft)/day with a stable terminus since at least 1950. The glacier's ice tongue began to break apart in 2000, leading to almost complete disintegration in 2003, while the retreat rate increased to over 30 m (98 ft)/day.

Oceans

The role of the oceans in global warming is a complex one. The oceans serve as a sink for carbon dioxide, taking up much that would otherwise remain in the atmosphere, but increased levels of CO_2 have led to ocean acidification. Furthermore, as the temperature of the oceans increases, they become less able to absorb excess CO_2. Global warming is projected to have a number of effects on the oceans. Ongoing effects include rising sea levels due to thermal expansion and melting of glaciers and ice sheets, and warming of the ocean surface, leading to increased temperature stratification. Other possible effects include large-scale changes in ocean circulation.

Sea Level Rise

IPCC reported that since 1961, global average sea level had risen at an average rate of 1.8 [1.3 to 2.3] mm/yr. Between 1993 and 2003, the rate increased above the previous period to 3.1 [2.4 to 3.8] mm/yr. IPCC were uncertain whether the increase in rate from 1993 to 2003 was due to natural variations in sea level over the time period, or whether it reflected an increase in the underlying long-term trend.

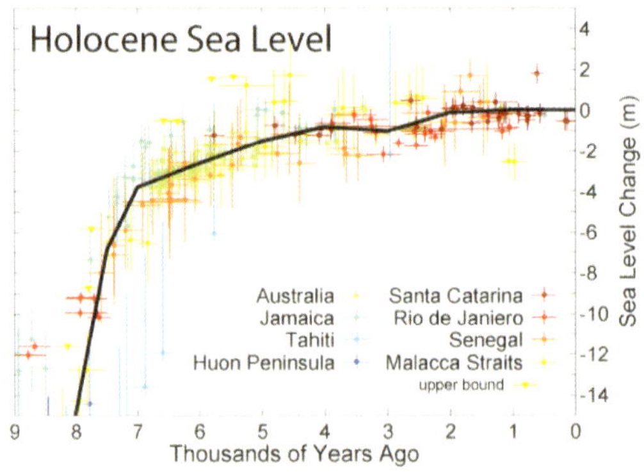

Sea level rise during the Holocene

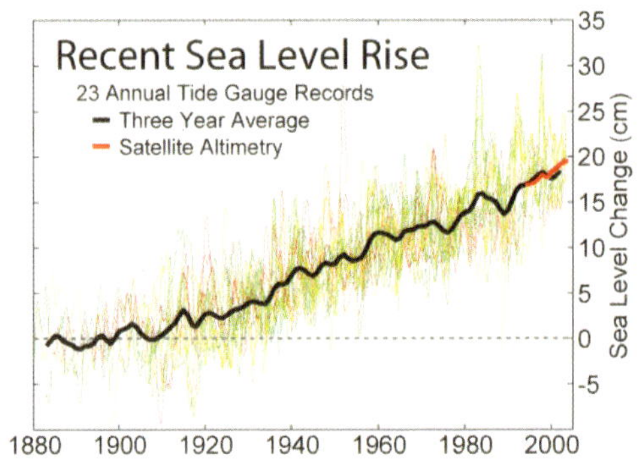

Sea level has been rising 0.2 cm/year, based on measurements of sea level rise from 23 long tide gauge records in geologically stable environments

IPCC projected sea level rise to the end of the 21st century using the SRES emission scenarios. Across the six SRES marker scenarios, sea level was projected to rise by 18 to 59 cm (7.1 to 23.2 inches). This projection was for the time period 2090–2099, with the increase in level relative to average sea levels over the 1980–1999 period. Due to a lack of scientific understanding, this sea level rise estimate does not include all of the possible contributions of ice sheets.

With increasing average global temperature, the water in the oceans expands in volume, and additional water enters them which had previously been locked up on land in glaciers and ice sheets. The Greenland and the Antarctic ice sheets are major ice masses, and at least the former of which may suffer irreversible decline. For most glaciers worldwide, an average volume loss of 60% until 2050 is predicted. Meanwhile, the estimated total ice melting rate over Greenland is 239 ± 23 cubic kilometres (57.3 ± 5.5 cu mi) per year, mostly from East Greenland. The Antarctic ice sheet, however, is expected to grow during the 21st century because of increased precipitation. Under the IPCC Special Report on Emission Scenario (SRES) A1B, by the mid-2090s global sea level will reach 0.22 to 0.44 m (8.7 to 17.3 in) above 1990 levels, and is currently rising at about 4 mm (0.16 in) per year.

Since 1900, the sea level has risen at an average of 1.7 mm (0.067 in) per year; since 1993, satellite altimetry from TOPEX/Poseidon indicates a rate of about 3 mm (0.12 in) per year.

The sea level has risen more than 120 metres (390 ft) since the Last Glacial Maximum about 20,000 years ago. The bulk of that occurred before 7000 years ago. Global temperature declined after the Holocene Climatic Optimum, causing a sea level lowering of 0.7 ± 0.1 m (27.6 ± 3.9 in) between 4000 and 2500 years before present. From 3000 years ago to the start of the 19th century, sea level was almost constant, with only minor fluctuations. However, the Medieval Warm Period may have caused some sea level rise; evidence has been found in the Pacific Ocean for a rise to perhaps 0.9 m (2 ft 11 in) above present level in 700 BP.

In a paper published in 2007, the climatologist James E. Hansen *et al.* claimed that ice at the poles does not melt in a gradual and linear fashion, but that another according to the geological record, the ice sheets can suddenly destabilize when a certain threshold is exceeded. In this paper Hansen *et al.* state:

> Our concern that BAU GHG scenarios would cause large sealevel rise this century (Hansen 2005) differs from estimates of IPCC (2001, 2007), which foresees little or no contribution to twentyfirst century sealevel rise from Greenland and Antarctica. However, the IPCC analyses and projections do not well account for the nonlinear physics of wet ice sheet disintegration, ice streams and eroding ice shelves, nor are they consistent with the palaeoclimate evidence we have presented for the absence of discernible lag between ice sheet forcing and sealevel rise.

Sea level rise due to the collapse of an ice sheet would be distributed nonuniformly across the globe. The loss of mass in the region around the ice sheet would decrease the gravitational potential there, reducing the amount of local sea level rise or even causing local sea level fall. The loss of the localized mass would also change the moment of inertia of the Earth, as flow in the Earth's mantle will require 10–15 thousand years to make up the mass deficit. This change in the moment of inertia results in true polar wander, in which the Earth's rotational axis remains fixed with respect to the sun, but the rigid sphere of the Earth rotates with respect to it. This changes the location of the equatorial bulge of the Earth and further affects the geoid, or global potential field. A 2009 study of the effects of collapse of the West Antarctic Ice Sheet shows the result of both of these effects. Instead of a global 5-meter sea level rise, western Antarctica would experience approximately 25 centimeters of sea level fall, while the United States, parts of Canada, and the Indian Ocean, would experience up to 6.5 meters of sea level rise.

A paper published in 2008 by a group of researchers at the University of Wisconsin led by Anders Carlson used the deglaciation of North America at 9000 years before present as an analogue to predict sea level rise of 1.3 meters in the next century, which is also much higher than the IPCC projections. However, models of glacial flow in the smaller present-day ice sheets show that a probable maximum value for sea level rise in the next century is 80 centimeters, based on limitations on how quickly ice can flow below the equilibrium line altitude and to the sea.

Temperature Rise and Ocean Heat Content

From 1961 to 2003, the global ocean temperature has risen by 0.10 °C from the surface to a depth of 700 m. There is variability both year-to-year and over longer time scales, with global ocean heat content observations showing high rates of warming for 1991 to 2003, but some cooling from 2003 to 2007. Nevertheless, there is a strong trend during the period of reliable measurements. Increasing heat content in the ocean is also consistent with sea level rise, which is occurring mostly as a result of thermal expansion of the ocean water as it warms.

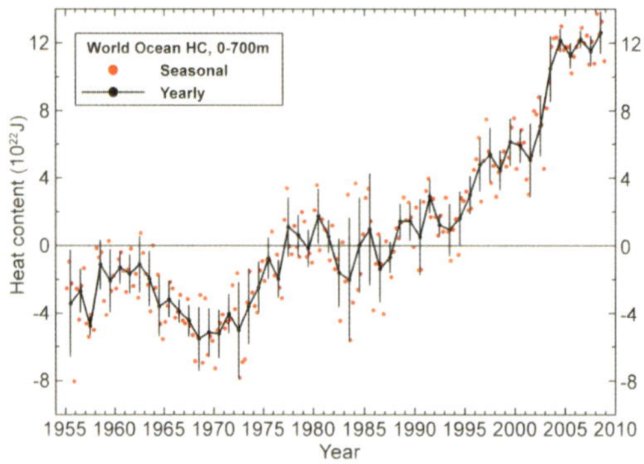

Time series of seasonal (red dots) and annual average (black line) global upper ocean heat content for the 0-700m layer between 1955 and 2008. The graph shows that ocean heat content has increased over this time period

The temperature of the Antarctic Southern Ocean rose by 0.17 °C (0.31 °F) between the 1950s and the 1980s, nearly twice the rate for the world's oceans as a whole. As well as having effects on ecosystems (e.g. by melting sea ice, affecting algae that grow on its underside), warming reduces the ocean's ability to absorb CO_2.

Acidification

Ocean acidification is an effect of rising concentrations of CO_2 in the atmosphere, and is not a direct consequence of global warming. The oceans soak up much of the CO_2 produced by living organisms, either as dissolved gas, or in the skeletons of tiny marine creatures that fall to the bottom to become chalk or limestone. Oceans currently absorb about one tonne of CO_2 per person per year. It is estimated that the oceans have absorbed around half of all CO_2 generated by human activities since 1800 (118 ± 19 petagrams of carbon from 1800 to 1994).

In water, CO_2 becomes a weak carbonic acid, and the increase in the greenhouse gas since the Industrial Revolution has already lowered the average pH (the laboratory measure of acidity) of seawater by 0.1 units, to 8.2. Predicted emissions could lower the pH by a further 0.5 by 2100, to a level probably not seen for hundreds of millennia and, critically, at a rate of change probably 100 times greater than at any time over this period.

There are concerns that increasing acidification could have a particularly detrimental effect on corals (16% of the world's coral reefs have died from bleaching caused by warm water in 1998, which coincidentally was, at the time, the warmest year ever recorded) and other marine organisms with calcium carbonate shells.

In November 2009, an article in *Science* by scientists at Canada's Department of Fisheries and Oceans reported they had found very low levels of the building blocks for the calcium chloride that forms plankton shells in the Beaufort Sea. Fiona McLaughlin, one of the DFO authors, asserted that the increasing acidification of the Arctic Ocean was close to the point it would start dissolving the walls of existing plankton: *"The Arctic ecosystem may be risk. In actual fact, they'll dissolve the shells."* Because cold water absorbs CO_2 more readily than warmer water the acidification is more severe in the polar regions. McLaughlin predicted the acidified water would travel to the North Atlantic within the next ten years.

Shutdown of Thermohaline Circulation

There is some speculation that global warming could, via a shutdown or slowdown of the thermohaline circulation, trigger localized cooling in the North Atlantic and lead to cooling, or lesser warming, in that region. This would affect in particular areas like Scandinavia and Britain that are warmed by the North Atlantic drift.

The chances of this near-term collapse of the circulation are unclear; there is some evidence for the short-term stability of the Gulf Stream and possible weakening of the North Atlantic drift. However, the degree of weakening, and whether it will be sufficient to shut down the circulation, is under debate. As yet, no cooling has been found in northern Europe or nearby seas. Lenton et al. found that "simulations clearly pass a THC tipping point this century".

IPCC concluded that a slowing of the Meridional Overturning Circulation would very likely occur this century. Due to global warming, temperatures across the Atlantic and Europe were still projected to increase.

Oxygen Depletion

The amount of oxygen dissolved in the oceans may decline, with adverse consequences for ocean life.

Sulfur Aerosols

Sulfur aerosols, especially stratospheric sulfur aerosols have a significant effect on climate. One source of such aerosols is the sulfur cycle, where plankton release gases such as DMS which eventually becomes oxidised to sulfur dioxide in the atmosphere. Disruption to the oceans as a result of ocean acidification or disruptions to the thermohaline circulation may result in disruption of the sulfur cycle, thus reducing its cooling effect on the planet through the creation of stratospheric sulfur aerosols.

Geology

Volcanoes

The retreat of glaciers and ice caps can cause increased volcanism. Reduction in ice cover reduces the confining pressure exerted on the volcano, increasing deviatoric stresses and potentially causing the volcano to erupt. This reduction of pressure can also cause decompression melting of material in the mantle, resulting in the generation of more magma. Researchers in Iceland have shown that the rate of volcanic rock production there following deglaciation (10,000 to 4500 years before present) was 20–30 times greater than that observed after 2900 years before present. While the original study addresses the first reason for increased volcanism (reduced confining pressure), scientists have more recently shown that these lavas have unusually high trace element concentrations, indicative of increased melting in the mantle. This work in Iceland has been corroborated by a study in California, in which scientists found a strong correlation between volcanism and periods of global deglaciation. The effects of current sea level rise could include increased crustal stress at the base of coastal volcanoes from a rise in the volcano's water table (and the associated saltwater intrusion), while the mass from extra water could activate dormant seismic faults around volcanoes. In addition, the wide-scale displacement of water from melting in places such as West Antarctica is likely to slightly alter the Earth's rotational period and may shift its axial tilt on the scale of hundreds of metres, inducing further crustal stress changes.

Current melting of ice is predicted to increase the size and frequency of volcanic eruptions. In particular, lateral collapse events at stratovolcanoes are likely to increase, and there are potential positive feedbacks between the removal of ice and magmatism.

Earthquakes

A numerical modeling study has demonstrated that seismicity increases during unloading, such as that due to the removal of ice.

Effects of Climate Change on Plant Biodiversity

Alpine flora at Logan Pass, Glacier National Park, in Montana, United States: Alpine plants are one group expected to be highly susceptible to the impacts of climate change

Environmental conditions play a key role in defining the function and distribution of plants, in combination with other factors. Changes in long term environmental conditions that can be collectively coined climate change are known to have had enormous impacts on current plant diversity patterns; further impacts are expected in the future. It is predicted that climate change will remain one of the major drivers of biodiversity patterns in the future.

Palaeo Context

Australian Rainforest: An ecosystem known to have significantly contracted in area over recent geological time as a result of climatic changes

The Earth has experienced a constantly changing climate in the time since plants first evolved. In comparison to the present day, this history has seen Earth as cooler, warmer, drier and wetter, and CO_2 (carbon dioxide) concentrations have been both higher and lower. These changes have been reflected by constantly shifting vegetation, for example forest communities dominating most areas in interglacial periods, and herbaceous communities dominating during glacial periods. It has been shown that past climatic change has been a major driver of the processes of speciation and extinction. The best known example of this is the Carboniferous Rainforest Collapse which occurred 350 million years ago. This event decimated amphibian populations and spurred on the evolution of reptiles.

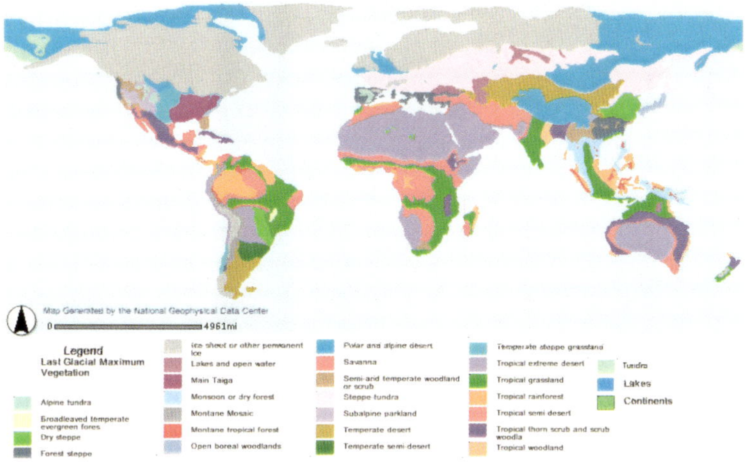

Map of global vegetation distributions during the last glacial maximum

Modern Context

There is significant current interest and research focus on the phenomenon of recent anthropogenic climate changes, or global warming. Focus is on identifying the current impacts of climate change on biodiversity, and predicting these effects into the future.

Changing climatic variables relevant to the function and distribution of plants include increasing CO_2 concentrations, increasing global temperatures, altered precipitation patterns, and changes in the pattern of 'extreme' weather events such as cyclones, fires or storms. Highly variable species distribution has resulted from different models with variable bioclimatic changes.

Because individual plants and therefore species can only function physiologically, and successfully complete their life cycles under specific environmental conditions (ideally within a subset of these), changes to climate are likely to have significant impacts on plants from the level of the individual right through to the level of the ecosystem or biome.

Effects of CO_2

CO_2 concentrations have been steadily rising for more than two centuries. Increases in atmospheric CO_2 concentration affect how plants photosynthesise, resulting in increases in plant water use efficiency, enhanced photosynthetic capacity and increased growth. Increased CO_2 has been implicated in 'vegetation thickening' which affects plant community structure and function. Depending on environment, there are differential responses to elevated atmospheric CO_2 between major 'functional types' of plant, such as C_3 and C_4 plants, or more or less woody species; which has the potential among other things to alter competition between these groups. Increased CO_2 can also lead to increased Carbon : Nitrogen ratios in the leaves of plants or in other aspects of leaf chemistry, possibly changing herbivore nutrition.

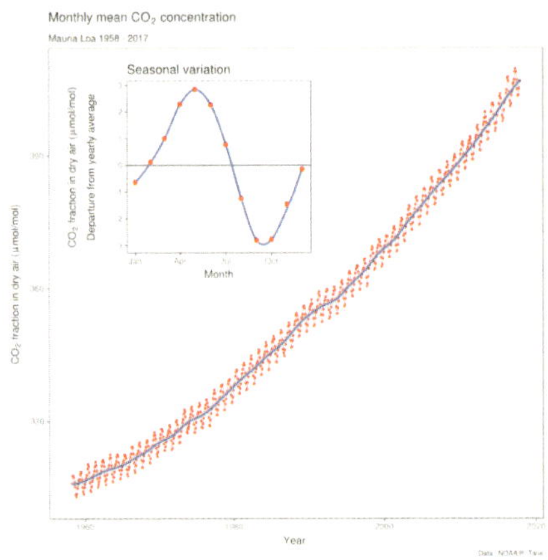

Recent increases in atmospheric CO_2

Effects of Temperature

Increases in temperature raise the rate of many physiological processes such as photosynthesis in plants, to an upper limit. These increases in photosynthesis and other physiological processes are driven by increased rates of chemical reactions and roughly a doubling of enzymatic product conversion rates for every 10 °C increase in temperature. Extreme temperatures can be harmful when beyond the physiological limits of a plant.

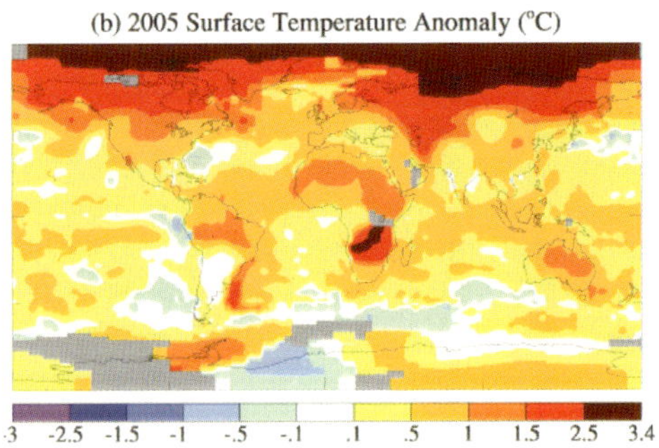

Global annual surface temperature anomaly, relative to 1951-1980 mean

One common hypothesis among scientists is that the warmer an area is, the higher the plant diversity. This hypothesis can be observed in nature, where higher plant biodiversity is often located at certain latitudes (which often correlates with a specific climate/temperature).

Effects of Water

As water supply is critical for plant growth, it plays a key role in determining the distribution of plants. Changes in precipitation are predicted to be less consistent than for temperature and more variable between regions, with predictions for some areas to become much wetter, and some much drier. This can cause a major change in some ecosystems which are dependent on water supply.

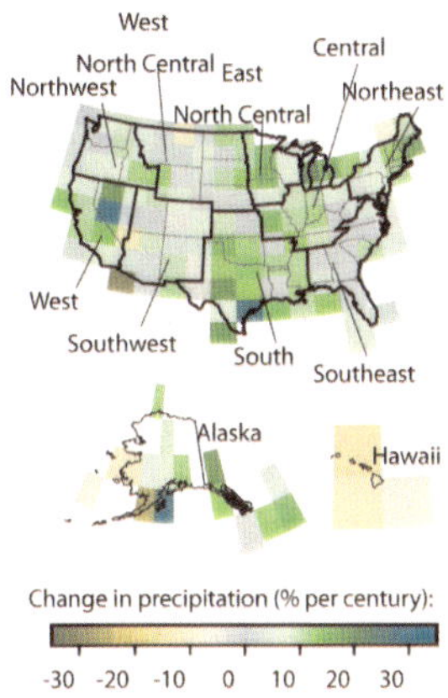

Precipitation trends in the United States, from the period 1901-2005. In some areas rainfall has increased in the last century, while some areas have dried

General Effects

Environmental variables act not in isolation, but in combination with other pressures such as habitat degradation, habitat loss, and the introduction of exotic species that can potentially be invasive. It is suggested that these other drivers of biodiversity change will act in synergy with climate change to increase the pressure on species to survive.

Direct Impacts of Climate Change

Changes in Distributions

Pine tree representing an elevational tree-limit rise of 105 m over the period 1915–1974. Nipfjället, Sweden

If climatic factors such as temperature and precipitation change in a region beyond the tolerance of a species phenotypic plasticity, then distribution changes of the species may be inevitable. There is already evidence that plant species are shifting their ranges in altitude and latitude as a response to changing regional climates. Yet it is difficult to predict how species ranges will change in response to climate and separate these changes from all the other man-made environmental changes such as eutrophication, acid rain and habitat destruction.

When compared to the reported past migration rates of plant species, the rapid pace of current change has the potential to not only alter species distributions, but also render many species as unable to follow the climate to which they are adapted. The environmental conditions required by some species, such as those in alpine regions may disappear altogether. The result of these changes is likely to be a rapid increase in extinction risk. Adaptation to new conditions may also be of great importance in the response of plants.

Predicting the extinction risk of plant species is not easy however. Estimations from particular periods of rapid climatic change in the past have shown relatively little species extinction in some regions, for example. Knowledge of how species may adapt or persist in the face of rapid change is still relatively limited.

Changes in the suitability of a habitat for a species drive distributional changes by not only changing the area that a species can physiologically tolerate, but how effectively it can compete with other plants within this area. Changes in community composition are therefore also an expected product of climate change.

Changes in Life-cycles (Phenology)

The timing of phenological events such as flowering are often related to environmental variables such as temperature. Changing environments are therefore expected to lead to changes in life cycle events, and these have been recorded for many species of plants. These changes have the potential to lead to the asynchrony between species, or to change competition between plants. Flowering times in British plants for example have changed, leading to annual plants flowering earlier than perennials, and insect pollinated plants flowering earlier than wind pollinated plants; with potential ecological consequences. A recently published study has used data recorded by the writer and naturalist Henry David Thoreau to confirm effects of climate change on the phenology of some species in the area of Concord, Massachusetts.

Indirect Impacts of Climate Change

All species are likely to be directly impacted by the changes in environmental conditions discussed above, and also indirectly through their interactions with other species. While direct impacts may be easier to predict and conceptualise, it is likely that indirect impacts are equally important in determining the response of plants to climate change. A species whose distribution changes as a direct result of climate change may 'invade' the range of another species or 'be invaded' for example, introducing a new competitive relationship or altering other processes such as carbon sequestration.

In Europe, the temperature and precipitation effects due to climate change can indirectly affect certain populations of people. The rise of temperatures and lack of precipitation results in different river floodplains, which reduce the populations of people sensitive to flood risk.

The range of a symbiotic fungi associated with plant roots may directly change as a result of altered climate, resulting in a change in the plant's distribution.

A new grass may spread into a region, altering the fire regime and greatly changing the species composition.

A pathogen or parasite may change its interactions with a plant, such as a pathogenic fungus becoming more common in an area where rainfall increases.

Increased temperatures may allow herbivores to expand further into alpine regions, significant impacting the composition of alpine herbfields.

Higher Level Changes

Species respond in very different ways to climate change. Variation in the distribution, phenology and abundance of species will lead to inevitable changes in the relative abundance of species and their interactions. These changes will flow on to affect the structure and function of ecosystems.

Challenges of Modeling Future Impacts

Accurate predictions of the future impacts of climate change on plant diversity are critical to the development of conservation strategies. These predictions have come largely from bioinformatic strategies, involving modeling individual species, groups of species such as 'functional types', communities, ecosystems or biomes. They can also involve modeling species observed environmental niches, or observed physiological processes.

Although useful, modeling has many limitations. Firstly, there is uncertainty about the future levels of greenhouse gas emissions driving climate change and considerable uncertainty in modeling how this will affect other aspects of climate such as local rainfall or temperatures. For most species the importance of specific climatic variables in defining distribution (e.g. minimum rainfall or maximum temperature) is unknown. It is also difficult to know which aspects of a particular climatic variable are most biologically relevant, such as average vs. maximum or minimum temperatures. Ecological processes such as interactions between species and dispersal rates and distances are also inherently complex, further complicating predictions.

Improvement of models is an active area of research, with new models attempting to take factors such as life-history traits of species or processes such as migration into account when predicting distribution changes; though possible trade-offs between regional accuracy and generality are recognised.

Climate change is also predicted to interact with other drivers of biodiversity change such as habitat destruction and fragmentation, or the introduction of foreign species. These threats may possibly act in synergy to increase extinction risk from that seen in periods of rapid climate change in the past.

Long-term Effects of Global Warming

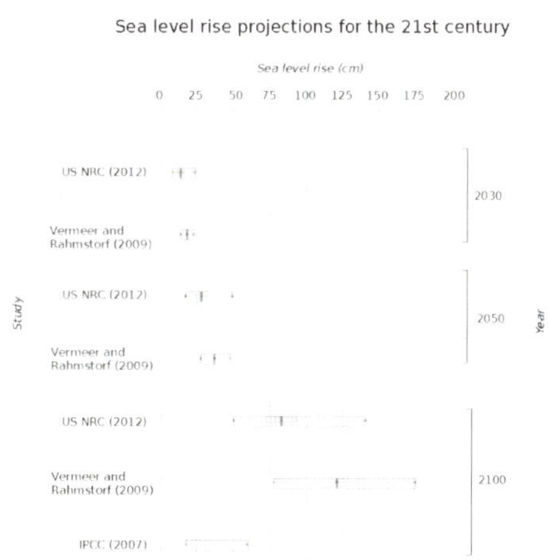

Sea level rise projections for the 21st century

There are expected to be various long-term effects of global warming. Most discussion and research, including that by the Intergovernmental Panel on Climate Change (IPCC) reports, concentrates on the effects of global warming up to 2100, with only an outline of the effects beyond this.

Ice loss and Sea Level Rise

Meltwater from melting ice sheets and glacier retreat contributes to a rise in the future sea level.

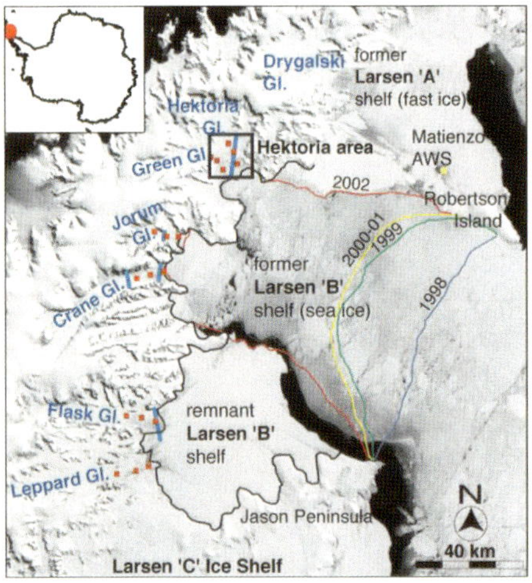

The collapse of Larsen B Shelf, showing the diminishing extent of the shelf from 1998 to 2002

Antarctica

Concern has been expressed about the stability of the West Antarctic ice sheet (WAIS). In 2002 Vaughan & Spouge argued that the "WAIS most likely will not collapse in the next few centuries." In an inaugural article for members of the National Academy of Sciences elected in 2005, Timothy Lenton and others suggest that a collapse of the West Antarctic ice sheet could occur within a millennium. Specifically, they state "Although the timescale is highly uncertain, a qualitative WAIS change could occur within this millennium, with collapse within 300 years being a worst-case scenario. Rapid sea-level rise (greater than 1 m per century) is more likely to come from the WAIS than from the Greenland ice sheet". A study in 2015 found that assuming cumulative fossil fuel emissions of 10 000 gigatonnes of carbon, the Antarctic Ice Sheet could melt completely over the following millennia, contributing 58 m to global sea-level rise, and 30 m within the first 1000 years.

Greenland

Greenland's ice sheet contains enough fresh water as ice to raise sea level worldwide by 7 metres (23 ft). Greenland may become warm enough by 2100 to begin an almost complete melt over more than 1,000 years. James E. Hansen suggests that inadequate attention is being given to this issue.

One study suggests it would take 3,000 years to completely melt the Greenland ice sheet. This figure was derived from the assumed levels of greenhouse gases over the duration of the experiment.

As the Greenland ice sheet loses mass from calving of icebergs as well as by melting of ice, any such processes tend to accelerate the loss of the ice sheet.

Millennial-timescale Events

Some long-term effects happen over thousands, not hundreds, of years.

Disruption to Thermohaline Circulation

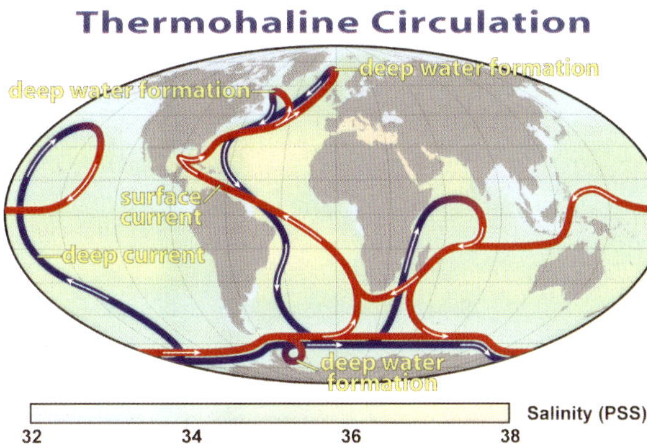

A schematic of the path of the thermohaline circulation. Blue paths represent deep-water currents, while red paths represent surface currents

Early work with simplified models suggested that global warming could cause a shutdown of the thermohaline circulation. This effect is not replicated in more sophisticated coupled ocean-atmosphere global climate models, which do not shutdown but do display varying degrees of slowdown.

This may take a considerable time to occur, as Knutti and Stocker found, again from a simplified model, when:" Thermohaline shutdowns can occur thousands of years after the warming has stopped."

Ocean Anoxia

One study suggests that the amount of oxygen dissolved in the oceans may decline, with adverse consequences for ocean life. This effect was determined using a model run of 100,000 years. The researchers predicted.

> Severe, long-term ocean oxygen depletion, as well as a great expansion of ocean oxygen-minimum zones for scenarios with high emissions or high climate sensitivity. We find that climate feedbacks within the Earth system amplify the strength and duration of global warming, ocean heating and oxygen depletion. Decreased oxygen solubility from surface-layer warming accounts for most of the enhanced oxygen depletion in the upper 500 m of the ocean. Possible weakening of ocean overturning and convection lead to further oxygen depletion, also in the deep ocean.

Other research by the Leibniz Institute for Marine Sciences predicts that minimum oxygen zones, particularly in the pacific are expanding, with an estimated total reduction of ocean oxygen levels from 1% to 7% by the end of the century.

Clathrate Decomposition

Methane clathrate, also called methane hydrate, is a form of water ice that contains a large amount of methane within its crystal structure. Extremely large deposits of methane clathrate have been found under sediments on the ocean floors of Earth (estimated at 3000–11,000 Gton C).

Worldwide distribution of confirmed or inferred offshore gas hydrate-bearing sediments

MacDonald suggests clathrate volumes are "about 11,000 Gt of carbon for ocean sediments, and about 400 Gt for sediments under permafrost regions". Buffett and Archer predict "eventual releases of 2000–4000 Gton C in response to a ~2000 Gton C anthropogenic carbon release", for which they suggest no timescale.

Archer considered the timescale and determined that "on longer time scales of 1–10 ky, there may be a positive feedback with ocean temperature, amplifying the long-term climate impact of anthropogenic CO_2 release."

Carbon Dioxide Levels in the Atmosphere

Stabilizing the global average temperature would require large reductions in CO_2 emissions, as well as reductions in emissions of other greenhouse gases such as methane and nitrous oxide. Emissions of CO_2 would need to be reduced by more than 80% relative to their peak level. Even if this were achieved, global average temperatures would remain close to their highest level for many centuries. As of 2016, emissions of CO_2 from burning fossil fuels had stopped increasing, but The Guardian reports they need to be "reduced to have a real impact on climate change". Meanwhile, this greenhouse gas continues to accumulate in the atmosphere. In that context, the New York Times reported that scientific installations analyzing oceanic air detected the excess carbon dioxide in the atmosphere "rose at the highest rate on record in 2015 and 2016." It has been suggested that this rise in CO_2 levels is the result of changing absorption patterns of the ocean and land surface in that they may have reached the limit of their ability to absorb carbon dioxide.

Long-term Return to Equilibrium

After the PETM peak there was an extended period of cooling or "climate whiplash".

While the warm surface waters of the oceans have limited ability to absorb anthropogenic carbon dioxide, the coldest surface waters near the poles (2–3% of ocean surfaces) can transfer significant amounts of carbon dioxide to deep-ocean reserves. Over a period of many centuries, this process and the process of calcium carbonate absorption of carbon dioxide on land and in the oceans will remove 60–80% of the excess carbon dioxide.

Igneous rock when exposed to a near surface environment absorbs carbon dioxide through a very slow weathering rate, but weathering increases in a warmer, higher rainfall climate, speeding the process. This geological weathering will absorb the remaining 20–40% of anthropogenic carbon dioxide over the period of tens of thousands to hundreds of thousands of years.

Effects of Global Warming on Humans

Flooding in the U.S. Midwest

Climate change has brought about possibly permanent alterations to Earth's geological, biological and ecological systems. These changes have led to the emergence of a not so large-scale environmental hazards to human health, such as extreme weather, ozone depletion, increased danger of wildland fires, loss of biodiversity, stresses to food-producing systems and the global spread of infectious diseases. The World Health Organization (WHO) estimates that 160,000 deaths, since 1950, are directly attributable to climate change.

To date, a neglected aspect of the climate change debate, much less research has been conducted on the impacts of climate change on health, food supply, economic growth, migration, security, societal change, and public goods, such as drinking water, than on the geophysical changes related to global warming. Human impacts can be both negative and positive. Climatic changes in Siberia, for instance, are expected to improve food production and local economic activity, at least in the short to medium term. Numerous studies suggest, however, that the current and future impacts of climate change on human society are and will continue to be overwhelmingly negative.

The majority of the adverse effects of climate change are experienced by poor and low-income communities around the world, who have much higher levels of vulnerability to environmental determinants of health, wealth and other factors, and much lower levels of capacity available for coping with environmental change. A report on the global human impact of climate change published by the Global Humanitarian Forum in 2009, estimated more than 300,000 deaths and about $125 billion in economic losses each year, and indicating that most climate change induced mortality is due to worsening floods and droughts in developing countries.

Key Vulnerabilities

Most of the key vulnerabilities to climate change are related to climate phenomena that exceed thresholds for adaptation; such as extreme weather events or abrupt climate change, as well as limited access to resources (financial, technical, human, institutional) to cope. In 2007, the IPCC published a report of key vulnerabilities of industry, settlements, and society to climate change.

This assessment included a level of confidence for each key vulnerability:

- Very high confidence: Interactions between climate change and urbanization: this is most notable in developing countries, where urbanization is often focused in vulnerable coastal areas.
- High confidence:
  - Interactions between climate change and global economic growth: Stresses due to climate change are not only linked to the impacts of climate change, but also to the impacts of climate change policies. For example, these policies might affect development paths by requiring high cost fuel choices.
  - Fixed physical infrastructures that are important in meeting human needs: These include infrastructures that are susceptible to damage from extreme weather events or sea level rise, and infrastructures that are already close to being inadequate.
- Medium confidence: Interactions with governmental and social cultural structures that already face other pressures, e.g., limited economic resources.

Health

Climate change poses a wide range of risks to population health – risks that will increase in future decades, often to critical levels, if global climate change continues on its current trajectory. The three main categories of health risks include: (i) direct-acting effects (e.g. due to heat waves, amplified air pollution, and physical weather disasters), (ii) impacts mediated via climate-related changes in ecological systems and relationships (e.g. crop yields, mosquito ecology, marine productivity), and (iii) the more diffuse (indirect) consequences relating to impoverishment, displacement, resource conflicts (e.g. water), and post-disaster mental health problems.

Climate change thus threatens to slow, halt or reverse international progress towards reducing child under-nutrition, deaths from diarrheal diseases and the spread of other infectious diseases. Climate change acts predominantly by exacerbating the existing, often enormous, health problems, especially in the poorer parts of the world. Current variations in weather conditions already have many adverse impacts on the health of poor people in developing nations, and these too are likely to be 'multiplied' by the added stresses of climate change.

A changing climate thus affects the prerequisites of population health: clean air and water, sufficient food, natural constraints on infectious disease agents, and the adequacy and security of shelter. A warmer and more variable climate leads to higher levels of some air pollutants. It increases the rates and ranges of transmission of infectious diseases through unclean water and contaminated food, and by affecting vector organisms (such as mosquitoes) and intermediate or reservoir host species that harbour the infectious agent (such as cattle, bats and rodents). Changes in temperature, rainfall and seasonality compromise agricultural production in many regions, including some of the least developed countries, thus jeopardising child health and growth and the overall health and functional capacity of adults. As warming proceeds, the severity (and perhaps frequency) of weather-related disasters will increase – and appears to have done so in a number of regions of the world over the past several decades. Therefore, in summary, global warming, together with resultant changes in food and water supplies, can indirectly cause increases in a range of adverse health outcomes, including malnutrition, diarrhea, injuries, cardiovascular and respiratory diseases, and water-borne and insect-transmitted diseases.

Health equity and climate change have a major impact on human health and quality of life, and are interlinked in a number of ways. The report of the WHO Commission on Social Determinants of Health points out that disadvantaged communities are likely to shoulder a disproportionate share of the burden of climate change because of their increased exposure and vulnerability to health threats. Over 90 percent of malaria and diarrhea deaths are borne by children aged 5 years or younger, mostly in developing countries. Other severely affected population groups include women, the elderly and people living in small island developing states and other coastal regions, mega-cities or mountainous areas.

Psychological Impacts

A 2011 article in the *American Psychologist* identified three classes of psychological impacts from global climate change:

- Direct - "Acute or traumatic effects of extreme weather events and a changed environment".
- Indirect - "Threats to emotional well-being based on observation of impacts and concern or uncertainty about future risks".
- Psychosocial – "Chronic social and community effects of heat, drought, migrations, and climate-related conflicts, and postdisaster adjustment" A psychological impact is shown through peoples behaviours and how they act towards the actual situation. The topic of climate change is very complex and difficult for people to understand, which effects how they act upon it. It is shown by Ranney and Clark(2016) that by informing people to make them understand the topic of climate science clearly, it promotes the change in behaviour towards mitigation of climate change.

Extreme Weather Events

This trend towards more variability and fluctuation is perhaps more important, in terms of its impact on human health, than that of a gradual and long-term trend towards higher average temperature. Infectious disease often accompanies extreme weather events, such as floods, earthquakes and drought. These local epidemics occur due to loss of infrastructure, such as hospitals and sanitation services, but also because of changes in local ecology and environment.

Diseases

Climate change may lead to dramatic increases in prevalence of a variety of infectious diseases. Beginning in the mid-'70s, there has been an "emergence, resurgence and redistribution of infectious diseases". Reasons for this are likely multicausal, dependent on a variety of social, environmental and climatic factors, however, many argue that the "volatility of infectious disease may be one of the earliest biological expressions of climate instability". Though many infectious diseases are affected by changes in climate, vector-borne diseases, such as malaria, dengue fever and leishmaniasis, present the strongest causal relationship. Observation and research detect a shift of pests and pathogens in the distribution away from the equator and towards Earth's poles.

Malaria

Increased precipitation like rain could increase the number of mosquitos indirectly by expanding larval habitat and food supply. Malaria kills approximately 300,000 children (under age 5) annually, poses an imminent threat through temperature increase . Models suggest, conservatively, that risk of malaria will increase 5-15% by 2100 due to climate change. In Africa alone, according to the MARA Project (Mapping Malaria Risk in Africa), there is a projected increase of 16–28% in person-month exposures to malaria by 2100.

Non-climatic Determinants

Sociodemographic factors include, but are not limited to: patterns of human migration and travel, effectiveness of public health and medical infrastructure in controlling and treating disease, the extent of anti-malarial drug resistance and the underlying health status of the population at hand. Environmental factors include: changes in land-use (e.g. deforestation), expansion of agricultural and water development projects (which tend to increase mosquito breeding habitat), and the overall trend towards urbanization (i.e. increased concentration of human hosts). Patz and Olson argue that these changes in landscape can alter local weather more than long term climate change. For example, the deforestation and cultivation of natural swamps in the African highlands has created conditions favourable for the survival of mosquito larvae, and has, in part, led to the increasing incidence of malaria. The effects of these non-climatic factors complicate things and make a direct causal relationship between climate change and malaria difficult to confirm. It is highly unlikely that climate exerts an isolated effect.

Environment

Climate change may dramatically impact habitat loss, for example, arid conditions may cause the deforestation of rainforests, as has occurred in the past.

Temperature

A sustained wet-bulb temperature exceeding 35° is a threshold at which the resilience of human systems is no longer able to adequately cool the skin. A study by NOAA from 2013 concluded that heat stress will reduce labor capacity considerably under current emissions scenarios. There is evidence to show that high temperatures can increase mortality rates among fetuses and children Although the main focus is often on the health impacts and risks of higher temperatures, it should be remembered that they also reduce learning and worker productivity, which can impact a country's economy and development.

Water

The freshwater resources that humans rely on are highly sensitive to variations in weather and climate. In 2007, the IPCC reported with high confidence that climate change has a net negative impact on water resources and freshwater ecosystems in all regions. The IPCC also found with very high confidence that arid and semi-arid areas are particularly exposed to freshwater impacts.

As the climate warms, it changes the nature of global rainfall, evaporation, snow, stream flow and other factors that affect water supply and quality. Specific impacts include:

- Warmer water temperatures affect water quality and accelerate water pollution.
- Sea level rise is projected to increase salt-water intrusion into groundwater in some regions. This reduces the amount of freshwater available for drinking and farming.
- In some areas, shrinking glaciers and snow deposits threaten the water supply. Areas that depend on melted water runoff will likely see that runoff depleted, with less flow in the late summer and spring peaks occurring earlier. This can affect the ability to irrigate crops. (This situation is particularly acute for irrigation in South America, for irrigation and drinking supplies in Central Asia, and for hydropower in Norway, the Alps, and the Pacific Northwest of North America.)
- Increased extreme weather means more water falls on hardened ground unable to absorb it, leading to flash floods instead of a replenishment of soil moisture or groundwater levels.
- Increased evaporation will reduce the effectiveness of reservoirs.
- At the same time, human demand for water will grow for the purposes of cooling and hydration.

Displacement and Migration

A refugee camp

Climate change causes displacement of people in several ways, the most obvious—and dramatic—being through the increased number and severity of weather-related disasters which destroy homes and habitats causing people to seek shelter or livelihoods elsewhere. Effects of climate change such as desertification and rising sea levels gradually erode livelihood and force communities to abandon traditional homelands for more accommodating environments. This is currently happening in areas of Africa's Sahel, the semi-arid belt that spans the continent just below its northern deserts. Deteriorating environments triggered by climate change can also lead to increased conflict over resources which in turn can displace people.

The IPCC has estimated that 150 million environmental migrants will exist by the year 2050, due mainly to the effects of coastal flooding, shoreline erosion and agricultural disruption. However, the IPCC also cautions that it is extremely difficult to measure the extent of environmental migration due to the complexity of the issue and a lack of data.

According to the Internal Displacement Monitoring Centre, more than 42 million people were displaced in Asia and the Pacific during 2010 and 2011, more than twice the population of Sri Lanka. This figure includes those displaced by storms, floods, and heat and cold waves. Still others were displaced drought and sea-level rise. Most of those compelled to leave their homes eventually returned when conditions improved, but an undetermined number became migrants, usually within their country, but also across national borders.

Asia and the Pacific is the global area most prone to natural disasters, both in terms of the absolute number of disasters and of populations affected. It is highly exposed to climate impacts, and is home to highly vulnerable population groups, who are disproportionately poor and marginalized. A recent Asian Development Bank report highlights "environmental hot spots" that are particular risk of flooding, cyclones, typhoons, and water stress.

Some Pacific Ocean island nations, such as Tuvalu, Kiribati, and the Maldives, are considering the eventual possibility of evacuation, as flood defense may become economically unrealistic. Tuvalu already has an ad hoc agreement with New Zealand to allow phased relocation. However, for some islanders relocation is not an option. They are not willing to leave their homes, land and families. Some simply don't know the threat that climate change has on their island and this is mainly down to the lack of awareness that climate change even exists. In Vutia on Viti Levu, Fiji's main island, half the respondents to a survey had not heard of climate change (Lata and Nuun 2012). Even where there is awareness many believe that it is a problem caused by developed countries and should therefore be solved by developed countries.

Governments have considered various approaches to reduce migration compelled by environmental conditions in at-risk communities, including programs of social protection, livelihoods development, basic urban infrastructure development, and disaster risk management. Some experts even support migration as an appropriate way for people to cope with environmental changes. However, this is controversial because migrants – particularly low-skilled ones – are among the most vulnerable people in society and are often denied basic protections and access to services.

Climate change is only one factor that may contribute to a household's decision to migrate; other factors may include poverty, population growth or employment options. For this reason, it is difficult to classify environmental migrants as actual "refugees" as legally defined by the UNHCR. Neither the UN Framework Convention on Climate Change nor its Kyoto Protocol, an international agreement on climate change, includes any provisions concerning specific assistance or protection for those who will be directly affected by climate change.

In small islands and megadeltas, inundation as a result of sea level rise is expected to threaten vital infrastructure and human settlements. This could lead to issues of statelessness for populations in countries such as the Maldives and Tuvalu and homelessness in countries with low-lying areas such as Bangladesh.

The World Bank predicts that a "severe hit" will spur conflict and migration across the Middle East, Central Asia, and Africa.

Security

Climate change has the potential to exacerbate existing tensions or create new ones — serving as a threat multiplier. It can be a catalyst for violent conflict and a threat to international security. A meta-analysis of over 50 quantitative studies that examine the link between climate and conflict found that "for each 1 standard deviation (1σ) change in climate toward warmer temperatures or more extreme rainfall, median estimates indicate that the frequency of interpersonal violence rises 4% and the frequency of intergroup conflict rises 14%". The IPCC has suggested that the disruption of environmental migration may serve to exacerbate conflicts, though they are less confident of the role of increased resource scarcity. Of course, climate change does not always lead to violence, and conflicts are often caused by multiple interconnected factors.

A variety of experts have warned that climate change may lead to increased conflict. The Military Advisory Board, a panel of retired U.S. generals and admirals, predicted that global warming will serve as a "threat multiplier" in already volatile regions. The Center for Strategic and International Studies and the Center for a New American Security, two Washington think tanks, have reported that flooding "has the potential to challenge regional and even national identities," leading to "armed conflict over resources." They indicate that the greatest threat would come from "large-scale migrations of people — both inside nations and across existing national borders." However, other researchers have been more skeptical: One study found no statistically meaningful relationship between climate and conflict using data from Europe between the years 1000 and 2000.

The link between climate change and security is a concern for authorities across the world, including United Nations Security Council and the G77 group of developing nations. Climate change's impact as a security threat is expected to hit developing nations particularly hard. In Britain, Foreign Secretary Margaret Beckett has argued that "An unstable climate will exacerbate some of the core drivers of conflict, such as migratory pressures and competition for resources." The links between the human impact of climate change and the threat of violence and armed conflict are particularly important because multiple destabilizing conditions are affected simultaneously.

Experts have suggested links to climate change in several major conflicts:

- War in Darfur, where sustained drought encouraged conflict between herders and farmers
- Syrian Civil War, preceded by the displacement of 1.5 million people due to drought-induced crop and livestock failure
- Islamist insurgency in Nigeria, which exploited natural resource shortages to fuel anti-government sentiment
- Somali Civil War, in which droughts and extreme high temperatures have been linked to violence

Additionally, researchers studying ancient climate patterns (paleoclimatology) have shown that long-term fluctuations of war frequency and population changes have followed cycles of temperature change since the preindustrial era. A 2016 study finds that "drought can contribute to sustaining conflict, especially for agriculturally dependent groups and politically excluded groups in very poor countries. These results suggest a reciprocal nature–society interaction in which violent conflict and environmental shock constitute a vicious circle, each phenomenon increasing the group's vulnerability to the other."

Social Impacts

The consequences of climate change and poverty are not distributed uniformly within communities. Individual and social factors such as gender, age, education, ethnicity, geography and language lead to differential vulnerability and capacity to adapt to the effects of climate change. Climate change effects such as hunger, poverty and diseases like diarrhea and malaria, disproportionately impact children; about 90 percent of malaria and diarrhea deaths are among young children. Children are also 14–44 percent more likely to die from environmental factors, again leaving them the most vulnerable.Those in urban areas will be affected by lower air quality and overcrowding, and will struggle the most to better their situation.

Social Effects of Extreme Weather

As the World Meteorological Organization explains, "recent increase in societal impact from tropical cyclones has largely been caused by rising concentrations of population and infrastructure in coastal regions." Pielke *et al.* (2008) normalized mainland U.S. hurricane damage from 1900 to 2005 to 2005 values and found no remaining trend of increasing absolute damage. The 1970s and 1980s were notable because of the extremely low amounts of damage compared to other decades. The decade 1996–2005 has the second most damage among the past 11 decades, with only the decade 1926–1935 surpassing its costs. The most damaging single storm is the 1926 Miami hurricane, with $157 billion of normalized damage.

The American *Insurance Journal* predicted that "catastrophe losses should be expected to double roughly every 10 years because of increases in construction costs, increases in the number of structures and changes in their characteristics." The Association of British Insurers has stated that limiting carbon emissions would avoid 80% of the projected additional annual cost of tropical cyclones by the 2080s. The cost is also increasing partly because of building in exposed areas such as coasts and floodplains. The ABI claims that reduction of the vulnerability to some inevitable effects of climate change, for example through more resilient buildings and improved flood defences, could also result in considerable cost-savings in the longterm.

Human Settlement

A major challenge for human settlements is sea-level rise, indicated by ongoing observation and research of rapid declines in ice-mass balance from both Greenland and Antarctica. Estimates for 2100 are at least twice as large as previously estimated by IPCC AR4, with an upper limit of about two meters. Depending on regional changes, increased precipitation patterns can cause more flooding or extended drought stresses water resources.

Coasts and Low-lying Areas

For historical reasons to do with trade, many of the world's largest and most prosperous cities are on the coast. In developing countries, the poorest often live on floodplains, because it is the only available space, or fertile agricultural land. These settlements often lack infrastructure such as dykes and early warning systems. Poorer communities also tend to lack the insurance, savings, or access to credit needed to recover from disasters.

In a journal paper, Nicholls and Tol (2006) considered the effects of sea level rise:

> The most vulnerable future worlds to sea-level rise appear to be the A2 and B2 [IPCC] scenarios, which primarily reflects differences in the socio-economic situation (coastal population, Gross Domestic Product (GDP) and GDP/capita), rather than the magnitude of sea-level rise. Small islands and deltaic settings stand out as being more vulnerable as shown in many earlier analyses. Collectively, these results suggest that human societies will have more choice in how they respond to sea-level rise than is often assumed. However, this conclusion needs to be tempered by recognition that we still do not understand these choices and significant impacts remain possible.

The IPCC reported that socioeconomic impacts of climate change in coastal and low-lying areas would be overwhelmingly adverse. The following impacts were projected with very high confidence:

- Coastal and low-lying areas would be exposed to increasing risks including coastal erosion due to climate change and sea level rise.
- By the 2080s, millions of people would experience floods every year due to sea level rise. The numbers affected were projected to be largest in the densely populated and low-lying mega-deltas of Asia and Africa; and smaller islands were judged to be especially vulnerable.

A study in the April 2007 issue of *Environment and Urbanization* reports that 634 million people live in coastal areas within 30 feet (9.1 m) of sea level. The study also reported that about two thirds of the world's cities with over five million people are located in these low-lying coastal areas.

Energy Sector

Oil, Coal and Natural Gas

Oil and natural gas infrastructure is vulnerable to the effects of climate change and the increased risk of disasters such as storm, cyclones, flooding and long-term increases in sea level. Minimising these risks by building in less disaster prone areas, can be expensive and impossible in countries with coastal locations or island states. All thermal power stations depend on water to cool them. Not only is there increased demand for fresh water, but climate change can increase the likelihood of drought and fresh water shortages. Another impact for thermal power plants, is that increasing the temperatures in which they operate reduces their efficiency and hence their output. The source of oil often comes from areas prone to high natural disaster risks; such as tropical storms, hurricanes, cyclones, and floods. An example is Hurricane Katrina's impact on oil extraction in the Gulf of Mexico, as it destroyed 126 oil and gas platforms and damaged 183 more.

However, previously pristine arctic areas will now be available for resource extraction

Nuclear

Climate change, along with extreme weather and natural disasters can affect nuclear power plants in a similar way to those using oil, coal, and natural gas. However, the impact of water shortages on nuclear power plants cooled by rivers will be greater than on other thermal power plants. This is because old reactor designs with water-cooled cores must run at lower internal temperatures and thus, paradoxically, must dump more heat to the environment to produce a given amount of electricity. This situation has forced some nuclear reactors to be shut down and will do so again unless the cooling systems of these plants are enhanced to provide more capacity. Nuclear power supply was diminished by low river flow rates and droughts, which meant rivers had reached the maximum temperatures for cooling. Such shutdowns happened in France during the 2003 and 2006 heat waves. During the heat waves, 17 reactors had to limit output or shut down. 77% of French electricity is produced by nuclear power; and in 2009 a similar situation created a 8GW shortage, and forced the French government to import electricity. Other Cases have been reported from Germany, where extreme temperatures have reduced nuclear power production 9 times due to high temperatures between 1979 and 2007. In particular:

- The Unterweser nuclear power plant reduced output by 90% between June and September 2003.
- The Isar nuclear power plant cut production by 60% for 14 days due to excess river temperatures and low stream flow in the river Isar in 2006.

Similar events have happened elsewhere in Europe during those same hot summers. Many scientists agree that if global warming continues, this disruption is likely to increase.

Hydroelectricity

Changes in the amount of river flow will correlate with the amount of energy produced by a dam. Lower river flows because of drought, climate change, or upstream dams and diversions, will reduce the amount of live storage in a reservoir; therefore reducing the amount of water that can be used for hydroelectricity. The result of diminished river flow can be a power shortage in areas that depend heavily on hydroelectric power. The risk of flow shortage may increase as a result of climate change. Studies from the Colorado River in the United States suggests that modest climate changes (such as a 2 degree change in Celsius that could result in a 10% decline in precipitation), might reduce river run-off by up to 40%. Brazil in particular, is vulnerable due to its having reliance on hydroelectricity as increasing temperatures, lower water flow, and alterations in the rainfall regime, could reduce total energy production by 7% annually by the end of the century.

Cost

The scientific evidence for links between global warming and the increasing cost of natural disasters due to weather events is weak, but, nevertheless, prominent mainstream environmental spokesmen such as Barack Obama and Al Gore have emphasized the possible connection. For the most part increased costs due to events such as Hurricane Sandy are due to increased exposure to loss resulting from building insured facilities in vulnerable locations. This information has been denounced by Paul Krugman and ThinkProgress as climate change denial.

Insurance

An industry directly affected by the risks of climate change is the insurance industry. According to a 2005 report from the Association of British Insurers, limiting carbon emissions could avoid 80% of the projected additional annual cost of tropical cyclones by the 2080s. A June 2004 report by the Association of British Insurers declared "Climate change is not a remote issue for future generations to deal with; it is, in various forms here already, impacting on insurers' businesses now." The report noted that weather-related risks for households and property were already increasing by 2–4% per year due to the changing weather conditions, and claims for storm and flood damages in the UK had doubled to over £6 billion over the period from 1998–2003 compared to the previous five years. The results are rising insurance premiums, and the risk that in some areas flood insurance will become unaffordable for those in the lower income brackets.

Financial institutions, including the world's two largest insurance companies: Munich Re and Swiss Re, warned in a 2002 study that "the increasing frequency of severe climatic events, coupled with social trends could cost almost 150 billion US$ each year in the next decade". These costs would burden customers, taxpayers, and the insurance industry, with increased costs related to insurance and disaster relief.

In the United States, insurance losses have also greatly increased. It has been shown that a 1% climb in annual precipitation can increase catastrophe loss by as much as 2.8%. Gross increases are mostly attributed to increased population and property values in vulnerable coastal areas; though there was also an increase in frequency of weather-related events like heavy rainfalls since the 1950s.

Transport

Roads, airport runways, railway lines and pipelines, (including oil pipelines, sewers, water mains etc.) may require increased maintenance and renewal as they become subject to greater temperature variation. Regions already adversely affected include areas of permafrost, which are subject to high levels of subsidence, resulting in buckling roads, sunken foundations, and severely cracked runways.

Tropical Cyclones and Climate Change

Tropical cyclones and climate change concerns how tropical cyclones have changed (in number, intensity, track or otherwise), and are expected to further change, under global warming. The topic receives considerable attention from climate scientists who study the connections between storms and climate, and notably since 2005 makes news during active storm seasons.

Hurricane Katrina, near the Gulf Coast

A hurricane is a tropical cyclone that occurs in the Atlantic Ocean and northeastern Pacific Ocean, a typhoon occurs in the northwestern Pacific Ocean, and a cyclone occurs in the south Pacific or Indian Ocean.

General

A 2010 review of related sciences found substantial uptake in the frequency of the most intense tropical cyclones, and increases in rainfall rates within 100 km of the storm centre of up to 20%.

A tropical cyclone's rainfall is primarily controlled by its actual environmental sea surface temperature (SST) - relative to the tropical mean SST, called the relative sea surface temperature. The relative SST has also been identified to more precise explain variations in Atlantic hurricane frequency, than absolute SST and tropical convection. The relative SST is defined as the SST in the TC environment minus the tropical (30°S-30°N) mean SST. Rainfall will expand outwards as the relative SST increases, associated with an expansion of a storm wind field. The largest tropical cyclones are observed in the western North Pacific tropics, where the largest values of relative SST and mid-tropospheric relative humidity are located. Flooding by TCs with global warming is projected primarily by increases in rainfall rate, rather than area extent.

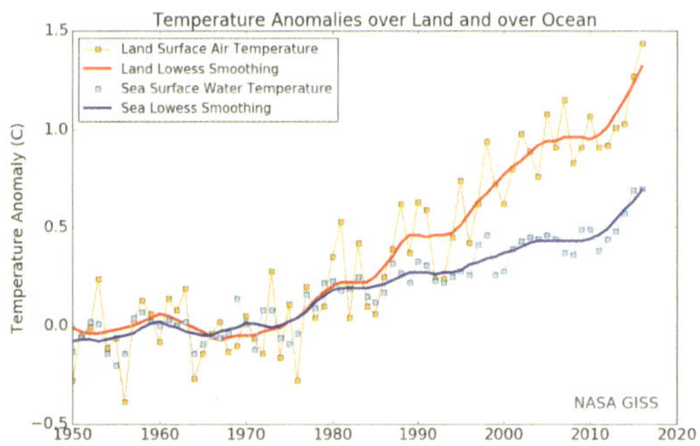

Annual (thin lines) and five-year lowess smooth (thick lines) for the temperature anomalies averaged over the Earth's land area and sea surface temperature anomalies (blue line) averaged over the part of the ocean that is free of ice at all times (open ocean)

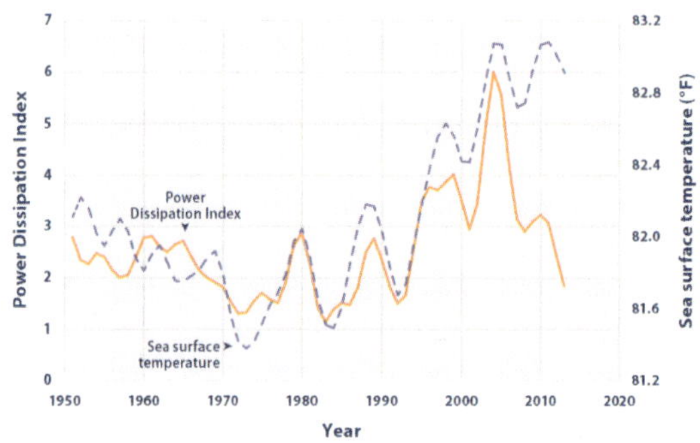

North Atlantic tropical cyclone activity according to the Power Dissipation Index, 1949–2015. Sea surface temperature is measured in different units, but the values have been plotted alongside the PDI to show how they compare. The lines have been smoothed using a five-year weighted average, plotted at the middle year. The most recent average (2011–2015) is plotted at 2013

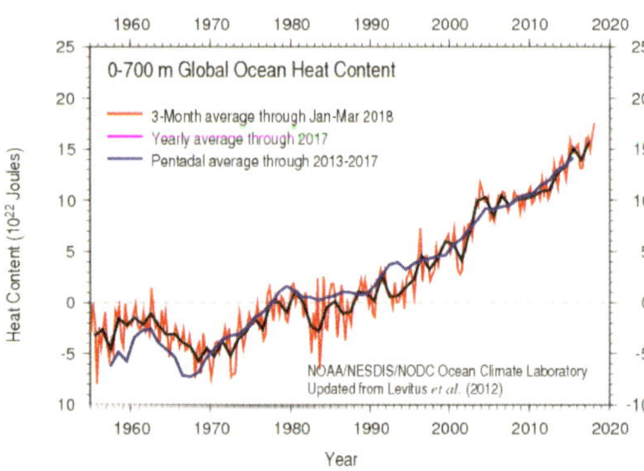

Global Heat Content (0–700 meters) layer

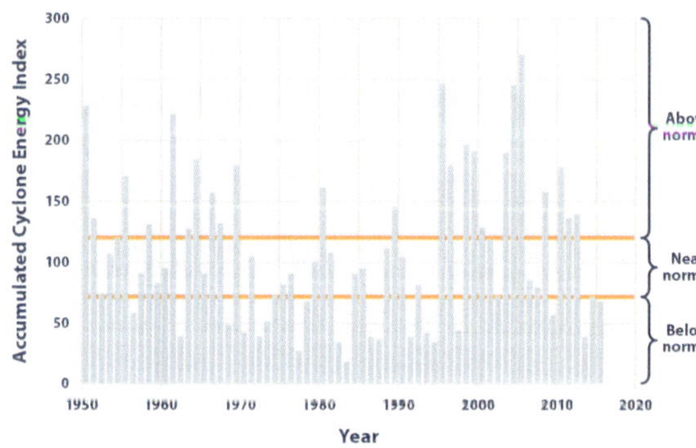

North Atlantic tropical cyclone activity according to the Accumulated Cyclone Energy Index, 1950–2015. For a global ACE graph visit this link

Most intense Atlantic hurricanes				
Rank	Hurricane	Season	Pressure	
			hPa	inHg
1	Wilma	2005	882	26.05
2	Gilbert	1988	888	26.23
3	"Labor Day"	1935	892	26.34
4	Rita	2005	895	26.43
5	Allen	1980	899	26.55
6	Camille	1969	900	26.58
7	Katrina	2005	902	26.64
8	Mitch	1998	905	26.73
	Dean	2007		
10	Maria	2017	908	26.81

Measurement

Based on satellite imagery, the Dvorak technique is the primary technique used to estimate globally the tropical cyclone intensity. Today measurement for a named system is also done with the ACE index (Accumulated Cyclone Energy), a wind energy index – the sum of the squares of the estimated six-hour maximum sustained wind speed in knots, to assess the storm duration and intensity. The Potential Intensity (PI) of tropical cyclones can be computed from observed data, primarily derived from vertical profiles of temperature, humidity and sea surface temperatures (SSTs). The convective available potential energy (CAPE), was computed from radiosonde stations in parts of the tropics from 1958 to 1997, but is considered to be of poor quality. The Power Dissipation Index (PDI) represents the total power dissipation for the North Atlantic and western North Pacific, and is strongly correlated with tropical SSTs. Various tropical cyclone scales exist to classify a system.

Historical Record

Since the satellite era, which began around 1970, trends are considered to be robust enough in regards to the connection of storms and sea surface temperatures. Agreement exists that there were active storm periods in the more distant past, but the sea surface temperature related Power Dissipation Index was not as high. Paleotempestology is the science of past tropical cyclone activity by means of geological proxies (flood sediment), or historical documentary records, such as shipwrecks or tree ring anomalies.

El Niño

El Niño (ENSO) shifts the region (warmer water, up and down welling at different locations, due to winds) in the Pacific and Atlantic where more storms form, resulting in nearly constant ACE values in any one basin. The El Niño event typically decreases hurricane formation in the Atlantic, and far western Pacific and Australian regions, but instead increases the odds in the central North and South Pacific and particular in the western North Pacific typhoon region.

When analyzing the correlation between extremely active tropical cyclones and El Niño events in the Eastern Pacific Ocean, it was concluded by Murakami et al. that strong El Niño events during hurricane seasons are not the only responsible factors and can most likely be attributed to greenhouse gases.

Public Perception

The tremendous destruction caused by recent Atlantic Ocean tropical cyclones, such as Hurricanes Katrina, Wilma, and Sandy caused a substantial upsurge in interest in the subject of climate change and hurricanes by news media and the wider public, and concerns that global climatic change may have played a significant role in those events. In 2005 and 2017, related polling of populations affected by hurricanes concluded in 2005 that 39 percent of Americans believed climate change helped to fuel the intensity of hurricanes, with 55 percent in September 2017.

The 2005 Atlantic hurricane season was the most active Atlantic hurricane season in recorded history, shattering numerous records. In response articles appeared in relation to climate change and hurricanes. *Time Magazine*, published an article titled, "Is Global Warming Fueling Katrina?". Shortly after the hurricane, former *Boston Globe* reporter Ross Gelbspan wrote an op-ed piece for the *Globe* titled, "Katrina's Real Name", declaring that the hurricane's "real name is global warming." Gelbspan went on to assert:

> "Although Katrina began as a relatively small hurricane that glanced off south Florida, it was supercharged with extraordinary intensity by the relatively blistering sea surface temperatures in the Gulf of Mexico."

Gelbspan did not single out Katrina from other recent storms in that regard; in the article he went on to attribute other major weather events over the preceding year to global warming, including a blizzard in Los Angeles, high winds in Scandinavia, wildfires in Spain, and a drought centered in Missouri. Britain's then deputy prime minister, John Prescott, has linked Katrina with global warming, and statements made shortly after the hurricane by Germany's environment minister, Jürgen Trittin, indicate he believes that global warming is responsible for an increase in the frequency of destructive natural events.

Scientists pointed out that the increase of CO_2 emissions contributes to warmer ocean waters and more moist air for rain. Because of sea level rise it is assumed that the storm surge of Hurricane Irma and other storms will cause greater flooding to vulnerable areas. Data collected by NASA showed that ocean surface temperatures in the path of Irma were above 30 °C (86 °F), capable of sustaining a Category 5 hurricane. Prior to affecting the U.S. mainland, Miami's mayor Tomás Regalado noted on Hurricane Irma, "This is the time to talk about climate change. This is the time that the president and the EPA and whoever makes decisions needs to talk about climate change." A day later the head of the EPA, Scott Pruitt said, "to discuss the cause and effect of these storms, there's the place (and time) to do that, it's not now." Following Irma's landfall, Donald Trump was asked about the connection between hurricanes and climate change, and stated that "We've had bigger storms than this." Richard Branson who was directly impacted by hurricane Irma noted, "hurricanes are the start of things to come. Look, climate change is real. Ninety-nine per cent of scientists know it's real. The whole world knows it's real except for maybe one person in the White House."

Three simultaneous hurricanes active – with Katia *(left)*, Irma *(center)*, and Jose *(right)*. All three were threatening land at the time

United Nations secretary general António Guterres citing the devastation from hurricanes noted in September, "The catastrophic Atlantic hurricane season has been made worse by climate change. Cutting carbon emissions must clearly be part of our response to the disaster. The rise in the surface temperature of the ocean has had an impact on weather patterns and we must do everything possible to bring it down."

The Associated Press looked at the yearly average Accumulated Cyclone Energy (ACE), which accounts for wind speed and storm duration to assess hurricane power of the past 30 years and found it to be 41 percent higher than the previous 30 years. They asked several experts about their opinion, James Kossin from NOAA "There's no question that the storms are stronger than they were 30 years ago." Climate scientist Stefan Rahmstorf of the Potsdam Institute for Climate Impact Research, "The only caveat being that the increase might be exaggerated somewhat because of undercounting early storms." Meteorologist Philip Klotzbach noted, "What's happening with hurricanes — the frequency, the duration, and the energy — is probably a combination of factors caused by both nature and man, a mish-mosh of everything." Kerry Emanuel who studies hurricanes, told the BBC, "The warming of the climate has increased the underlying probabilities of very heavy rain events like happened in Harvey and very high category hurricanes like Irma. It is just not sensible to say either storm was caused by climate change, but the underlying probabilities are going up."

Research

A review by Walsh et al. 2015 found good understanding of the links between climate and tropical cyclones on various timescales, with significant observed trends in the Atlantic for frequencies and intensities in the past few decades. However, these trends are still heavily studied. The authors note that the theory for maximum tropical cyclone intensity is considered well established, but with cyclone genesis still elusive. Most climate models predict fewer tropical cyclones globally, and more of the strongest storms, with increased rainfall rates. Additional sea level rise will likely increase storm surge threats. A 2017 study looked at compounding effects from floods, storm surge, and terrestrial flooding (rivers), and projects an increase due to global warming.

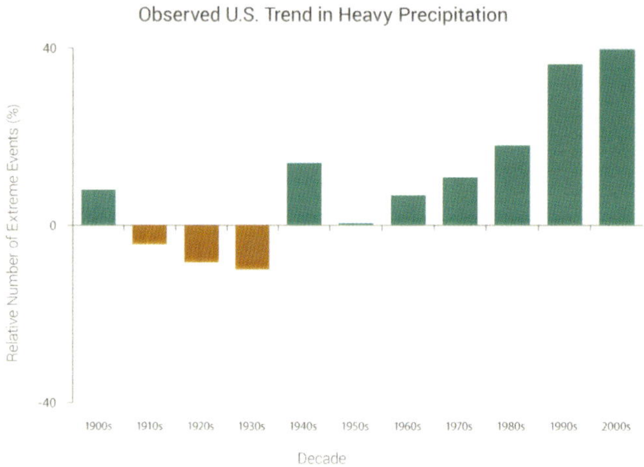

Warmer air can contain more water vapor than cooler air. Global analyses show that the amount of water vapor in the atmosphere has in fact increased due to human-caused warming. This extra moisture is available to storm systems, resulting in heavier rainfalls

Studies on increased rainfall rates are largely based on the Clausius–Clapeyron relation, which yields ≈7% increase in water vapor in the atmosphere per 1 °C warming, and the duration of storms. Storm tracks indicate a global poleward expansion of the latitude at which the maximum intensity of tropical cyclones occur. Some model based research predicts a significant increase (+90%) in the frequency of very intense (category 4 and 5) hurricanes, with average climate change projections.

The National Climate Assessment from 2014 states:

> The recent increases in activity are linked, in part, to higher sea surface temperatures in the region that Atlantic hurricanes form in and move through. Numerous factors have been shown to influence these local sea surface temperatures, including natural variability, human-induced emissions of heat-trapping gases, and particulate pollution. Quantifying the relative contributions of natural and human-caused factors is an active focus of research.

The World Meteorological Organization stated 2017 that the quantity of rainfall from Hurricane Harvey had very likely been increased by climate change. The relationship between climate change and the frequency of hurricanes (or tropical cyclones) is still unclear, and is the subject of continued research.

Greenhouse gases are affecting hurricane frequency and severity. Research conducted by Murakami et al. following the 2015 hurricane season in the eastern and central Pacific Ocean where a record number of tropical cyclones and three simultaneous category-4 hurricanes occurred, concludes that greenhouse gas forcing enhances subtropical Pacific warming which they project will increase the frequency of extremely active tropical cyclones in this area.

Hurricanes

A 2011 study linked increased activity of intense hurricanes in the North Atlantic with a northward shift and amplification of convective activities from the African easterly waves (AEWs). A

2014 study investigated the response of AEWs to high emissions scenarios, and found increases in regional temperature gradients, convergence and uplift along the Intertropical Front of Africa, resulting in strengthening of the African easterly waves, affecting the climate over West Africa and the larger Atlantic basin.

Studies conducted in 2008 and 2016 looked at the duration of The Atlantic hurricane season, and found it may be getting longer, particular south of 30°N and east of 75°W, or the tendency toward more early- and late-season storms, correlated to warming sea surface temperatures. However, uncertainty is still high, and one study found no trend, another mixed results.

Typhoons

Research based on records from Japan and Hawaii indicate that typhoons in the north-west Pacific intensified by 12–15% on average since 1977. The observed strongest typhoons doubled, or tripled in some regions, the intensity of particular landfalling systems is most pronounced. This uptick in storm intensity affects coastal populations in China, Japan, Korea and the Philippines, and has been attributed to warming ocean waters. The authors noted that it is not yet clear to what extent global warming caused the increased water temperatures, but observations are consistent with what the IPCC projects for warming of sea surface temperatures.

Policy Response

Given the scientific consensus that tropical cyclones have been strengthening as a response to climate change, there have been numerous recommended in peer-reviewed journal articles on how best to respond. There has been grave concern that in the immediate aftermath of a tropical cyclone, the focus on applying resources towards immediate relief to those affected has and will continue to divert attention from more long-term solutions. This is further exacerbated in lower-income communities and countries as they suffer most from the consequences of tropical cyclones. Despite this, in order to maintain long term stability and resilience, the implementation of long term goals cannot be understated.

Policymakers have recommended certain proposed initiatives by the scientific and economic communities, and certain nations and supranational communities have already taken the initiative to either implement these or have created a plan of action to implement them in the future. One active policy decision that would assist in mitigating risks of flooding would be reforestation of inland areas in order to strengthen the soil of the communities and reduce coastal inundation. It is also recommended that local schools, churches, and other local community infrastructure be permanently equipped to become cyclone shelters at a moment's notice such that residents are not required to remain in their own housing for longer than absolutely necessary.

Specific national and supranational decisions have already been made and are being implemented in this regard. The Framework for Resilient Development in the Pacific (FRDP) has been instituted to strengthen and better coordinate disaster response and climate change adaptation among nations and communities in the region. Specific nations such as Tonga and the Cook Islands in the Southern Pacific under this regime have developed a Joint National Action Plan on Climate Change and Disaster Risk Management (JNAP) to coordinate and execute responses to the rising risk for climate change. These countries have identified the most vulnerable areas of their nations, generated national and supranational policies to be implemented, and provided specific goals and timelines to achieve these goals. These actions to be implemented include reforestation, building of levees and dams, creation of early warning systems, reinforcing existing communication infrastructure, finding new sources of fresh water, promoting and subsidizing the proliferation renewable energy, improving irrigation techniques to promote sustainable agriculture, increase public education efforts on sustainable measures, and lobbying internationally for the increased use of renewable energy sources.

United States

In the United States, there have been several concrete initiatives taken to better prepare for the strengthening of hurricanes in the foreseeable future.

North Carolina

North Carolina has been focusing on environmental issues since the 1970s when rising environmental awareness pushed the state to take advantage of the federal funds made available to states to deal with potential natural disasters in the Coastal Zone Management Act of 1972. In the interim, North Carolina has identified nine coastal areas which are most likely to suffer most in the event of hurricane activity. The state required that all municipalities and counties in these areas to prepare comprehensive action plans to be implemented and carried out in the event of a major natural disaster. The localities are further required to periodically and constantly update these action plans to keep them up to date. The state requires these plans to include both long term goals to be achieved such as reforestation, building of levees, finding new fresh water sources, etc. as well as short term goals such as preparing local emergency shelters, increasing funding towards research in how best to mitigate the harm of hurricanes, stocking of emergency supplies, etc.

South Carolina

Hurricane Hugo near peak intensity

In South Carolina, the state has been actively monitoring and taking an interest in hurricane aftermath management since 1977 when it passed its first coastal management act. The act was administered by the Coastal Council in the state's executive branch until 1993 when it was incorporated into the Department of Health and Environmental Control. From its inception, the South Carolina model has been management of local flora and fauna habitats to strengthen the coastal areas' resilience to flooding and storm surge from hurricanes. In 1988, the state significantly strengthened its coastal laws to deal with the increasing coastal erosion and oceanfront development. These actions came one year before the landfall of Hurricane Hugo in 1989, which devastated the state, and since then South Carolina has continued to focus on increased preparedness for ensuing hurricanes. The Coastal Council's authority was amended by the legislature to provide and further enforce additional guidelines to shoreline development, which the act required be done only as a last resort. The state has also mandated planning and zoning provisions for local communities which are vulnerable to hurricane impacts, and it has required relocation of certain residents in the coastal areas to inland part of the states to prevent further loss or damage to human life.

Maine

The state of Maine has a long history of coastal management in all forms dating back to the 1960s, which was then eventually codified into written law in 1978 after the "Blizzard of 1978". The state passed various laws to mandate the management and preparation of state authorities to prepare for further natural disasters, including the potential landfalls of future hurricanes. The legislature allocated funds for several state executive agencies to take an active role in the regulation of coastal development and residence. Maine also utilized already allocated federal funds to better manage its coastal areas. These agencies and funds were utilized to take several precautionary measures such as building sand dunes to serve as temporary levees, construction of seawalls to fortify the coast against storm surges and flooding, reforestation of the state to strengthen the soil, and the banning of construction in certain extremely flood-prone areas.

Climate Change and Ecosystems

Future climate change is expected to affect particular ecosystems, including tundra, mangroves, coral reefs, and caves.

Impacts

Unchecked global warming could affect most terrestrial ecoregions. Increasing global temperature means that ecosystems will change; some species are being forced out of their habitats (possibly to extinction) because of changing conditions, while others are flourishing. Secondary effects of global warming, such as lessened snow cover, rising sea levels, and weather changes, may influence not only human activities but also the ecosystem.

For the IPCC Fourth Assessment Report, experts assessed the literature on the impacts of climate change on ecosystems. Rosenzweig *et al.* (2007) concluded that over the last three decades, human-induced warming had likely had a discernible influence on many physical and biological systems. Schneider *et al.* (2007) concluded, with very high confidence, that regional temperature trends had already affected species and ecosystems around the world. With high confidence, they concluded that climate change would result in the extinction of many species and a reduction in the diversity of ecosystems.

- Terrestrial ecosystems and biodiversity: With a warming of 3 °C, relative to 1990 levels, it is likely that global terrestrial vegetation would become a net source of carbon. With high confidence, concluded that a global mean temperature increase of around 4 °C (above the 1990-2000 level) by 2100 would lead to major extinctions around the globe.
- Marine ecosystems and biodiversity: With very high confidence, concluded that a warming of 2 °C above 1990 levels would result in mass mortality of coral reefs globally. In addition, several studies dealing with planktonic organisms and modelling have shown that temperature plays a transcendental role in marine microbial food webs, which may have a deep influence on the biological carbon pump of marine planktonic pelagic and mesopelagic ecosystems.
- Freshwater ecosystems: Above about a 4 °C increase in global mean temperature by 2100 (relative to 1990-2000), concluded, with high confidence, that many freshwater species would become extinct.

Studying the association between Earth climate and extinctions over the past 520 million years, scientists from the University of York write, "The global temperatures predicted for the coming centuries may trigger a new 'mass extinction event', where over 50 per cent of animal and plant species would be wiped out."

Many of the species at risk are Arctic and Antarctic fauna such as polar bears and emperor penguins. In the Arctic, the waters of Hudson Bay are ice-free for three weeks longer than they were thirty years ago, affecting polar bears, which prefer to hunt on sea ice. Species that rely on cold weather conditions such as gyrfalcons, and snowy owls that prey on lemmings that use the cold winter to their advantage may be hit hard. Marine invertebrates enjoy peak growth at the temperatures they have adapted to, regardless of how cold these may be, and cold-blooded animals found at greater latitudes and altitudes generally grow faster to compensate for the short growing season. Warmer-than-ideal conditions result in higher metabolism and consequent reductions in body size despite increased foraging, which in turn elevates the risk of predation. Indeed, even a slight increase in temperature during development impairs growth efficiency and survival rate in rainbow trout.

Rising temperatures are beginning to have a noticeable impact on birds, and butterflies have shifted their ranges northward by 200 km in Europe and North America. Plants lag behind, and larger animals' migration is slowed down by cities and roads. In Britain, spring butterflies are appearing an average of 6 days earlier than two decades ago.

A 2002 article in *Nature* surveyed the scientific literature to find recent changes in range or seasonal behaviour by plant and animal species. Of species showing recent change, 4 out of 5 shifted their ranges towards the poles or higher altitudes, creating "refugee species". Frogs were breeding, flowers blossoming and birds migrating an average 2.3 days earlier each decade; butterflies, birds and plants moving towards the poles by 6.1 km per decade. A 2005 study concludes human activity is the cause of the temperature rise and resultant changing species behaviour, and links these effects with the predictions of climate models to provide validation for them. Scientists have observed that Antarctic hair grass is colonizing areas of Antarctica where previously their survival range was limited.

Mechanistic studies have documented extinctions due to recent climate change: McLaughlin *et al.* documented two populations of Bay checkerspot butterfly being threatened by precipitation change. Parmesan states, "Few studies have been conducted at a scale that encompasses an entire species" and McLaughlin *et al.* agreed "few mechanistic studies have linked extinctions to recent climate change." Daniel Botkin and other authors in one study believe that projected rates of extinction are overestimated.

Many species of freshwater and saltwater plants and animals are dependent on glacier-fed waters to ensure a cold water habitat that they have adapted to. Some species of freshwater fish need cold water to survive and to reproduce, and this is especially true with salmon and cutthroat trout. Reduced glacier runoff can lead to insufficient stream flow to allow these species to thrive. Ocean krill, a cornerstone species, prefer cold water and are the primary food source for aquatic mammals such as the blue whale. Alterations to the ocean currents, due to increased freshwater inputs from glacier melt, and the potential alterations to thermohaline circulation of the worlds oceans, may affect existing fisheries upon which humans depend as well.

The *white lemuroid possum*, only found in the Daintree mountain forests of northern Queensland, may be the first mammal species to be driven extinct by global warming in Australia. In 2008, the white possum has not been seen in over three years. The possums cannot survive extended temperatures over 30 °C (86 °F), which occurred in 2005.

A 27-year study of the largest colony of Magellanic penguins in the world, published in 2014, found that extreme weather caused by climate change is responsible for killing 7% of penguin chicks per year on average, and in some years studied climate change accounted for up to 50% of all chick deaths. Since 1987, the number of breeding pairs in the colony has reduced by 24%.

Climate change is leading to a mismatch between the snow camouflage of arctic animals such as snowshoe hares with the increasingly snow-free landscape.

Forests

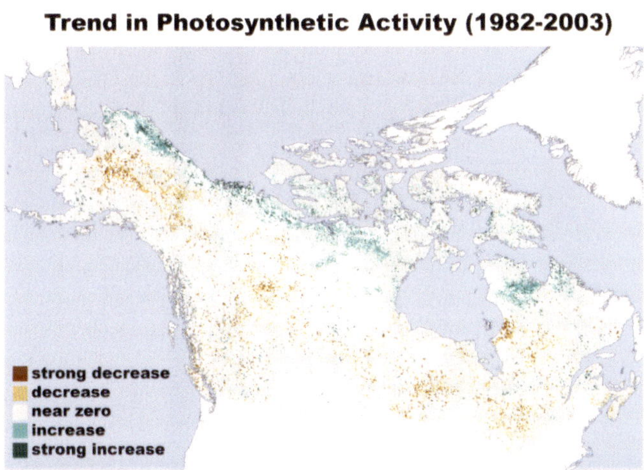

Change in Photosynthetic Activity in Northern Forests 1982-2003; NASA Earth Observatory

Pine forests in British Columbia have been devastated by a pine beetle infestation, which has expanded unhindered since 1998 at least in part due to the lack of severe winters since that time;

a few days of extreme cold kill most mountain pine beetles and have kept outbreaks in the past naturally contained. The infestation, which (by November 2008) has killed about half of the province's lodgepole pines (33 million acres or 135,000 km^2) is an order of magnitude larger than any previously recorded outbreak. One reason for unprecedented host tree mortality may be due to that the mountain pine beetles have higher reproductive success in lodgepole pine trees growing in areas where the trees have not experienced frequent beetle epidemics, which includes much of the current outbreak area. In 2007 the outbreak spread, via unusually strong winds, over the continental divide to Alberta. An epidemic also started, be it at a lower rate, in 1999 in Colorado, Wyoming, and Montana. The United States forest service predicts that between 2011 and 2013 virtually all 5 million acres (20,000 km^2) of Colorado's lodgepole pine trees over five inches (127 mm) in diameter will be lost.

As the northern forests are a carbon sink, while dead forests are a major carbon source, the loss of such large areas of forest has a positive feedback on global warming. In the worst years, the carbon emission due to beetle infestation of forests in British Columbia alone approaches that of an average year of forest fires in all of Canada or five years worth of emissions from that country's transportation sources.

Besides the immediate ecological and economic impact, the huge dead forests provide a fire risk. Even many healthy forests appear to face an increased risk of forest fires because of warming climates. The 10-year average of boreal forest burned in North America, after several decades of around 10,000 km^2 (2.5 million acres), has increased steadily since 1970 to more than 28,000 km^2 (7 million acres) annually. Though this change may be due in part to changes in forest management practices, in the western U.S., since 1986, longer, warmer summers have resulted in a fourfold increase of major wildfires and a sixfold increase in the area of forest burned, compared to the period from 1970 to 1986. A similar increase in wildfire activity has been reported in Canada from 1920 to 1999.

Forest fires in Indonesia have dramatically increased since 1997 as well. These fires are often actively started to clear forest for agriculture. They can set fire to the large peat bogs in the region and the CO_2 released by these peat bog fires has been estimated, in an average year, to be 15% of the quantity of CO_2 produced by fossil fuel combustion.

Mountains

Mountains cover approximately 25 percent of earth's surface and provide a home to more than one-tenth of global human population. Changes in global climate pose a number of potential risks to mountain habitats. Researchers expect that over time, climate change will affect mountain and lowland ecosystems, the frequency and intensity of forest fires, the diversity of wildlife, and the distribution of water.

Studies suggest that a warmer climate in the United States would cause lower-elevation habitats to expand into the higher alpine zone. Such a shift would encroach on the rare alpine meadows and other high-altitude habitats. High-elevation plants and animals have limited space available for new habitat as they move higher on the mountains in order to adapt to long term changes in regional climate.

Changes in climate will also affect the depth of the mountains snowpacks and glaciers. Any changes in their seasonal melting can have powerful impacts on areas that rely on freshwater runoff from mountains. Rising temperature may cause snow to melt earlier and faster in the spring and shift the timing and distribution of runoff. These changes could affect the availability of freshwater for natural systems and human uses.

Ecological Productivity

- According to a paper by Smith and Hitz (2003:66), it is reasonable to assume that the relationship between increased global mean temperature and ecosystem productivity is parabolic. Higher carbon dioxide concentrations will favourably affect plant growth and demand for water. Higher temperatures could initially be favourable for plant growth. Eventually, increased growth would peak then decline.
- According to IPCC (2007:11), a global average temperature increase exceeding 1.5–2.5 °C (relative to the period 1980–99), would likely have a predominantly negative impact on ecosystem goods and services, e.g., water and food supply.
- Research done by the Swiss Canopy Crane Project suggests that slow-growing trees only are stimulated in growth for a short period under higher CO_2 levels, while faster growing plants like liana benefit in the long term. In general, but especially in rainforests, this means that liana become the prevalent species; and because they decompose much faster than trees their carbon content is more quickly returned to the atmosphere. Slow growing trees incorporate atmospheric carbon for decades.

Species Migration

In 2010, a gray whale was found in the Mediterranean Sea, even though the species had not been seen in the North Atlantic Ocean since the 18th century. The whale is thought to have migrated from the Pacific Ocean via the Arctic. Climate Change & European Marine Ecosystem Research (CLAMER) has also reported that the Neodenticula seminae alga has been found in the North Atlantic, where it had gone extinct nearly 800,000 years ago. The alga has drifted from the Pacific Ocean through the Arctic, following the reduction in polar ice.

In the Siberian subarctic, species migration is contributing to another warming albedo-feedback, as needle-shedding larch trees are being replaced with dark-foliage evergreen conifers which can absorb some of the solar radiation that previously reflected off the snowpack beneath the forest canopy. It has been projected many fish species will migrate towards the North and South poles as a result of climate change, and that many species of fish near the Equator will go extinct as a result of global warming.

Migratory birds are especially at risk for endangerment due to the extreme dependability on temperature and air pressure for migration, foraging, growth, and reproduction. Much research has been done on the effects of climate change on birds, both for future predictions and for conservation. The species said to be most at risk for endangerment or extinction are populations that are not of conservation concern. It is predicted that a 3.5 degree increase in surface temperature will occur by year 2100, which could result in between 600 and 900 extinctions, which mainly will occur in the tropical environments.

Agriculture

Droughts have been occurring more frequently because of global warming and they are expected to become more frequent and intense in Africa, southern Europe, the Middle East, most of the Americas, Australia, and Southeast Asia. Their impacts are aggravated because of increased water demand, population growth, urban expansion, and environmental protection efforts in many areas. Droughts result in crop failures and the loss of pasture grazing land for livestock.

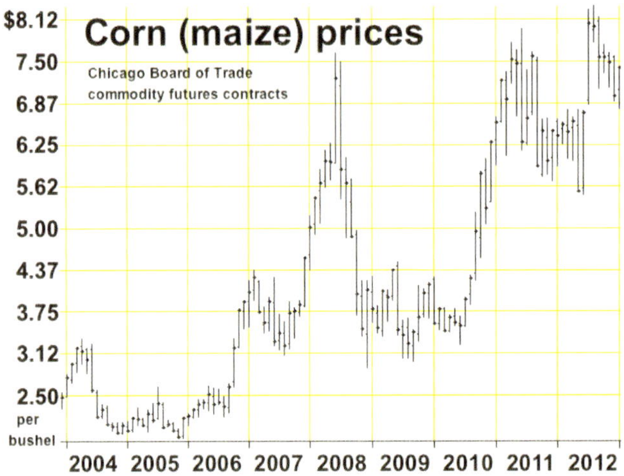

Price of corn in North America, in U.S. dollars per bushel, 2004-2012

Droughts are becoming more frequent and intense in arid and semiarid western North America as temperatures have been rising, advancing the timing and magnitude of spring snow melt floods and reducing river flow volume in summer. Direct effects of climate change include increased heat and water stress, altered crop phenology, and disrupted symbiotic interactions. These effects may be exacerbated by climate changes in river flow, and the combined effects are likely to reduce the abundance of native trees in favor of non-native herbaceous and drought-tolerant competitors, reduce the habitat quality for many native animals, and slow litter decomposition and nutrient cycling. Climate change effects on human water demand and irrigation may intensify these effects. By 2012, North American corn prices had risen to a record $8.34 per bushel in August, leaving 20 of the 211 U.S. ethanol fuel plants idle.

Climate Change and Poverty

In an ever-progressing world with an increasing demand for energy and technology, it is difficult to avoid climate change and its impacts on societies both locally and globally. Climate change affects social development factors, such as, poverty, infrastructure, technology, security, and economics across the globe. Although climate change affects everything we see around us, the interrelation between climate change and social vulnerability and inequality is particularly evident in impoverished communities. In particular, impoverished communities experience reductions in safe drinking water as well as food security as a result of climate change (OECD 2013). These typically rural, isolated communities do not exhibit sufficient financial and technical capacities to manage the risks associated with climate change (climate risk) (Skoufias 2012). Energy development and policy alteration could adjust the severity of climate change impacts; this is being tested now, as renewable energy sources develop.

The Intergovernmental Panel on Climate Change (IPCC) Fifth Assessment Report projects that there is likely to be at least a 0.4–1.6 °C increase in global mean surface temperature by the period of 2046–2065 and likely a sea level rise of 0.17–0.32 meters by this time due to recent trends relative to 1986–2005. The report claims that "over the period 1993 to 2010, global mean sea level rise is, with high confidence, consistent with the sum of the observed contributions from ocean thermal expansion due to warming...the sum of these contributions is 2.8 mm/yr". Glaciers and ice sheets are clearly melting now more than they have been in recent history. Not only that, but "the global ocean will continue to warm during the 21st century. Heat will penetrate from the surface to the deep ocean and affect ocean circulation".

The impacts that climate change has had on development in recent years are significant. According to McMichael, an International Professor of Development Sociology at Cornell University, the "development theory" signifies the market as "the vehicle for social change", for the market allocates resources. Privatization then follows the market influence to enable or prevent development. In consideration of climate change, the market has most likely not yet invested enough in alternative energy research to reduce adverse climate change affects and enable social change in low-income communities around the world. To provide an example, during the Cold War, global capitalist expansion occurred, in order to secure markets and resources for the Western world. In the case of climate change, some, like McMichael, believe that the market has not yet invested enough in alternative energy research to reduce adverse climate change affects.

Those in poverty have a higher chance of experiencing the ill-effects of climate change due to increased exposure and vulnerability. Vulnerability represents the degree to which a system is susceptible to, or unable to cope with, adverse effects of climate change including climate variability and environmental extremes. Also, a lack of capacity available for coping with environmental change is experienced in lower-income communities. According to the United Nations Development Programme, developing countries suffer 99% of the casualties attributable to climate change.

There are several ways to measure poverty, and there have been considerable development of definitions of poverty and the standards that regulate that aspect. The classical view of poverty used to be the economic indicator, which was the most dominant indicator; it used to view poverty as a poor quality of life experienced by most stressed and vulnerable people on the national or international level. However, according to Robinin Leichenko, "in the recent years, the nonmonetary indicators have incorporated in the definition of poverty along the economic indicator and added to the consideration the voice, the power, the human rights, and physiological well-being."

Climate change raises some climate ethics issues, as the least 50 developed countries of the world account for an imbalanced 1% contribution to the worldwide emissions of greenhouse gasses which are theorized to be attributable to global warming. The issue of distributive justice questions how to fairly share the benefits and burdens of climate change policy options. Many of the policy tools often employed to solve environmental problems, such as cost-benefit analysis, usually do not adequately handle these issues because they often ignore questions of just distribution and the effects on human rights.

The solution for climate change must empower global development by improving the livelihoods and alleviating the poverty off the poor, stressed, and vulnerable people in the third world countries.

This is crucial, because 80% of humanity around the world lives in poverty; 50% of those in poverty live in abject poverty (less than $2.50 a day).

The focus should be on the poor and struggling countries that are located in the southern hemisphere to receive guidance and help from the developed and rich countries located in the northern hemisphere to imply sustainable environmental policies, in order to avoid a catastrophic biological demise that could damage human well-being on a global scale.

Social Development Factors

Poverty

"Energy has strong links with poverty reduction – through household income, health, education, gender, and the environment".

The carbon footprint of the world's most poverty-stricken billion people is only about 3 percent of the world's total footprint, yet death rates are about 500 times greater in the regions where these individuals reside than in the Global North. Moreover, it is estimated that climatic disasters will cause affects to 375 million people a year by 2015, and "up to a billion people will be forced out of their homes by 2050 due to climate change".

The cycle of poverty is an endless cycle that has plenty of causes, such as historical, societal, economic, political, orthodoxy, geographical, wars, and colonialism factors. The cycle of poverty exacerbates the potential negative impacts of climate change. This phenomenon is defined when poor families become trapped in poverty for at least three generations and obtain limited or no access to resources, causing them to be disadvantaged and unable to break the cycle; this cycle is referred to as the poverty trap. In well-off countries, coping with climate change has largely been a matter of adjusting thermostats, dealing with longer, hotter summers, and observing seasonal shifts; for those in poverty, weather-related disasters, unproductive harvests, or even family members falling ill can facilitate crippling economic shocks. Aside from these economic shocks, widespread famine, drought, and potential humanistic shocks can affect an entire nation. High levels of poverty and low levels of human development limit capacity of poor households to manage climate risks. With limited access to formal insurance, low incomes, and meager assets, poor households must cope with climate-related shocks under highly constrained conditions. 'according to the world bank, climate change is likely to reduce agricultural productivity, especially in tropical regions'.

Impoverished communities tend to be more dependent on climate-sensitive sectors and natural resources for survival, so climate change poses an extreme threat on the livelihood, food security, and health of the poor; women are particularly vulnerable. The Kyoto Protocol established the Clean Development Mechanism, supposedly to provide "global benefits from carbon sequestration as well as sustainable development benefits to developing countries.

With climate change, availability of fresh water will reduce dramatically along areas with large coastal populations, especially those that are less economically stable, like some regions of Latin America. Approximately 22 million people faced water scarcity at the end of the twentieth century, and according to projections configured by the IPCC, the number of people likely to endure issues in water sanitation by 2050 is 79 to 178 million.

Poverty is a shameful disease of the modern society, for which new remedies are required. The attitude toward poverty in the past was to remove the need for begging and panhandling, by finding employment for the poor and directing them. It is statistically proved that poor households in third world countries do not consume the required minimum balanced diet, and more females than males are out of school. Some of the third world countries have adopted interventions aimed at mitigating the poverty, including free education, subsidies for farmers & fishers, scholarships, and food supplies to the needy people. However, the solution for poverty problem requires multifaceted measures, which have the capacity to break the generational poverty cycle and have transformational effect.

The UN-Energy corporation was formed at the request of the World Summit on Sustainable Development (WSSD), and released a report which argues that today's climate-changing-world "entrenches poverty, constrains the delivery of social services, limits opportunities for women, and erodes environmental sustainability at the local, national and global levels" (UN 2005). According to the report, 2.4 billion individuals rely on biomass for cooking, while 1.6 billion people worldwide are without electricity. Consequently, there is an explicit connection between energy access and poverty reduction, often referred to energy poverty (UN 2005). Services such as heating, lighting, cooking, and transportation facilitate socioeconomic development, for they yield social benefits as well as employment and income (UN 2005). Integrated assessment modelling (IAMs) is utilized to quantify the current socioeconomic dimensions of climate change, and further predict future effects on nations.

Reasons for Helping

Those of us who are better-off would be stupid not to help the poor. Not only is this the morally right thing to do, but the purist of well being of our future generations requires that we help impoverished people in far lands. The rich world's response to West Africa's Ebola was tardy. This disease has been known for decades, but medical research to prevent and treat it has been very limited: "It only kills poor Africans, so who would pay for research?" A similarly sluggish response to an emerging HIV/AIDS happened 30 years ago permitted this disease to become pandemic, killing millions in Europe, US, and other rich countries. And now, we have Zika virus. Have we learned anything? A problem in one place can become a problem everywhere in this small densely populated and highly connected planet. Unfortunately, health is not the only example of the need of rich countries to help the poor countries.

Implications of Poverty on the National Security of the Western Countries

Poor people from Latin America vow to grow and transport cocaine to the United States, as their only economic opportunity they have in their hands. That caused large parts of Central America to become destabilized and it spilled the narco-violence over to the US. In Middle East, religious ideologies legitimate violence at home and abroad. So, it is not a matter of the question "should rich countries help poor countries or not?", but it should be "what are the best ways for rich countries to help the poor countries?"

The role of Rich Countries in the Northern Hemisphere toward the Poor Countries in the Southern Hemisphere

The role of rich nations will come as to support domestic processes of state formation, helping governments to work more effectively rather than transferring aids. Also, the levels of carbon emissions produced by factories, farms, transport, and houses have created a no-return point of global warming. The development process has to be redefined and if the world's population of 7 billion (10 billion in the futures) is to survive and have reasonable lives. Living well means better material conditions for 3 billion poor people with better access to food, clean water, shelter, sanitation, health services, energy, and other goods & services. However, it will also mean living differently for elites and middle classes around the world. The extravagant consumption and the energy-profligate lifestyle that rich nations live in cannot continue forever. That cannot continue in a world that is sustainable and socially just. So, helping the poorer countries is no longer about the change.

The mix of various reasons for rich nations in the northern hemisphere to help poor nations in the southern hemisphere range from "it's our moral duty" to the venal "we can look good and make a fast buck out of this." Thus, rich nations use combinations of altruism and self-interest objectives to justify the support for the distant and needy people to taxpayers.

Areas of Development

Infrastructure and Landscape

The potential effects of climate change and the security of infrastructure will have the most direct effect on the poverty cycle. Areas of infrastructure effects will include water systems, housing and settlements, transport networks, utilities, and industry. Infrastructure designers can contribute in three areas for improving living environment for the poor, in building design, in settlement planning and design as well as in urban planning. The National Research Council has identified five climate changes of particular importance to infrastructure and factors that should be taken into consideration when designing future structures. These factors include: increases in very hot days and heat waves, increases in Arctic temperatures, rising sea levels, increases in intense precipitation events, and increases in hurricane intensity. The UK Climate Impacts Programme established the Building Knowledge for a Changing Climate program, which works to strategize urban design in order to adapt to and diminish the effects of climate change. Under this program, architects, such as the American architect Michael Graves, consider that buildings, depending on their locations, are more vulnerable or more susceptible to flooding impacts. Green infrastructure, which absorbs rainwater, could alleviate some of these flood concerns, particularly in cities with little green space. Accordingly, transportation decision makers continually make short- and long-term investment decisions that affect how infrastructure will respond to climate change.

Nevertheless, there are certain economically-viable solutions that can be implemented in urban settings, such as green roofs. Green roofs may "keep surface temperatures below the baseline level for all time periods and emissions scenarios". Another strategy is to maintain green space as often as possible, and integrate drought-resistant plantings. Vegetation such as this can reduce heat that accumulates in buildings and thus diminish the need for air conditioning, which contributes to "both the greenhouse gas emissions as well as the intensification of the urban heat island through waste heat".

Economist Ross Garnaut examines the impacts and potential solutions of climate change in his review: the Garnaut Climate Change Review. Garnaut speaks of the effects associated with deforestation, or the destruction of vegetation, concentrating on the region of Australia. He argues that Australia should lead the way in making carbon capture storage viable, perhaps through biosequestration.

Technology

Two renewable technologies – solar water heaters and biodiesel – have already helped improve the welfare of homes in South Africa. Improved technologies, such as these, also provide job opportunities and skills development, which could contribute to poverty alleviation.

Other technologies, such as those involved in the carbon-capture of flue gas from power plants are prominent. With this technology, carbon dioxide is injected into oil reservoirs "to increase the mobility of the oil, and, therefore, the productivity of the reservoir," while minimizing emissions. At the end of the twentieth century, capture plants utilized chemical absorption processes using the solvent monoethanolamine (MEA). Other processes include membrane separation, cryogenic fractionation, and adsorption using molecular sieves (Herzog). Another option for disposing of carbon dioxide is underground storage of carbon in geological formations (Herzog). For example, mined salt domes or rock caverns typically have a large storage capacity. The ocean may have the greatest potential for storage of carbon dioxide – injection may occur using dry ice, injecting liquid carbon dioxide, creating a dense carbon dioxide-seawater mixture, etc. (Herzog). Although, greater research is necessary, concerning the physical-chemical interactions between seawater and carbon dioxide, ocean circulation and mixing, biological impacts, and ocean engineering (Herzog).

Security

The concept of human security and the effects that climate change may have on it will become increasingly important as the changes become more apparent. Some effects are already evident and will become very clear in the human and climatic short run (2007–2020). They will increase and others will manifest themselves in the medium term (2021–2050); whilst in the long run (2051–2100), they will all be active and interacting strongly with other major trends. There is the potential for the end of the petroleum economy for many producing and consuming nations, possible financial and economic crisis, a larger population of humans, and a much more urbanized humanity – far in excess of the 50% now living in small to very large cities. All these processes will be accompanied by redistribution of population nationally and internationally. Such redistributions typically have significant gender dimensions; for example, extreme event impacts can lead to male out migration in search of work, culminating in an increase in women-headed households – a group often considered particularly vulnerable. Indeed, the effects of climate change on impoverished women and children is crucial in that women and children in particular, have unequal human capabilities.

Economics

William D. Nordhaus, a Sterling Professor of Economics at Yale University, analyzes the DICE model (Dynamic Integrated model of Climate and the Economy) in order to determine the uncertainties and known benefits of several climate change policy approaches, including "no controls, economic optimization, geoengineering, stabilization of emission and climate, and a ten-year delay in undertaking climate change policies". He uncovered that climate change is inevitable due to the increasing buildup of greenhouse gas emissions, yet can be slowed with efficient policy, in place of no controls or a ten-year delay. Efficient policy will reap economic benefits in comparison to no or little policy action.

Martin L. Weitzman, a Professor with the Department of Economics at Harvard University, explains that "economic consequences of fat-tailed structural uncertainty of climate change can readily outweigh the effects of discounting in climate-change policy analysis".

Energy Development

Current Energy

"Current energy services fail to meet the needs of the poor".

Countries such as China have limited energy demand through aggressive and strictly-enforced energy-efficiency programs. The main components of these programs are close oversight of industry, issuing financial incentives, distribution of information through education and other venues, and research and development.

Alternatives to Current Energy

One way to promote environmental sustainability of energy supply is to introduce renewable energy.

"The Brazilian ethanol program provides fuel for more than 5 million cars each year. It has created 720,000 jobs directly and 200,000 indirectly in rural areas, curbed city air pollution and avoided 6 to 10 million tons of carbon emission per year since 1980".

Empowerment and Limitations

Vulnerability

Vulnerability to water stress is determined almost exclusively by population and development pressure. Vulnerability to agricultural practices depends on increasing water demands coupled with unsustainable water use.

Application of Energy in Society

Agricultural Production and Food Security

There has been considerable research comparing the interrelated processes of climate change and agriculture. Climate change will affect rainfall, temperature, and water availability for agriculture in vulnerable areas. Climate change could affect agriculture in several ways including productivity, agricultural practices, environmental effects, and distribution of rural space. Additional number affected by malnutrition could rise to 600 million by 2080. Climate change could worsen the prevalence of hunger through direct negative effects on production and indirect impacts on purchasing powers.

Countries, such as Indonesia and Mexico, are at the forefront of climate change, for they rely on crops such as rice and corn for income.

Water Insecurity

Of the 3 billion growth in population projected worldwide by the mid-21st century, the majority will be born in countries already experiencing water shortages. As the overall climate of the earth warms, changes in the nature of global rainfall, evaporation, snow, and runoff flows will be affected. Safe water sources are essential for survival within a community. Manifestations of the projected water crisis include inadequate access to safe drinking water for about 884 million people as well as inadequate access to water for sanitation and water disposal for 2.5 billion people.

Rising Sea Levels and Exposure to Climate Disasters

Sea levels could rise rapidly with accelerated ice sheet disintegration. Sea level rise brings risk to coastal groundwater aquifers, which increases vulnerability to saline intrusion. For example, with the compounding effects of sea level rise and overexploitation of aquifers, saline intrusion has greatly hindered vital groundwater resources in Lima, Peru, and has hindered much of the city's water supply. Global temperature increases of 3–4 degrees C could result in 330 million people being permanently or temporarily displaced through flooding Frequent flooding has occurred in northeast Argentina, Bolivia, southern Brazil, Ecuador, northwest Mexico, Paraguay, northwest Peru, and Uruguay. Warming seas will also fuel more intense tropical storms. The changes in temperature of water will cause the autopurification rate at which streams degrade organic loads to fluctuate, which will result in issues of water quality. Of particular concern are parts of Mexico, like the Balsas, South Pacific, and southern frontier basins are highly vulnerable to hurricanes and flooding. Extreme droughts are also the result of climate change. La Nina-related droughts have affected farming practices in western Argentina and in central Chile, while El Nino-related droughts have diminished the circulation of the Cauca River in Colombia. More than half a billion children live in zones of "extremely high flood occurrence severity" and 160 million live in zones of high or extremely high drought severity. With such high numbers affected already the rising sea levels and extreme weather will only continue to worsen especially for those already impacted.

Ecosystems and Biodiversity

Climate change is already transforming ecological systems. Around one-half of the world's coral reef systems have suffered bleaching as a result of warming seas. In addition, the direct human pressures that might be experienced include overfishing which could lead to resource depletion, nutrient and chemical pollution and poor land use practices such as deforestation and dredging. Also, climate change may increase the amount of arable land in high-latitude regions by reduction of the amount of frozen lands. A 2005 study reports that temperature in Siberia has increased three degree Celsius in average since 1960, which is reportedly more than in other areas of the world.

By 2050, over one million plant and animal species will grow extinct due to climate change. The 2005 Millennium Ecosystem Assessment (MEA) predicts that by 2100, climate change will be the dominant cause of biodiversity loss.

Forests release carbon dioxide into the atmosphere when cut down or burnt. Land use changes, particularly deforestation in tropical regions—where forests tend to be very rich in biodiversity—are responsible for roughly 18 percent of human-driven carbon dioxide emissions.

Human Health

The effects of climate change on human health can be organized into two categories: direct (physical health, trauma, illness etc.) and indirect (environmental changes, resource stress, psychological).

Exposure to multiple climate threats, in any given population, can lead to a myriad of health consequences. Climate change affects the essential components of maintaining good health: clean air, clean water, sufficient food, and adequate shelter. The effects of a changing climate will be widespread and vary through different populations. The report of the WHO Commission on Social Determinants of Health points out that disadvantaged communities are likely to shoulder a disproportionate share of the burden of climate change because of their increased exposure and vulnerability to health threats. Vulnerable populations, such as the poor, are at a higher risk of poor health. Agriculture is a major source of income in poorer countries, however, it is also the most susceptible to the negative impacts of climate change. Other health risks include an increase in vector borne illnesses, diarrheal diseases, malnutrition, and psychological consequences. Environmental consequences will result in extreme weather conditions, which could lead to a decrease in labor productivity and an increase in the frequency and severity of natural disasters. The Intergovernmental Panel on Climate Change (IPCC) reports that overall human health will be influenced more heavily as time goes on; this is due to a chain of events that start at environmental and social well-being and ends at population health of individuals at the top of the food chain.

Temperature Related Death and Disease

The second highest health risk is an increase in temperature-related illnesses and deaths related to prolonged heat waves and humidity. These effects will disproportionately affect children, the elderly, and those living in poverty. Rising global temperatures will extend seasons and increase the geographic range of vectors, specifically mosquito-borne disease such as malaria and dengue fever exposing new populations to disease. Due to increased carbon dioxide emissions, the temperature of the lower troposphere rise with elevation. This means that geographic ranges of disease carrying insects will spread vertically and horizontally. Over 90 percent of malaria and diarrhea deaths are children 5 years and younger, mostly in developing countries. Other severely affected population groups include women, the elderly, and people living in small island developing states and other coastal regions, mega-cities or mountainous areas. According to researchers from the University College Institute for Global Health, health effects of climate change are likely to deepen in both low- and middle-income communities. Perhaps the most detrimental health impacts, especially for poorer communities, are illness that debilitate, but don't cause death. These diseases will cause a decrease in labor productivity, resulting in less income. This cyclical chain of events is why those living in poverty are at a higher risk.

Psychological Impacts

Psychological impacts are considered an indirect cost of climate change as a result of physical injury and community stress. There are also both direct and indirect types of psychological effects: direct would be considered physical ailments, displacement, and disruptions to food supplies and indirect includes poverty related stress, community well-being, and isolation/alienation. An increase in climate and weather related disasters could lead to a higher rate of mental health issues. Studies have shown that flooding and extended droughts lead to an escalation in the amount of anxiety, depression, and post traumatic stress disorder cases.

Populations living in poverty and those who already have chronic mental health illnesses are particularly vulnerable to the psychological risks resulting from a changing climate. Although children are more susceptible to natural disasters, in the long term they are more resilient and the effects may fade. First responders and emergency personnel are also at a higher risk of psychological consequences.

Proposed Policy Solutions

Mitigation Efforts

Climate change mitigation is the action to decrease the intensity of radiative forcing in order to reduce the potential effects of global warming. Most often, mitigation efforts involve reductions in the concentrations of greenhouse gases, either by reducing their source or by increasing their sinks. Several international institutions, such as the Green Climate Fund, are financing projects to mitigate greenhouse gas while also reducing poverty. Various United Nations agencies, such as UNIDO have been tackling issues such as the link between industrial energy efficiency, economic development and poverty reduction. Aid agencies have also supported research into gaining a better understanding of how energy can impact rural poverty, especially women.

However, the cash distribution will perpetuate the problem than easing it, and these stipends don't generate investable capital in the form of accumulation of surplus income, because the poor tend to spend the money on the basic needs of life.

Adaptation Efforts

Adaptation to global warming involves actions to tolerate the effects of global warming. Collaborative research from the Institute of Development Studies draws links between adaptation and poverty to help develop an agenda for pro-poor adaptation that can inform climate-resilient poverty reduction. Adaptation to climate change will be "ineffective and inequitable if it fails to learn and build upon an understanding of the multidimensional and differentiated nature of poverty and vulnerability". Poorer countries tend to be more seriously affected by climate change, yet have reduced assets and capacities with which to adapt. This has led to more activities to integrate adaptation within development and poverty reduction programs. The rise of adaptation as a development issue has been influenced by concerns around minimizing threats to progress on poverty reduction, notably the Millennium Development Goals, and by the injustice of impacts that are felt hardest by those who have done least to contribute to the problem, framing adaptation as an equity and human rights issue.

The idea that development can be achieved via foreign aid has proved its failure. Governments who have improved their populations' well-being in recent years have achieved this through engaging with international trade, international markets, and globalization. In addition, civil societies of different countries like the women's movement, the environment movement, development and human rights, faith-based organizations, diaspora communities, etc. have played an important role in the development. On the other hand, the countries that have made little development in the process had high level in political instability and violent conflict.

Future Development

Policies and programs can help facilitate new technologies that will increase environmentally-sustainable energy production, but diminish the harmful effects of typical consumption. "Priorities include pricing energy to account for environmental costs, removing subsidies that increase harmful emissions, adopting incentives for beneficial new technologies during their market scale up stage, and promulgating regulatory standards for energy efficiency". These objectives will demand the cooperation of all stakeholders.

Developed countries have started levying new taxes in order to help aid the global poverty-reduction and adaptation efforts related to climate change. For example, France has imposed taxes on international flights to aid HIV/AIDS in Africa, and a fifth of the money from the EU's emissions-trading scheme has contributed to climate change-related improvements.

Rural areas would benefit from the removal of market barriers on modern fuels. Decentralizing decision-making and energy investments may also improve energy services in rural areas.

Proposed Policy Challenges

Most difficult policy challenge is related to distribution. While this is a potential catastrophic risk for the entire globe, the short and medium-term distribution of the costs and benefits will be far from uniform. Distribution challenge is made particularly difficult because those who have largely caused the problem – richer nations – are not going to be those who suffer the most in the short term. It is the poorest who did not and still are not contributing significantly to green house gas emissions that are the most vulnerable.

Developed countries have started levying new taxes in order to help aid the global poverty-reduction and adaptation efforts related to climate change. For example, France has imposed taxes on international flights to aid HIV/AIDS in Africa, and a fifth of the money from the EU's emissions-trading scheme has contributed to climate change-related improvements.

Also, decentralizing decision-making and energy saving may improve services in rural areas.

References

- Wallace, John M. and Peter V. Hobbs. Atmospheric Science; An Introductory Survey. Elsevier. Second Edition, 2006. ISBN 978-0-12-732951-2
- Retallack, Gregory J. (2001). "Cenozoic Expansion of Grasslands and Climatic Cooling". The Journal of Geology. 109 (4): 407–426. Bibcode:2001JG....109..407R. doi:10.1086/320791
- Marger (2008). Social Inequality: Patterns and Processes, 4th edition. McGraw Hill publishing. ISBN 0-07-352815-3
- Miles, M. G.; Grainger, R. G.; Highwood, E. J. (2004). "The significance of volcanic eruption strength and frequency for climate" (PDF). Quarterly Journal of the Royal Meteorological Society. 130 (602): 2361–2376. doi:10.1256/qj.30.60
- Bridging the Emissions Gap: A UNEP Synthesis Report (PDF), Nairobi, Kenya: United Nations Environment Programme (UNEP), November 2011, ISBN 978-92-807-3229-0 UNEP Stock Number: DEW/1470/NA
- Jacob, Daniel (1999). Introduction to atmospheric chemistry. Princeton University Press. pp. 25–26. ISBN 0-691-00185-5. Archived from the original on 2 September 2011
- David Archer (2009). The Long Thaw: How Humans Are Changing the Next 100,000 Years of Earth's Climate. Princeton University Press. p. 109. ISBN 978-0-691-13654-7

- Schneeberger, Christian; et al. (2003). "Modelling changes in the mass balance of glaciers of the northern hemisphere for a transient 2×CO_2 scenario". Journal of Hydrology. 282 (1–4): 145–163. Bibcode:-2003JHyd..282..145S. doi:10.1016/S0022-1694(03)00260-9
- Langford, Bill & Lewis, Greg. "Hadley Cell Expansion in Today's Climate and Paleoclimates" (PDF). Retrieved 19 October2014
- Botkin DB; et al. (2007). "Forecasting the effects of global warming on biodiversity". BioScience. 57 (3): 227–36. doi:10.1641/B570306
- America›s Climate Choices. Washington, D.C.: The National Academies Press. 2011. p. 15. ISBN 978-0-309-14585-5
- New, M., Todd, M., Hulme, M. and Jones, P. (December 2001). "Review: Precipitation measurements and trends in the twentieth century". International Journal of Climatology. 21 (15): 1889–1922. Bibcode:2001IJCli..21.1889N. doi:10.1002/joc.680
- Vaara, Miska (1 January 2003), Use of ozone depleting substances in laboratories, TemaNord, p. 170, ISBN 92-893-0884-2, Archived from the original on 6 August 2011
- "Glacial Lake Outburst Flood Monitoring and Early Warning System". United Nations Environment Programme. Archived from the original on 2006-07-17. Retrieved 2007-12-12

Chapter 5

Ozone Layer and Ozone Depletion

The ozone layer is the region in the stratosphere of the Earth, which absorbs most of the ultraviolet radiation coming from the Sun. Ozone depletion refers to the steady decrease in stratospheric ozone. This chapter has been carefully written to provide an easy understanding of the ozone layer and the phenomenon of ozone layer depletion.

Ozone Layer

Ozone layer is a vital component of our atmosphere. Instratosphere, it acts as a protective layer as it absorbs the harmful ultraviolet radiation and does not allow it to reach the earth's surface. Therefore, the presence of the ozone layer is essential factor in the life environment. Ozone is a very reactive molecule and with the help of a catalyst it is easily reduced to the more stable oxygen. The ozone destroying catalysts are natural as well as manmade. The primary cause of ozone depletion is the trace gases mainly chlorine and bromine, from anthropogenic sources.The depletion of ozone layer was discovered for the first time by Farman in 1985. It was established that, ozone 'hole' has occurred in the ozone layer over Antarctica. In 1987, the Montreal protocol set legally binding controls on the production and consumption of gases associated with ozone depletion. The outcome of this response is decline in the total abundance of CFCs in the atmosphere and trend of gradual healing of ozone layer.

The ozone layer was discovered in 1913 by the French physicists Charles Fabry and Henri Buisson. Measurements of the sun showed that the radiation sent out from its surface and reaching the ground on Earth is usually consistent with the spectrum of a black body with a temperature in the range of 5,500–6,000 K (5,227 to 5,727 °C), except that there was no radiation below a wavelength of about 310 nm at the ultraviolet end of the spectrum. It was deduced that the missing radiation was being absorbed by something in the atmosphere. Eventually the spectrum of the missing radiation was matched to only one known chemical, ozone. Its properties were explored in detail by the British meteorologist G. M. B. Dobson, who developed a simple spectrophotometer (the Dobsonmeter) that could be used to measure stratospheric ozone from the ground. Between 1928 and 1958, Dobson established a worldwide network of ozone monitoring stations, which continue to operate to this day. The "Dobson unit", a convenient measure of the amount of ozone overhead, is named in his honor.

The ozone layer absorbs 97 to 99 percent of the Sun's medium-frequency ultraviolet light (from about 200 nm to 315 nm wavelength), which otherwise would potentially damage exposed life forms near the surface.

In 1976 atmospheric research revealed that the ozone layer was being depleted by chemicals released by industry, mainly chlorofluorocarbons (CFCs). Concerns that increased UV radiation due to ozone depletion threatened life on Earth led to bans on the chemicals, and the latest evidence is that ozone depletion has slowed or stopped. The United Nations General Assembly has designated September 16 as the International Day for the Preservation of the Ozone Layer.

Venus also has a thin ozone layer at an altitude of 100 kilometers from the planet's surface.

Sources

The photochemical mechanisms that give rise to the ozone layer were discovered by the British physicist Sydney Chapman in 1930. Ozone in the Earth's stratosphere is created by ultraviolet light striking ordinary oxygen molecules containing two oxygen atoms (O_2), splitting them into individual oxygen atoms (atomic oxygen); the atomic oxygen then combines with unbroken O_2 to create ozone, O_3. The ozone molecule is unstable (although, in the stratosphere, long-lived) and when ultraviolet light hits ozone it splits into a molecule of O_2 and an individual atom of oxygen, a continuing process called the ozone-oxygen cycle. Chemically, this can be described as:

$$O_2 + h\nu_{uv} \rightarrow 2O$$

$$O + O_2 \leftrightarrow O_3$$

About 90 percent of the ozone in the atmosphere is contained in the stratosphere. Ozone concentrations are greatest between about 20 and 40 kilometres (66,000 and 131,000 ft), where they range from about 2 to 8 parts per million. If all of the ozone were compressed to the pressure of the air at sea level, it would be only 3 millimetres ($\frac{1}{8}$ inch) thick.

Ultraviolet light

Although the concentration of the ozone in the ozone layer is very small, it is vitally important to life because it absorbs biologically harmful ultraviolet (UV) radiation coming from the sun. Extremely short or vacuum UV (10–100 nm) is screened out by nitrogen. UV radiation capable of penetrating nitrogen is divided into three categories, based on its wavelength; these are referred to as UV-A (400–315 nm), UV-B (315–280 nm), and UV-C (280–100 nm).

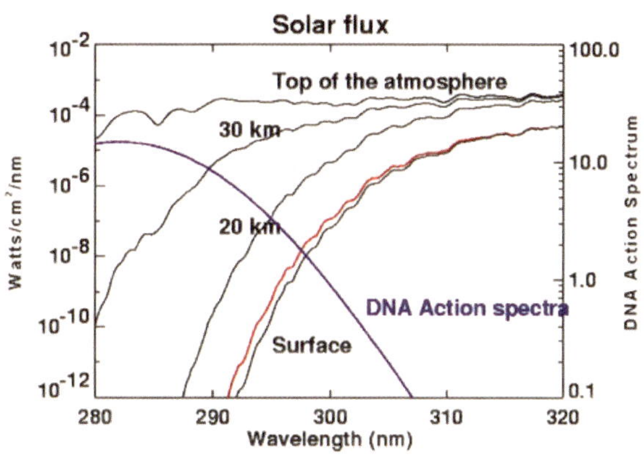

UV-B energy levels at several altitudes. Blue line shows DNA sensitivity. Red line shows surface energy level with 10 percent decrease in ozone

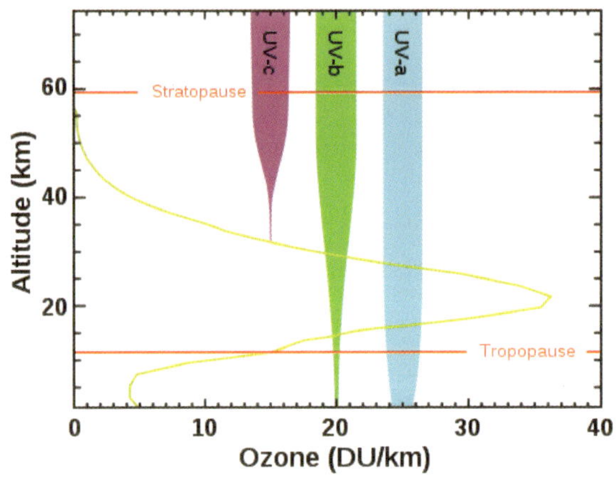

Levels of ozone at various altitudes and blocking of different bands of ultraviolet radiation. Essentially all UVC (100–280 nm) is blocked by dioxygen (from 100–200 nm) or else by ozone (200–280 nm) in the atmosphere

The shorter portion of the UV-C band and the more energetic UV above this band causes the formation of the ozone layer, when single oxygen atoms produced by UV photolysis of dioxygen (below 240 nm) react with more dioxygen. The ozone layer also blocks most, but not quite all, of the sunburn-producing UV-B (280–315 nm) band, which lies in the wavelengths longer than UV-C. The band of UV closest to visible light, UV-A (315–400 nm), is hardly affected by ozone, and most of it reaches the ground. UV-A does not primarily cause skin reddening, but there is evidence that it causes long-term skin damage.

UV-C, which is very harmful to all living things, is entirely screened out by a combination of dioxygen (< 200 nm) and ozone (> about 200 nm) by around 35 kilometres (115,000 ft) altitude. UV-B radiation can be harmful to the skin and is the main cause of sunburn; excessive exposure can also cause cataracts, immune system suppression, and genetic damage, resulting in problems such as skin cancer. The ozone layer (which absorbs from about 200 nm to 310 nm with a maximal absorption at about 250 nm) is very effective at screening out UV-B; for radiation with a wavelength of 290 nm, the intensity at the top of the atmosphere is 350 million times stronger than at the Earth's surface. Nevertheless, some UV-B, particularly at its longest wavelengths, reaches the surface, and is important for the skin's production of vitamin D.

Ozone is transparent to most UV-A, so most of this longer-wavelength UV radiation reaches the surface, and it constitutes most of the UV reaching the Earth. This type of UV radiation is significantly less harmful to DNA, although it may still potentially cause physical damage, premature aging of the skin, indirect genetic damage, and skin cancer.

Distribution in the Stratosphere

The thickness of the ozone layer—that is, the total amount of ozone in a column overhead—varies by a large factor worldwide, being in general smaller near the equator and larger towards the poles. It also varies with season, being in general thickest during the autumn and thinner during the early spring. The reasons for this latitude and seasonal dependence are complicated, which involve in atmospheric circulation patterns as well as solar intensity. But in general, bitterly cold temperatures mixed with sunlight assist in maximum ozone depletion, which occurs in early spring.

Since stratospheric ozone is produced by solar UV radiation, one might expect to find the highest ozone levels over the tropics and the lowest over polar regions. The same argument would lead one to expect the highest ozone levels in the summer and the lowest in the winter. The observed behavior is very different: most of the ozone is found in the mid-to-high latitudes of the northern and southern hemispheres, and the highest levels are found in the spring, not summer, and the lowest in the autumn, not winter in the northern hemisphere. During winter, the ozone layer actually increases in depth. This puzzle is explained by the prevailing stratospheric wind patterns, known as the Brewer-Dobson circulation. While most of the ozone is indeed created over the tropics, the stratospheric circulation then transports it poleward and downward to the lower stratosphere of the high latitudes. However, owing to the ozone hole phenomenon, the lowest amounts of column ozone found anywhere in the world are over the Antarctic in the southern spring period of September and October and to a lesser extent over the Arctic in the northern spring period of March, April, and May.

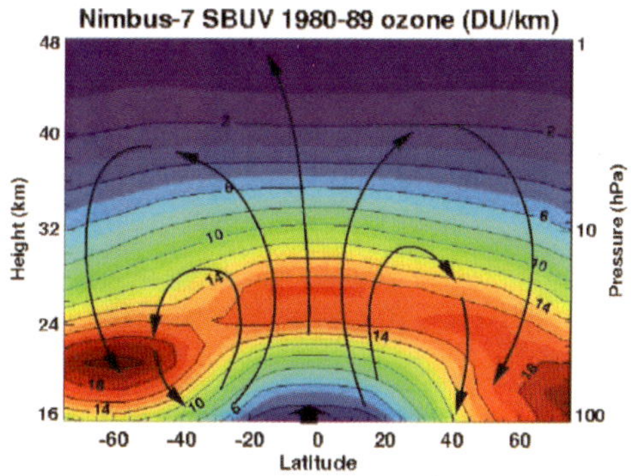

Brewer-Dobson circulation in the ozone layer

The ozone layer is higher in altitude in the tropics, and lower in altitude outside the tropics, especially in the polar regions. This altitude variation of ozone results from the slow circulation that lifts the ozone-poor air out of the troposphere into the stratosphere. As this air slowly rises in the tropics, ozone is produced as the sun overhead photolyzes oxygen molecules. As this slow circulation levels off and flows towards the mid-latitudes, it carries the ozone-rich air from the tropical middle stratosphere to the lower stratosphere middle and high latitudes . The high ozone concentrations at high latitudes are due to the accumulation of ozone at lower altitudes.

The Brewer-Dobson circulation moves very slowly. The time needed to lift an air parcel by 1 km in the lower tropical stratosphere is about 2 months (18 m per day). However, horizontal poleward transport in the lower stratosphere is much faster and amounts to approximately 100 km per day in the northern hemisphere whilst it is only half as much in the southern hemisphere (~51 km per day). Even though ozone in the lower tropical stratosphere is produced at a very slow rate, the lifting circulation is so slow that ozone can build up to relatively high levels by the time it reaches 26 kilometres (16 mi).

Ozone amounts over the continental United States (25°N to 49°N) are highest in the northern spring (April and May). These ozone amounts fall over the course of the summer to their lowest amounts in October, and then rise again over the course of the winter. Again, wind transport of ozone is principally responsible for the seasonal changes of these higher latitude ozone patterns.

The total column amount of ozone generally increases as we move from the tropics to higher latitudes in both hemispheres. However, the overall column amounts are greater in the northern hemisphere high latitudes than in the southern hemisphere high latitudes. In addition, while the highest amounts of column ozone over the Arctic occur in the northern spring (March–April), the opposite is true over the Antarctic, where the lowest amounts of column ozone occur in the southern spring (September–October).

Depletion

The ozone layer can be depleted by free radical catalysts, including nitric oxide (NO), nitrous oxide (N_2O), hydroxyl (OH), atomic chlorine (Cl), and atomic bromine (Br). While there are natural sources for all of these species, the concentrations of chlorine and bromine increased markedly in recent decades because of the release of large quantities of man-made organohalogen compounds, especially chlorofluorocarbons (CFCs) and bromofluorocarbons. These highly stable compounds are capable of surviving the rise to the stratosphere, where Cl and Br radicals are liberated by the action of ultraviolet light. Each radical is then free to initiate and catalyze a chain reaction capable of breaking down over 100,000 ozone molecules. By 2009, nitrous oxide was the largest ozone-depleting substance (ODS) emitted through human activities.

The breakdown of ozone in the stratosphere results in reduced absorption of ultraviolet radiation. Consequently, unabsorbed and dangerous ultraviolet radiation is able to reach the Earth's surface at a higher intensity. Ozone levels have dropped by a worldwide average of about 4 percent since the late 1970s. For approximately 5 percent of the Earth's surface, around the north and south poles, much larger seasonal declines have been seen, and are described as "ozone holes". The discovery of the annual depletion of ozone above the Antarctic was first announced by Joe Farman, Brian Gardiner and Jonathan Shanklin, in a paper which appeared in *Nature* on May 16, 1985.

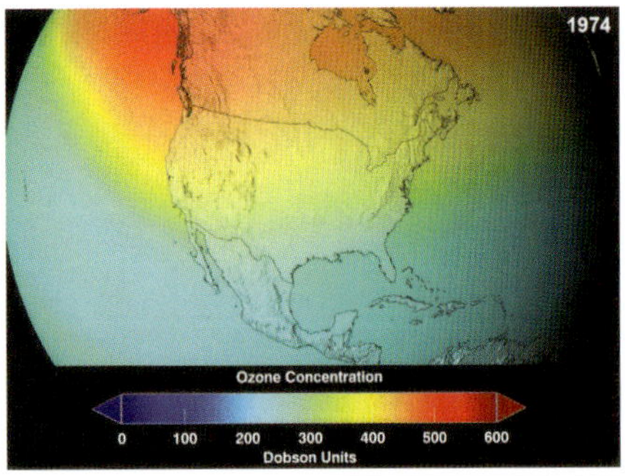

NASA projections of stratospheric ozone concentrations if chlorofluorocarbons had not been banned

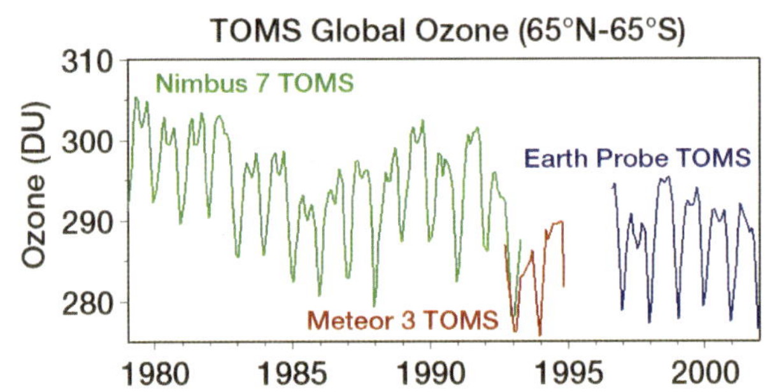

Levels of atmospheric ozone measured by satellite show clear seasonal variations and appear to verify their decline over time

Regulation

To support successful regulation attempts, the ozone case was communicated to lay persons "with easy-to-understand bridging metaphors derived from the popular culture" and related to "immediate risks with everyday relevance". The specific metaphors used in the discussion (ozone shield, ozone hole) proved quite useful and, compared to global climate change, the ozone case was much more seen as a "hot issue" and imminent risk. Lay people were cautious about a depletion of the ozone layer and the risks of skin cancer.

In 1978, the United States, Canada and Norway enacted bans on CFC-containing aerosol sprays that damage the ozone layer. The European Community rejected an analogous proposal to do the same. In the U.S., chlorofluorocarbons continued to be used in other applications, such as refrigeration and industrial cleaning, until after the discovery of the Antarctic ozone hole in 1985. After negotiation of an international treaty (the Montreal Protocol), CFC production was capped at 1986 levels with commitments to long-term reductions. Since that time, the treaty was amended to ban CFC production after 1995 in the developed countries, and later in developing countries. Today, all of the world's 197 countries have signed the treaty. Beginning January 1, 1996, only recycled and stockpiled CFCs were available for use in developed countries like the US. This production phaseout was possible because of efforts to ensure that there would be substitute chemicals and technologies for all ODS uses.

On August 2, 2003, scientists announced that the global depletion of the ozone layer may be slowing down because of the international regulation of ozone-depleting substances. In a study organized by the American Geophysical Union, three satellites and three ground stations confirmed that the upper-atmosphere ozone-depletion rate slowed down significantly during the previous decade. Some breakdown can be expected to continue because of ODSs used by nations which have not banned them, and because of gases which are already in the stratosphere. Some ODSs, including CFCs, have very long atmospheric lifetimes, ranging from 50 to over 100 years. It has been estimated that the ozone layer will recover to 1980 levels near the middle of the 21st century. A gradual trend toward "healing" was reported in 2016.

Compounds containing C–H bonds (such as hydrochlorofluorocarbons, or HCFCs) have been designed to replace CFCs in certain applications. These replacement compounds are more reactive and less likely to survive long enough in the atmosphere to reach the stratosphere where they could affect the ozone layer. While being less damaging than CFCs, HCFCs can have a negative impact on the ozone layer, so they are also being phased out. These in turn are being replaced by hydrofluorocarbons (HFCs) and other compounds that do not destroy stratospheric ozone at all.

The residual effects of CFCs accumulating within the atmosphere lead to a concentration gradient between the atmosphere and the ocean. This organohalogen compound is able to dissolve into the ocean's surface waters and is able to act as a time-dependent tracer. This tracer helps scientists study ocean circulation by tracing biological, physical and chemical pathways.

Implications for Astronomy

As ozone in the atmosphere prevents most energetic ultraviolet radiation reaching the surface of the Earth, astronomical data in these wavelengths have to be gathered from satellites orbiting above the atmosphere and ozone layer. Most of the light from young hot stars is in the ultraviolet and so study of these wavelengths is important for studying the origins of galaxies. The Galaxy Evolution Explorer, GALEX, is an orbiting ultraviolet space telescope launched on April 28, 2003, which operated until early 2012.

This GALEX image of the Cygnus Loop nebula could not have been taken from the surface of the Earth because the ozone layer blocks the ultra-violet radiation emitted by the nebula

Ozone Depletion

Ozone is a very reactive molecule and with the help of a catalyst it is easily reduced to the more stable oxygen. The ozone destroying catalysts are natural as well as manmade. Ozone can be destroyed by a number of free radical catalysts such as (i) OH (hydroxyl) -the odd hydrogen atoms and OH come from the dissociation of water vapour, molecular hydrogen and methane (CH_4); (ii) nitric oxide –ozone is destroyed in stratosphere due to the presence nitrogen oxides (NOx, i.e. NO_2 and NO). The source gas of the NOx is nitrous oxide (N_2O) - it is produced by combustion, supersonic jets and fertilizer use; (iii) Chlorine and (iv) Bromine. Anthropogenic activities have sharply increased the levels of chlorine and bromine.

A single chlorine atom has the capability to interact with thousands of ozone molecules in catalytic cycle before it is finally removed. On a per atom basis, the efficiency of bromine is far more than chlorine to destroy ozone, but its abundance is much less in the atmosphere. Therefore, both chlorine and bromine are the major contributors in overall ozone destruction. The increase in abundance of chlorine and bromine during the decades of 1970-90 is reflected in the observed decrease of stratospheric ozone over the Antarctica and other parts. A simplified illustration is given to show ozone destruction below:

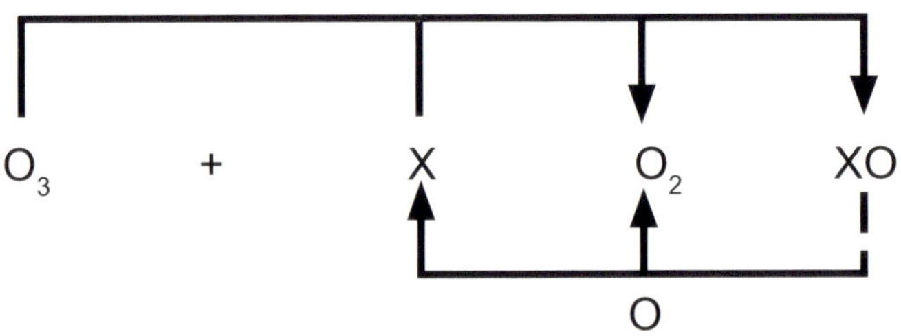

Ozone Destruction Process (where X is any ozone-destroying species, e.g. OH, NO, CR, Br)

The depletion of ozone in the stratosphere due to human activities is a serious global environmental problem. The major offending chemicals are Freons, synthetic compounds containing carbon, fluorine and chlorine atoms. Compounds of this class are also known as *halocarbons*. The popular alternate name for these is*chlorofluorocarbons*, or CFCs. They are versatile compounds that are chemically stable, odorless, nontoxic, noncorrosive, and inexpensive to produce. CFCs, carbon tetrachloride (used in fire extinguishers and solvents), alternates of CFCs - hydrochlorofluorocarbons (HCFCs) and methyl chloroform are the chief chlorine containing gases. These gases are used in many applications such as – refrigeration, air conditioning, foam blowing, aerosol propellants and cleaning of metals and electronic components. Methyl chloroform is used in industries for cold cleaning, vapor degreasing, chemical processing, adhesives and aerosols.

The most important source halogen gases for bromine are the halons and methyl bromide. Halons are used to extinguish fires. Methyl bromide, used as an agricultural fumigant, is another significant source of bromine to the atmosphere. Methane and nitrogen oxide, which react in stratosphere to form water vapour and reactive hydrogen and nitrogen oxides, respectively have ozone depletion capabilities.The emission of oxides of nitrogen by supersonic jet planes is also destructive for ozone layer.

Table: Common Chloroflurocarbons and Halons

Compound (chemical formula)	Ozone Depletion Potentiala	Atmospheric Lifetime (Years)	Major Uses
CFC-11 (CFCl3)	1.0	64	Rigid and flexible foams, refrigration
CFC-12 (CF2Cl3)	1.0	108	Air conditioning, refrigeration, rigid foam
CFC-113 (C2F3Cl3)	0.8	88	Solvent
Halon-1211(CF3BrCl)	3.0	25	Portable fire extinguishers
Halon-1301 (CF3Br)	10.0	110	Total flooding fire extinguisher systems
HCFC-22 (HCCIF2)	0.05	22	Air conditioning

There are a few halogen source gases present in the stratosphere that have large natural sources. Methyl chloride contributes about 17 per cent of the chlorine currently into the stratosphere. Likewise, about 30 per cent of the bromine is contributed by natural source methyl bromide. These gases are emitted by oceanic and terrestrial ecosystems. Sunspot cycle shows that global total ozone levels vary by 1 to 2 per cent between the maximum and minimum.Volcanoes can emit some chlorine containing gases. In addition, volcanic sulphur and particulate matter emissions reduce solar transmission and ozone abundance. Sunspots and volcanic eruptions bring only short term changes in ozone abundance. Therefore, the main causes of ozone depletion are anthropogenic.

There was no concern about the negative consequences of nitrous oxide and CFCs until three scientists, Paul Crutzen, F.S. Rowland, and Mario Molina, studied the relationship. In 1970, Crutzen highlighted that emissions of nitrous oxide (N_2O), from increased use of fertilizers and supersonic aircrafts could deplete ozone in the stratosphere, where due to photochemical process it is converted into nitric oxide (NO). In 1974, Rowland and Molina concluded that, like N_2O, the CFCs under ultraviolet radiation in stratosphere would be dissociated and released chlorine atoms will destroy ozone layer. This way they alerted the world through their research work that CFCs were probably reducing the average concentration of ozone in the stratosphere.

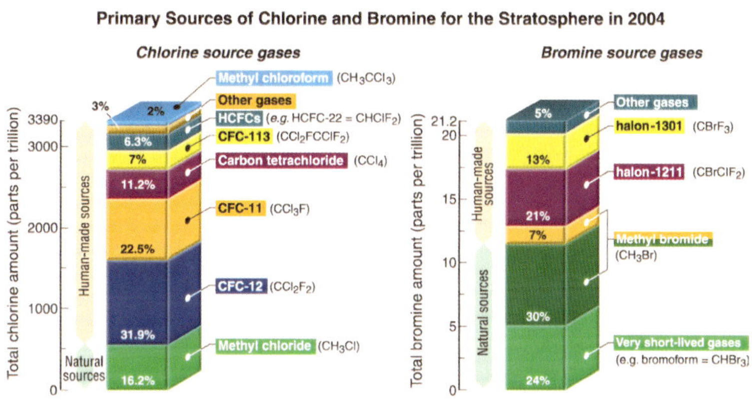

Natural and Anthropogenic Sources of Chlorine and Bromine in Stratosphere

Later on, when in 1995 ozone hole formation was reported, Dr Rowland expressed his frustration and stated, "What's the use of having developed a science well enough to make predictions, if in the end all we are willing to do is stand around and wait for them come true. Unfortunately, this means that if a disaster is in the making in the stratosphere we are probably not going to avoid it." In 1995, scholarsCrutzen, Rowland and Molina, were awarded the Nobel Prize in chemistry for their pioneering research work on ozone depleting substances.

They discovered that molecules of ozone depleting substances like CFCs drift upward and reach the ozone layer in stratosphere. In stratosphere, CFCs in photochemical process absorb ultraviolet radiation and get decomposed, releasing chlorine. These released chlorine atoms in a complex series of reactions attack molecules of ozone. In a series of chain reactions large numbers of ozone molecules are converted into ordinary oxygen molecules. The chlorine atoms basically interact with ozone molecules by acquiring one oxygen atom to constitute chlorine monoxide (CIO) and leaving behind an oxygen molecule (O_2). As the molecule of chlorine monoxide encounters a single oxygen atom, the oxygen "breaks up" the chlorine monoxide, acquires its oxygen atom and chlorine is released back to indulge in further destruction of ozone in stratosphere. These reactions are chain reactions. In this manner, through chain reactions each chlorine atom that reaches the stratosphere is able to destroy thousands of molecules of ozone resulting into ozone depletion. The simplified reactions are following:

(i) $O_3 + Cl \rightarrow ClO + O_2$ (chlorine acting as the catalyst)

$O + ClO \rightarrow O_2 + Cl$

(ii) $O_3 + NO \rightarrow NO_2 + O_2$ (with nitrous oxide acting as the catalyst)

$O + NO_2 \rightarrow O_2 + NO$

The main cause of ozone depletion and the ozone hole is man-made chemicals, especially man-made halocarbon refrigerants, solvents, propellants, and foam-blowing agents (chlorofluorocarbon (CFCs), HCFCs, halons), referred to as ozone-depleting substances (ODS). These compounds are transported into the stratosphere by the winds after being emitted at the surface. Once in the stratosphere, they release halogen atoms through photodissociation, which catalyze the breakdown of ozone (O_3) into oxygen (O_2). Both types of ozone depletion were observed to increase as emissions of halocarbons increased.

Ozone depletion and the ozone hole have generated worldwide concern over increased cancer risks and other negative effects. The ozone layer prevents most harmful UVB wavelengths of ultraviolet light (UV light) from passing through the Earth's atmosphere. These wavelengths cause skin cancer, sunburn, and cataracts, which were projected to increase dramatically as a result of thinning ozone, as well as harming plants and animals. These concerns led to the adoption of the Montreal Protocol in 1987, which bans the production of CFCs, halons, and other ozone-depleting chemicals.

The ban came into effect in 1989. Ozone levels stabilized by the mid-1990s and began to recover in the 2000s. Recovery is projected to continue over the next century, and the ozone hole is expected to reach pre-1980 levels by around 2075. The Montreal Protocol is considered the most successful international environmental agreement to date.

Ozone Cycle Overview

Three forms (or allotropes) of oxygen are involved in the ozone-oxygen cycle: oxygen atoms (O or atomic oxygen), oxygen gas (O_2 or diatomic oxygen), and ozone gas (O_3 or triatomic oxygen). Ozone is formed in the stratosphere when oxygen molecules photodissociate after intaking ultraviolet photons. This converts a single O_2 into two atomic oxygen radicals. The atomic oxygen radicals then combine with separate O_2 molecules to create two O_3 molecules. These ozone molecules absorb ultraviolet (UV) light, following which ozone splits into a molecule of O_2 and an oxygen atom. The oxygen atom then joins up with an oxygen molecule to regenerate ozone. This is a continuing process that terminates when an oxygen atom recombines with an ozone molecule to make two O_2 molecules.

$$O + O_3 \rightarrow 2 \quad O_2$$

The total amount of ozone in the stratosphere is determined by a balance between photochemical production and recombination.

Ozone can be destroyed by a number of free radical catalysts; the most important are the hydroxyl radical (OH·), nitric oxide radical (NO·), chlorine radical (Cl·) and bromine radical (Br·). The dot is a notation to indicate that each species has an unpaired electron and is thus extremely reactive. All of these have both natural and man-made sources; at the present time, most of the OH· and NO· in the stratosphere is naturally occurring, but human activity has drastically increased the levels of chlorine and bromine. These elements are found in stable organic compounds, especially chlorofluorocarbons, which can travel to the stratosphere without being destroyed in the troposphere due to their low reactivity. Once in the stratosphere, the Cl and Br atoms are released from the parent compounds by the action of ultraviolet light, e.g.

$$CFCl_3 + \text{electromagnetic radiation} \rightarrow Cl\cdot + \cdot CFCl_2$$

Ozone is a highly reactive molecule that easily reduces to the more stable oxygen form with the assistance of a catalyst. Cl and Br atoms destroy ozone molecules through a variety of catalytic cycles. In the simplest example of such a cycle, a chlorine atom reacts with an ozone molecule (O_3), taking an oxygen atom to form chlorine monoxide (ClO) and leaving an oxygen molecule (O_2). The ClO can react with a second molecule of ozone, releasing the chlorine atom and yielding two molecules of oxygen. The chemical shorthand for these gas-phase reactions is:

- $Cl\cdot + O_3 \rightarrow ClO + O_2$

  A chlorine atom removes an oxygen atom from an ozone molecule to make a ClO molecule

- $ClO + O_3 \rightarrow Cl\cdot + 2\ O_2$

This ClO can also remove an oxygen atom from another ozone molecule; the chlorine is free to repeat this two-step cycle

The overall effect is a decrease in the amount of ozone, though the rate of these processes can be decreased by the effects of null cycles. More complicated mechanisms have also been discovered that lead to ozone destruction in the lower stratosphere.

A single chlorine atom would continuously destroy ozone (thus a catalyst) for up to two years (the time scale for transport back down to the troposphere) were it not for reactions that remove them from this cycle by forming reservoir species such as hydrogen chloride (HCl) and chlorine nitrate ($ClONO_2$). Bromine is even more efficient than chlorine at destroying ozone on a per atom basis, but there is much less bromine in the atmosphere at present. Both chlorine and bromine contribute significantly to overall ozone depletion. Laboratory studies have also shown that fluorine and iodine atoms participate in analogous catalytic cycles. However, fluorine atoms react rapidly with water and methane to form strongly bound HF in the Earth's stratosphere, while organic molecules containing iodine react so rapidly in the lower atmosphere that they do not reach the stratosphere in significant quantities.

A single chlorine atom is able to react with an average of 100,000 ozone molecules before it is removed from the catalytic cycle. This fact plus the amount of chlorine released into the atmosphere yearly by chlorofluorocarbons (CFCs) and hydrochlorofluorocarbons (HCFCs) demonstrates the danger of CFCs and HCFCs to the environment.

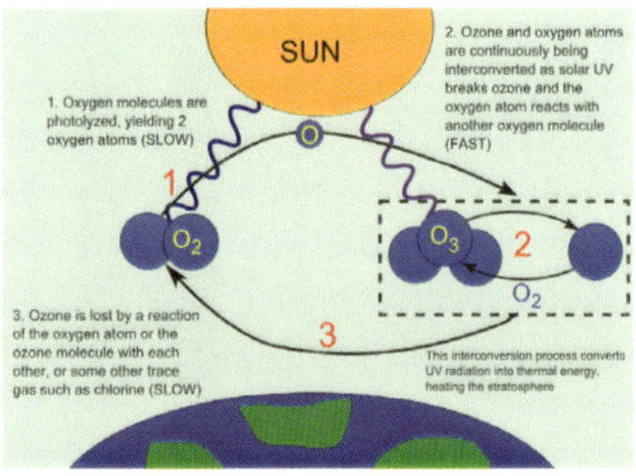

The ozone cycle

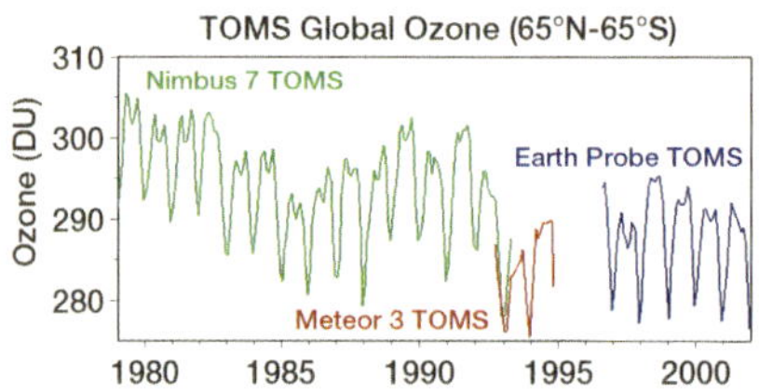

Global monthly average total ozone amount

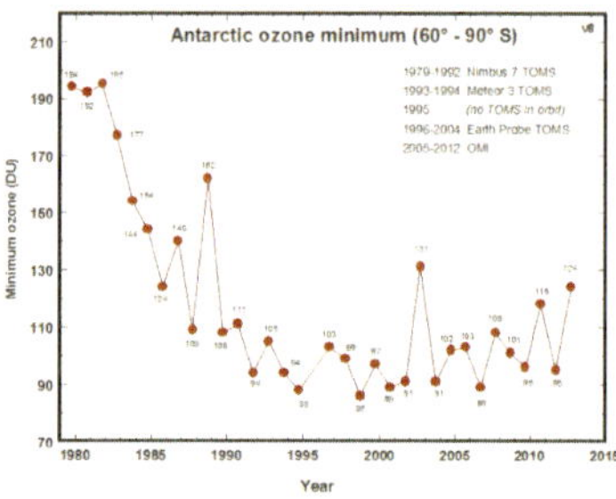

Lowest value of ozone measured by TOMS each year in the ozone hole

Observations on Ozone Layer Depletion

The ozone hole is usually measured by reduction in the total *column ozone* above a point on the Earth's surface. This is normally expressed in Dobson units; abbreviated as "DU". The most prominent decrease in ozone has been in the lower stratosphere. Marked decreases in column ozone in the Antarctic spring and early summer compared to the early 1970s and before have been observed using instruments such as the Total Ozone Mapping Spectrometer (TOMS).

Reductions of up to 70 percent in the ozone column observed in the austral (southern hemispheric) spring over Antarctica and first reported in 1985 (Farman et al.) are continuing. Antarctic total column ozone in September and October have continued to be 40–50 percent lower than pre-ozone-hole values since the 1990s. A gradual trend toward "healing" was reported in 2016. In 2017, NASA announced that the ozone hole was the weakest since 1988 because of warm stratospheric conditions. It is expected to recover around 2070.

The amount lost is more variable year-to-year in the Arctic than in the Antarctic. The greatest Arctic declines are in the winter and spring, reaching up to 30 percent when the stratosphere is coldest.

Reactions that take place on polar stratospheric clouds (PSCs) play an important role in enhancing ozone depletion. PSCs form more readily in the extreme cold of the Arctic and Antarctic stratosphere. This is why ozone holes first formed, and are deeper, over Antarctica. Early models failed to take PSCs into account and predicted a gradual global depletion, which is why the sudden Antarctic ozone hole was such a surprise to many scientists.

It is more accurate to speak of ozone depletion in middle latitudes rather than holes. Total column ozone declined below pre-1980 values between 1980 and 1996 for mid-latitudes. In the northern mid-latitudes, it then increased from the minimum value by about two percent from 1996 to 2009 as regulations took effect and the amount of chlorine in the stratosphere decreased. In the Southern Hemisphere's mid-latitudes, total ozone remained constant over that time period. There are no significant trends in the tropics, largely because halogen-containing compounds have not had time to break down and release chlorine and bromine atoms at tropical latitudes.

Large volcanic eruptions have been shown to have substantial albeit uneven ozone-depleting effects, as observed with the 1991 eruption of Mt. Pinotubo in the Philippines.

Ozone depletion also explains much of the observed reduction in stratospheric and upper tropospheric temperatures. The source of the warmth of the stratosphere is the absorption of UV radiation by ozone, hence reduced ozone leads to cooling. Some stratospheric cooling is also predicted from increases in greenhouse gases such as CO_2 and CFCs themselves; however, the ozone-induced cooling appears to be dominant.

Predictions of ozone levels remain difficult, but the precision of models' predictions of observed values and the agreement among different modeling techniques have increased steadily. The World Meteorological Organization Global Ozone Research and Monitoring Project—Report No. 44 comes out strongly in favor of the Montreal Protocol, but notes that a UNEP 1994 Assessment overestimated ozone loss for the 1994–1997 period.

Chemicals in the Atmosphere

CFCs and Related Compounds in the Atmosphere

Chlorofluorocarbons (CFCs) and other halogenated ozone depleting substances (ODS) are mainly responsible for man-made chemical ozone depletion. The total amount of effective halogens (chlorine and bromine) in the stratosphere can be calculated and are known as the equivalent effective stratospheric chlorine (EESC).

CFCs were invented by Thomas Midgley, Jr. in the 1930s. They were used in air conditioning and cooling units, as aerosol spray propellants prior to the 1970s, and in the cleaning processes of delicate electronic equipment. They also occur as by-products of some chemical processes. No significant natural sources have ever been identified for these compounds—their presence in the atmosphere is due almost entirely to human manufacture. As mentioned above, when such ozone-depleting chemicals reach the stratosphere, they are dissociated by ultraviolet light to release chlorine atoms. The chlorine atoms act as a catalyst, and each can break down tens of thousands of ozone molecules before being removed from the stratosphere. Given the longevity of CFC molecules, recovery times are measured in decades. It is calculated that a CFC molecule takes an average of about five to seven years to go from the ground level up to the upper atmosphere, and it can stay there for about a century, destroying up to one hundred thousand ozone molecules during that time.

1,1,1-Trichloro-2,2,2-trifluoroethane, also known as CFC-113a, is one of four man-made chemicals newly discovered in the atmosphere by a team at the University of East Anglia. CFC-113a is the only known CFC whose abundance in the atmosphere is still growing. Its source remains a mystery, but illegal manufacturing is suspected by some. CFC-113a seems to have been accumulating unabated since 1960. Between 2010 and 2012, emissions of the gas jumped by 45 percent.

Computer Modeling

Scientists have attributed ozone depletion to the increase of man-made (anthropogenic) halogen compounds from CFCs by combining observational data with computer models. These complex chemistry transport models (e.g. SLIMCAT, CLaMS—Chemical Lagrangian Model of the Stratosphere) work by combining measurements of chemicals and meteorological fields with chemical reaction rate constants. They identify key chemical reactions and transport processes that bring CFC photolysis products into contact with ozone.

Ozone Hole and its Causes

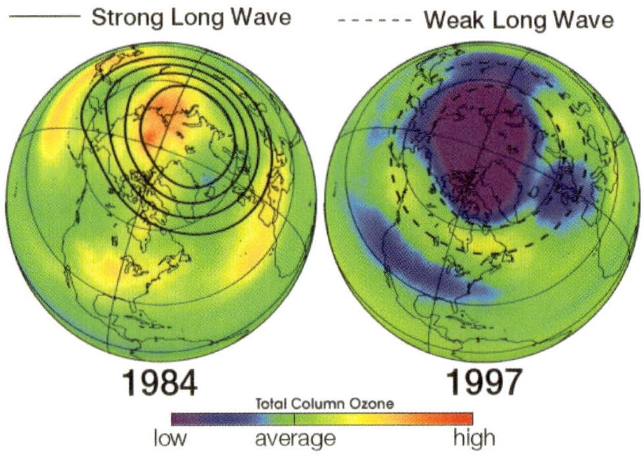

Ozone hole in North America during 1984 (abnormally warm reducing ozone depletion) and 1997 (abnormally cold resulting in increased seasonal depletion)

The Antarctic ozone hole is an area of the Antarctic stratosphere in which the recent ozone levels have dropped to as low as 33 percent of their pre-1975 values. The ozone hole occurs during the Antarctic spring, from September to early December, as strong westerly winds start to circulate around the continent and create an atmospheric container. Within this polar vortex, over 50 percent of the lower stratospheric ozone is destroyed during the Antarctic spring.

As explained above, the primary cause of ozone depletion is the presence of chlorine-containing source gases (primarily CFCs and related halocarbons). In the presence of UV light, these gases dissociate, releasing chlorine atoms, which then go on to catalyze ozone destruction. The Cl-catalyzed ozone depletion can take place in the gas phase, but it is dramatically enhanced in the presence of polar stratospheric clouds (PSCs).

These polar stratospheric clouds (PSC) form during winter, in the extreme cold. Polar winters are dark, consisting of 3 months without solar radiation (sunlight). The lack of sunlight contributes to a decrease in temperature and the polar vortex traps and chills air. Temperatures hover around or below −80 °C. These low temperatures form cloud particles. There are three types of PSC clouds—nitric acid trihydrate clouds, slowly cooling water-ice clouds, and rapid cooling water-ice (nacerous) clouds—provide surfaces for chemical reactions whose products will, in the spring lead to ozone destruction.

The photochemical processes involved are complex but well understood. The key observation is that, ordinarily, most of the chlorine in the stratosphere resides in "reservoir" compounds, primarily chlorine nitrate ($ClONO_2$) as well as stable end products such as HCl. The formation of end products essentially remove Cl from the ozone depletion process. The former sequester Cl, which can be later made available via absorption of light at shorter wavelengths than 400 nm. During the Antarctic winter and spring, however, reactions on the surface of the polar stratospheric cloud particles convert these "reservoir" compounds into reactive free radicals (Cl and ClO). The process by which the clouds remove NO_2 from the stratosphere by converting it to nitric acid in the PSC particles, which then are lost by sedimentation is called denitrification. This prevents newly formed ClO from being converted back into $ClONO_2$.

The role of sunlight in ozone depletion is the reason why the Antarctic ozone depletion is greatest during spring. During winter, even though PSCs are at their most abundant, there is no light over the pole to drive chemical reactions. During the spring, however, the sun comes out, providing energy to drive photochemical reactions and melt the polar stratospheric clouds, releasing considerable ClO, which drives the hole mechanism. Further warming temperatures near the end of spring break up the vortex around mid-December. As warm, ozone and NO_2-rich air flows in from lower latitudes, the PSCs are destroyed, the enhanced ozone depletion process shuts down, and the ozone hole closes.

Most of the ozone that is destroyed is in the lower stratosphere, in contrast to the much smaller ozone depletion through homogeneous gas phase reactions, which occurs primarily in the upper stratosphere.

Interest in Ozone Layer Depletion

Public misconceptions and misunderstandings of complex issues like the ozone depletion are common. The limited scientific knowledge of the public led to a confusion with global warming or the perception of global warming as a subset of the "ozone hole". In the beginning, classical green NGOs refrained from using CFC depletion for campaigning, as they assumed the topic was too complicated. They became active much later, e.g. in Greenpeace's support for a CFC-free fridge produced by the former East German company VEB dkk Scharfenstein.

The metaphors used in the CFC discussion (ozone shield, ozone hole) are not "exact" in the scientific sense. The "ozone hole" is more of a *depression*, less "a hole in the windshield". The ozone does not disappear through the layer, nor is there a uniform "thinning" of the ozone layer. However they resonated better with non-scientists and their concerns. The ozone hole was seen as a "hot issue" and imminent risk as lay people feared severe personal consequences such as skin cancer, cataracts, damage to plants, and reduction of plankton populations in the ocean's photic zone. Not only on the policy level, ozone regulation compared to climate change fared much better in public opinion. Americans voluntarily switched away from aerosol sprays before legislation was enforced, while climate change failed to achieve comparable concern and public action. The sudden recognition in 1985 that there was a substantial "hole" was widely reported in the press. The especially rapid ozone depletion in Antarctica had previously been dismissed as a measurement error. Scientific consensus was established after regulation.

While the Antarctic ozone hole has a relatively small effect on global ozone, the hole has generated a great deal of public interest because:

- Many have worried that ozone holes might start appearing over other areas of the globe, though to date the only other large-scale depletion is a smaller ozone "dimple" observed during the Arctic spring around the North Pole. Ozone at middle latitudes has declined, but by a much smaller extent (about 4–5 percent decrease).
- If stratospheric conditions become more severe (cooler temperatures, more clouds, more active chlorine), global ozone may decrease at a greater pace. Standard global warming theory predicts that the stratosphere will cool.
- When the Antarctic ozone hole breaks up each year, the ozone-depleted air drifts out into nearby regions. Decreases in the ozone level of up to 10 percent have been reported in New Zealand in the month following the breakup of the Antarctic ozone hole, with ultraviolet-B radiation intensities increasing by more than 15 percent since the 1970s.

Consequences of Ozone Layer Depletion

Since the ozone layer absorbs UVB ultraviolet light from the sun, ozone layer depletion increases surface UVB levels (all else equal), which could lead to damage, including increase in skin cancer. This was the reason for the Montreal Protocol. Although decreases in stratospheric ozone are well-tied to CFCs and to increases in surface UVB, there is no direct observational evidence linking ozone depletion to higher incidence of skin cancer and eye damage in human beings. This is partly because UVA, which has also been implicated in some forms of skin cancer, is not absorbed by ozone, and because it is nearly impossible to control statistics for lifestyle changes over time.

Increased UV

Ozone, while a minority constituent in Earth's atmosphere, is responsible for most of the absorption of UVB radiation. The amount of UVB radiation that penetrates through the ozone layer decreases exponentially with the slant-path thickness and density of the layer. When stratospheric ozone levels decrease, higher levels of UVB reach the Earth's surface. UV-driven phenolic formation in tree rings has dated the start of ozone depletion in northern latitudes to the late 1700s.

In October 2008, the Ecuadorian Space Agency published a report called HIPERION. The study used ground instruments in Ecuador and the last 28 years' data from 12 satellites of several countries, and found that the UV radiation reaching equatorial latitudes was far greater than expected, with the UV Index climbing as high as 24 in Quito; the WHO considers 11 as an extreme index and a great risk to health. The report concluded that depleted ozone levels around the mid latitudes of the planet are already endangering large populations in these areas. Later, the CONIDA, the Peruvian Space Agency, published its own study, which yielded almost the same findings as the Ecuadorian study.

Biological Effects

The main public concern regarding the ozone hole has been the effects of increased surface UV radiation on human health. So far, ozone depletion in most locations has been typically a few percent and, as noted above, no direct evidence of health damage is available in most latitudes. If the high levels of depletion seen in the ozone hole were to be common across the globe, the effects could be substantially more dramatic. As the ozone hole over Antarctica has in some instances grown so large as to affect parts of Australia, New Zealand, Chile, Argentina, and South Africa, environmentalists have been concerned that the increase in surface UV could be significant.

Ozone depletion would magnify all of the effects of UV on human health, both positive (including production of Vitamin D) and negative (including sunburn, skin cancer, and cataracts). In addition, increased surface UV leads to increased tropospheric ozone, which is a health risk to humans.

Basal and Squamous Cell Carcinomas

The most common forms of skin cancer in humans, basal and squamous cell carcinomas, have been strongly linked to UVB exposure. The mechanism by which UVB induces these cancers is well understood—absorption of UVB radiation causes the pyrimidine bases in the DNA molecule to form dimers, resulting in transcription errors when the DNA replicates. These cancers are relatively mild and rarely fatal, although the treatment of squamous cell carcinoma sometimes requires extensive reconstructive surgery. By combining epidemiological data with results of animal studies, scientists have estimated that every one percent decrease in long-term stratospheric ozone would increase the incidence of these cancers by two percent.

Malignant Melanoma

Another form of skin cancer, malignant melanoma, is much less common but far more dangerous, being lethal in about 15–20 percent of the cases diagnosed. The relationship between malignant melanoma and ultraviolet exposure is not yet fully understood, but it appears that both UVB and UVA are involved. Because of this uncertainty, it is difficult to estimate the effect of ozone depletion on melanoma incidence. One study showed that a 10 percent increase in UVB radiation was associated with a 19 percent increase in melanomas for men and 16 percent for women. A study of people in Punta Arenas, at the southern tip of Chile, showed a 56 percent increase in melanoma and a 46 percent increase in nonmelanoma skin cancer over a period of seven years, along with decreased ozone and increased UVB levels.

Cortical Cataracts

Epidemiological studies suggest an association between ocular cortical cataracts and UVB exposure, using crude approximations of exposure and various cataract assessment techniques. A detailed assessment of ocular exposure to UVB was carried out in a study on Chesapeake Bay Watermen, where increases in average annual ocular exposure were associated with increasing risk of cortical opacity. In this highly exposed group of predominantly white males, the evidence linking cortical opacities to sunlight exposure was the strongest to date. Based on these results, ozone depletion is predicted to cause hundreds of thousands of additional cataracts by 2050.

Increased Tropospheric Ozone

Increased surface UV leads to increased tropospheric ozone. Ground-level ozone is generally recognized to be a health risk, as ozone is toxic due to its strong oxidant properties. The risks are particularly high for young children, the elderly, and those with asthma or other respiratory difficulties. At this time, ozone at ground level is produced mainly by the action of UV radiation on combustion gases from vehicle exhausts.

Increased Production of Vitamin D

Vitamin D is produced in the skin by ultraviolet light. Thus, higher UVB exposure raises human vitamin D in those deficient in it. Recent research (primarily since the Montreal Protocol) shows that many humans have less than optimal vitamin D levels. In particular, in the U.S. population, the lowest quarter of vitamin D (<17.8 ng/ml) were found using information from the National Health and Nutrition Examination Survey to be associated with an increase in all-cause mortality in the general population. While blood level of Vitamin D in excess of 100 ng/ml appear to raise blood calcium excessively and to be associated with higher mortality, the body has mechanisms that prevent sunlight from producing Vitamin D in excess of the body's requirements.

Effects on Non-human Animals

A November 2010 report by scientists at the Institute of Zoology in London found that whales off the coast of California have shown a sharp rise in sun damage, and these scientists "fear that the thinning ozone layer is to blame". The study photographed and took skin biopsies from over 150 whales in the Gulf of California and found "widespread evidence of epidermal damage commonly associated with acute and severe sunburn", having cells that form when the DNA is damaged by UV radiation. The findings suggest "rising UV levels as a result of ozone depletion are to blame for the observed skin damage, in the same way that human skin cancer rates have been on the increase in recent decades."

Effects on crops

An increase of UV radiation would be expected to affect crops. A number of economically important species of plants, such as rice, depend on cyanobacteria residing on their roots for the retention of nitrogen. Cyanobacteria are sensitive to UV radiation and would be affected by its increase. "Despite mechanisms to reduce or repair the effects of increased ultraviolet radiation, plants have a limited ability to adapt to increased levels of UVB, therefore plant growth can be directly affected by UVB radiation."

Public Policy

The full extent of the damage that CFCs have caused to the ozone layer is not known and will not be known for decades; however, marked decreases in column ozone have already been observed. The Montreal and Vienna conventions were installed long before a scientific consensus was established or important uncertainties in the science field were being resolved. The ozone case was understood comparably well by lay persons as e.g. *Ozone shield* or *ozone hole* were useful "easy-to-understand bridging metaphors". Americans voluntarily switched away from aerosol sprays, resulting in a 50 percent sales loss even before legislation was enforced.

After a 1976 report by the United States National Academy of Sciences concluded that credible scientific evidence supported the ozone depletion hypothesis a few countries, including the United States, Canada, Sweden, Denmark, and Norway, moved to eliminate the use of CFCs in aerosol spray cans. At the time this was widely regarded as a first step towards a more comprehensive regulation policy, but progress in this direction slowed in subsequent years, due to a combination of political factors (continued resistance from the halocarbon industry and a general change in attitude towards environmental regulation during the first two years of the Reagan administration) and scientific developments (subsequent National Academy assessments that indicated that the first estimates of the magnitude of ozone depletion had been overly large). A critical DuPont manufacturing patent for Freon was set to expire in 1979. The United States banned the use of CFCs in aerosol cans in 1978. The European Community rejected proposals to ban CFCs in aerosol sprays, and in the U.S., CFCs continued to be used as refrigerants and for cleaning circuit boards. Worldwide CFC production fell sharply after the U.S. aerosol ban, but by 1986 had returned nearly to its 1976 level. In 1993, DuPont Canada closed its CFC facility.

The U.S. Government's attitude began to change again in 1983, when William Ruckelshaus replaced Anne M. Burford as Administrator of the United States Environmental Protection Agency. Under Ruckelshaus and his successor, Lee Thomas, the EPA pushed for an international approach to halocarbon regulations. In 1985 20 nations, including most of the major CFC producers, signed the Vienna Convention for the Protection of the Ozone Layer, which established a framework for negotiating international regulations on ozone-depleting substances. That same year, the discovery of the Antarctic ozone hole was announced, causing a revival in public attention to the issue. In 1987, representatives from 43 nations signed the Montreal Protocol. Meanwhile, the halocarbon industry shifted its position and started supporting a protocol to limit CFC production. However, this shift was uneven with DuPont acting more quickly than their European counterparts. DuPont may have feared court action related to increased skin cancer especially as the EPA had published a study in 1986 claiming that an additional 40 million cases and 800,000 cancer deaths were to be expected in the U.S. in the next 88 years. The EU shifted its position as well after Germany gave up its defence of the CFC industry and started supporting moves towards regulation. Government and industry in France and the UK tried to defend their CFC producing industries even after the Montreal Protocol had been signed.

At Montreal, the participants agreed to freeze production of CFCs at 1986 levels and to reduce production by 50 percent by 1999. After a series of scientific expeditions to the Antarctic produced convincing evidence that the ozone hole was indeed caused by chlorine and bromine from manmade organohalogens, the Montreal Protocol was strengthened at a 1990 meeting in London. The participants agreed to phase out CFCs and halons entirely (aside from a very small amount marked for certain "essential" uses, such as asthma inhalers) by 2000 in non-Article 5 countries and by 2010 in Article 5 (less developed) signatories. At a 1992 meeting in Copenhagen, the phase-out date was moved up to 1996. At the same meeting, methyl bromide (MeBr), a fumigant used primarily in agricultural production, was added to the list of controlled substances. For all substances controlled under the protocol, phaseout schedules were delayed for less developed ('Article 5(1)') countries, and phaseout in these countries was supported by transfers of expertise, technology, and money from non-Article 5(1) Parties to the Protocol. Additionally, exemptions from the agreed schedules could be applied for under the Essential Use Exemption (EUE) process for substances other than methyl bromide and under the Critical Use Exemption (CUE) process for methyl bromide.

Civil society including especially NGOs, played critical roles at all stages of policy development leading up to the Vienna Conference, the Montreal Protocol, and in assessing compliance afterwards. The major companies claimed that no alternatives to HFC existed. An ozone-safe hydrocarbon refrigerant was developed at a Hamburg technological institute in Germany, and in 1992 came to the attention of the non-governmental organization (NGO) Greenpeace. Greenpeace was given the patent, called it "Greenfreeze," and left the patent as open source. The NGO then worked successfully first with a small and struggling company to market an appliance beginning in Europe, then Asia and later Latin America, receiving a 1997 UNEP award. By 1995, Germany had already made CFC refrigerators illegal. Since 2004, corporations like Coca-Cola, Carlsberg, and IKEA have been forming a coalition to promote the ozone-safe Greenfreeze units. Production spread to companies like Electrolux, Bosch, and LG, with sales reaching some 300 million refrigerators by 2008. In Latin America, a domestic Argentinian company began Greenfreeze production in 2003, while the giant Bosch in Brazil began a year later. By 2013 it was being used by some 700 million refrigerators, making up about 40 percent of the market. In the U.S., however, change has been much slower. To some extent, CFCs were being replaced by the less damaging hydrochlorofluorocarbons (HCFCs), although concerns remain regarding HCFCs also. In some applications, hydrofluorocarbons (HFCs) were being used to replace CFCs. HFCs, which contain no chlorine or bromine, do not contribute at all to ozone depletion although they are potent greenhouse gases. The best known of these compounds is probably HFC-134a (R-134a), which in the United States has largely replaced CFC-12 (R-12) in automobile air conditioners. In laboratory analytics (a former "essential" use) the ozone depleting substances can be replaced with various other solvents. Chemical companies like Du Pont, whose representatives even disparaged Greenfreeze as "that German technology," maneuvered the EPA to block the technology in the U.S. until 2011. Ben & Jerry's of Unilever and General Electric, spurred by Greenpeace, had expressed formal interest in 2008 which figured in the EPA's final approval.

More recently, policy experts have advocated for efforts to link ozone protection efforts to climate protection efforts. Many ODS are also greenhouse gases, some thousands of times more powerful agents of radiative forcing than carbon dioxide over the short and medium term. Thus policies protecting the ozone layer have had benefits in mitigating climate change. In fact, the reduction of the radiative forcing due to ODS probably masked the true level of climate change effects of other GHGs, and was responsible for the "slow down" of global warming from the mid-90s. Policy decisions in one arena affect the costs and effectiveness of environmental improvements in the other.

ODS requirements in the Marine Industry

The IMO has amended MARPOL Annex VI Regulation 12 regarding ozone depleting substances. As from July 1, 2010, all vessels where MARPOL Annex VI is applicable should have a list of equipment using ozone depleting substances. The list should include name of ODS, type and location of equipment, quantity in kg and date. All changes since that date should be recorded in an ODS Record book on board recording all intended or unintended releases to the atmosphere. Furthermore, new ODS supply or landing to shore facilities should be recorded as well.

Prospects of Ozone Depletion

Since the adoption and strengthening of the Montreal Protocol has led to reductions in the emissions of CFCs, atmospheric concentrations of the most-significant compounds have been declining. These substances are being gradually removed from the atmosphere; since peaking in 1994, the Effective Equivalent Chlorine (EECl) level in the atmosphere had dropped about 10 percent by 2008. The decrease in ozone-depleting chemicals has also been significantly affected by a decrease in bromine-containing chemicals. The data suggest that substantial natural sources exist for atmospheric methyl bromide (CH_3Br). The phase-out of CFCs means that nitrous oxide (N_2O), which is not covered by the Montreal Protocol, has become the most highly emitted ozone-depleting substance and is expected to remain so throughout the 21st century.

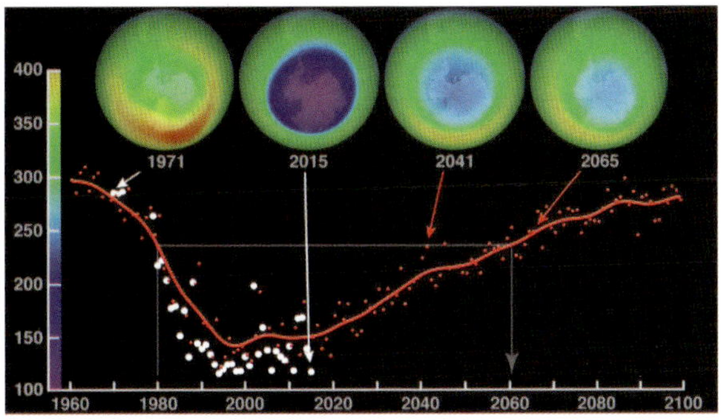

Ozone levels stabilized in the 1990s following the Montreal Protocol, and have started to recover. They are projected to reach pre-1980 levels before 2075

A 2005 IPCC review of ozone observations and model calculations concluded that the global amount of ozone has now approximately stabilized. Although considerable variability is expected from year to year, including in polar regions where depletion is largest, the ozone layer is expected to begin to recover in coming decades due to declining ozone-depleting substance concentrations, assuming full compliance with the Montreal Protocol.

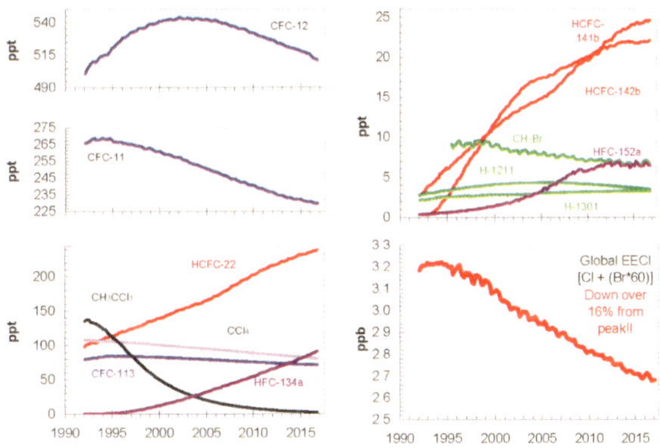

Ozone-depleting gas trends

The Antarctic ozone hole is expected to continue for decades. Ozone concentrations in the lower stratosphere over Antarctica will increase by 5–10 percent by 2020 and return to pre-1980 levels by about 2060–2075. This is 10–25 years later than predicted in earlier assessments, because of revised estimates of atmospheric concentrations of ozone-depleting substances, including a larger predicted future usage in developing countries. Another factor that may prolong ozone depletion is the drawdown of nitrogen oxides from above the stratosphere due to changing wind patterns. A gradual trend toward "healing" was reported in 2016.

Ozone Depletion and Climate Change

Ozone depletion and climate change, or Ozone hole and global warming in more popular terms, are environmental challenges whose connections have been explored and which have been compared and contrasted, for example in terms of global regulation, in various studies and books.

There is widespread scientific interest in better regulation of climate change, ozone depletion and air pollution, as in general the human relationship with the biosphere is deemed of major historiographical and political significance. Already by 1994 the legal debates about respective regulation regimes on climate change, ozone depletion and air pollution were being dubbed "monumental" and a combined synopsis provided.

There are some parallels between atmospheric chemistry and anthropogenic emissions in the discussions which have taken place and the regulatory attempts which have been made. Most important is that the gases causing both problems have long lifetimes after emission to the atmosphere, thus causing problems which are difficult to reverse. However, the Vienna Convention for the Protection of the Ozone Layer and the Montreal Protocol that amended it are seen as success stories, while the Kyoto Protocol on anthropogenic climate change has largely failed. Currently efforts are being undertaken to assess the reasons and to use synergies, for example with regard to data reporting and policy design and further exchanging of information. While the general public tends to see global warming as a subset of ozone depletion, in fact ozone and chemicals such as chlorofluorocarbons (CFCs) and other halocarbons, which are held responsible for ozone depletion, are important greenhouse gasses. Furthermore, natural levels of ozone in both the stratosphere and troposphere have a warming effect.

Policy Approach

There are both links and major differences between ozone depletion and global warming and the way the two challenges have been handled. While in the case of atmospheric ozone depletion, in a situation of high uncertainty and against strong resistance, climate change regulation attempts at the international level such as the Kyoto Protocol have failed to reduce global emissions. The Vienna Convention for the Protection of the Ozone Layer and the Montreal Protocol were both originally signed by only some member states of the United Nations (43 nations in the case of the Montreal Protocol in 1986) while Kyoto attempted to create a worldwide agreement from scratch. Expert consensus concerning CFCs in the form of the Scientific Assessment of Ozone Depletion was reached long after the first regulatory steps were taken, and as of 29 December 2012, all countries in the United Nations plus the Cook Islands, the Holy See, Niue and the supranational European Union had ratified the original Montreal Protocol. These countries have also ratified the London, Copenhagen, and Montreal amendments to the Protocol. As of 15 April 2014, the Beijing amendments had not been ratified by two state parties.

After the Vienna Convention, the halocarbon industry shifted its position and started supporting a protocol to limit CFC production. US manufacturer DuPont acted more quickly than their European counterparts. The EU shifted its position as well after Germany, which has a substantial chemical industry, gave up its defence of the CFC industry and started supporting more regulation. Government and industry in France and the UK had tried to defend their CFC-producing industries even after the Montreal Protocol had been signed.

The Vienna Convention was installed before a scientific consensus on the ozone hole was established. On the contrary, until the 1980s the EU, NASA, NAS, UNEP, WMO and the British government had issued scientific reports with divergent conclusions. Sir Robert (Bob) Watson, Director of the Science Division at NASA, played a crucial role in the process of reaching a unified assessment.

Policy and Consensus

Bob Watson, successfully united the international science community in 1985 to act on the problem with the ozone hole, before a consensus existed.

Sir Robert (Bob) Watson played an important role in both cases

Aant Elzinga wrote in 1996 about the consensus, that the Intergovernmental Panel on Climate Change has tried in the prior two reports a global consensus approach to climate action. Stephen Schneider and Paul N. Edwards, noted in 1997, that after the IPCC Second Assessment Report, the lobby group Global Climate Coalition and a few self-proclaimed "contrarian" scientists tried to discredit the conclusions of the report. They pointed out that the goal of the IPCC is to fairly represent the complete range of credible scientific opinion and if possible a consensus view.

In 2007, Reiner Grundmann compared climate actions in Europe and the United States, he interpreted the inaction besides existing consensus, and noted, *Political agenda that drove US climate change policy. The high visibility of sceptical scientists in the media resonates with this*, and wrote that Germany started ambitious goals, reduced emissions, because 'balanced reporting' led to a bias in climate change coverage in advantage of sceptical arguments in the U.S., but not so much in Germany. Additionally, Grundmann pointed out that after warnings from scientists in 1986 the German Parliament commissioned the *Enquetekommission 'Vorsorge zum Schutz der Erdatmosphäre'* (Precaution for the Protection of the Earth's Atmosphere), to assess the situation, consisting of scientists, politicians and representatives of interest groups. Three years later the report made an impact with the assessment of the state of the art in climate research, an assessment of the threat of climate change itself as well as suggestions for clear emissions reduction targets, even though he argues there was no consensus, and attributed the success of the report to strong precautionary action, and that no scientific outsiders or climate change skeptics were involved.

A linear model of policy-making, based on a position that "the more knowledge we have, the better the political response will be", was not applied in the ozone case. On the contrary, the CFC regulation process focused more on managing ignorance and uncertainties as a basis of political decision making, as the relationships between science, public (lack of) understanding and policy were better taken into account. In the meantime, such a player in the IPCC process as Michael

Oppenheimer conceded some limitations of the IPCC consensus approach and asked for concurring, smaller assessments of special problems instead of repetitions of the large-scale approach every six years. It has become more important to provide a broader exploration of uncertainties. Others also see mixed blessings in the drive for consensus within the IPCC process and have asked for dissenting or minority positions to be included or for statements about uncertainties to be improved.

Public Opinion

The two atmospheric problems have achieved significantly different levels of understanding by the public, including both the basic science and policy issues. People have limited scientific knowledge about global warming and tend to confuse it with or see it as a subset of the ozone hole. Not only on the policy level, ozone regulation fared much better than climate change in public opinion. Americans voluntarily switched away from aerosol sprays before legislation was enforced, while climate change has failed in achieving a broader scientific comprehension and in raising comparable concern.

The metaphors used in the CFC discussion (ozone shield, ozone hole) resonated better with non-scientists and their concerns. The ozone case was communicated to lay persons "with easy-to-understand bridging metaphors derived from the popular culture" and related to "immediate risks with everyday relevance", while the public opinion on climate change sees no imminent danger. The ozone hole was much more seen as a "hot issue" and imminent risk compared to global climate change, as lay people feared a depletion of the ozone layer (ozone shield) risked increasing severe consequences such skin cancer, cataracts, damage to plants, and reduction of plankton populations in the ocean's photic zone. This was not the case with global warming.

Personal Risk Assessment and Knowledge

Sheldon Ungar, a Canadian sociologist, assumes that while the quantity of specialized knowledge is exploding, in contrast scientific ignorance among lay people is the norm and even increasing. Public opinion failed to tie climate change to concrete events which could be used as a threshold or beacon to signify immediate danger. Scientific predictions of a temperature rise of 2 °C (4 °F) to 3 °C (5 °F) over several decades do not resonate with people, for example in North America, who experience similar swings during a single day. As scientists define global warming as a problem of the future, a liability in the "attention economy", pessimistic outlooks in general and the attribution of extreme weather to climate change have often been discredited or ridiculed in the public arena (compare the Gore effect). Even when James Hansen tried to use the 1988–89 North American drought as a call to action, scientists kept stating, in line with the IPCC findings, that even extreme weather is not climate. While the greenhouse effect, *per se*, is essential for life on earth, the case was quite different with the ozone hole and other metaphors about ozone depletion. The scientific assessment of the ozone problem also had large uncertainties; both the ozone content of the upper atmosphere and its depletion are complicated to measure and the link between ozone depletion and rates of enhanced skin cancer is rather weak. But the metaphors used in the discussion (ozone shield, ozone hole) resonated better with lay people and their concerns.

> The idea of rays penetrating a damaged "shield" meshes nicely with abiding and resonant cultural motifs, including "Hollywood affinities." These range from the shields on the Starship Enterprise to Star Wars... It is these pre-scientific bridging metaphors built

> around the penetration of a deteriorating shield that render the ozone problem relatively simple. That the ozone threat can be linked with Darth Vader means that it is encompassed in common sense understandings that are deeply ingrained and widely shared.
>
> —*Sheldon Ungar*

The CFC regulation attempts at the end of the 1980s profited from those easy to grasp metaphors and the personal risk assumptions taken from them. The fate of celebrities like President Ronald Reagan, who had skin cancer removal from his nose in 1985 and 1987, was also of high importance. In case of the public opinion on climate change, no imminent danger is perceived.

Cost-benefit Assessments and Industry Policy

Cass Sunstein and others have compared the differing approach of the United States to the Montreal Protocol, which it accepted, and the Kyoto Protocol, which it rejected. Sunstein assumes that the cost-benefit assessments of climate change action for the US were instrumental in the US' withdrawal from participation in Kyoto. Daniel Magraw, also a lawyer, considers governmental motivations besides relative costs and benefits as being of higher importance. Peter Orszag and Terry Dinan took an insurance perspective and assume that an assessment which predicted dire consequences of climate change would be more of a motivation for the US to change its stance on global warming and adopting regulation measurements.

The US chemical company DuPont had already lost some of their zeal in defending their products after a strategic manufacturing patent for Freon was set to expire in 1979. A citizen boycott of spray cans gained importance in parallel. Not by chance, the United States banned the use of CFCs in aerosol cans in 1978.

Government and industry in France and the UK tried to defend their CFC-producing industries even after the Montreal Protocol had been signed. The European Community rejected proposals to ban CFCs in aerosol sprays for a long time. The EU shifted its position after Germany, which also has a large chemical industry, gave up its defence of the CFC industry and started supporting moves towards regulation. After regulation was more and more enforced, DuPont acted faster than their European counterparts as they may have feared court action related to increased skin cancer, especially as the EPA had published a study in 1986 claiming that an additional 40 million cases and 800,000 cancer deaths were to be expected in the US in the next 88 years. The identification and marketing of a 100% ozone-safe hydrocarbon refrigerant called "Greenfreeze" by the NGO Greenpeace in the early 1990s had a rapid significant impact in major markets of Europe and Asia. The climate change protocols were less successful. In the case of Kyoto, then secretary of the environment Angela Merkel, prevented a possible failure by suggesting to use 1990 as starting date for emission reduction. In so far the demise of the Eastern European heavy industry allowed for a high commitment, but actual emissions kept on growing on a global scale.

Science Background

There are various links between the two fields of human-atmospheric interaction. Policy experts have advocated for a closer linking of ozone protection and climate protection efforts.

Drew Shindell has used climate models to assess both climate change and ozone depletion. In his view, while research up to now has been more about the impact of CFC emissions on stratospheric ozone, the future will be more about the interaction between climate change and ozone feedback. Ozone is a greenhouse gas itself. Many ozone-depleting substances are also greenhouse gases, some agents of radiative forcing are thousands of times more powerful than carbon dioxide over the short and medium term. The increases in concentrations of these chemicals have produced 0.34 ± 0.03 W/m² of radiative forcing, corresponding to about 14% of the total radiative forcing from increases in the concentrations of well-mixed greenhouse gases. Already the natural ozone variability in the stratosphere seems to be closely correlated with the 11-year solar cycle of irradiance changes and has, via a dynamic coupling between the stratosphere and troposphere, a significant impact on climate.

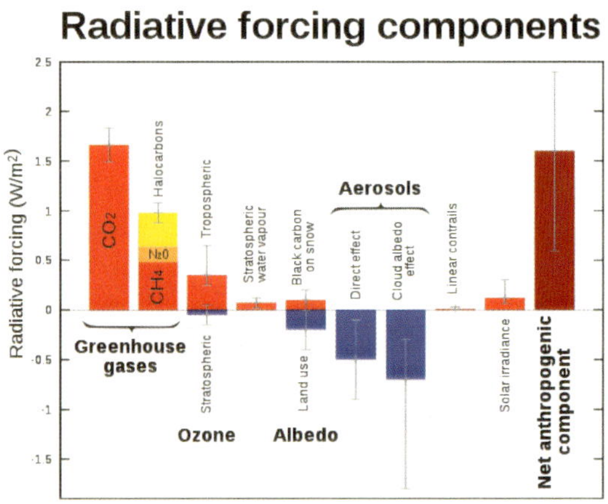

Radiative forcing from various greenhouse gases

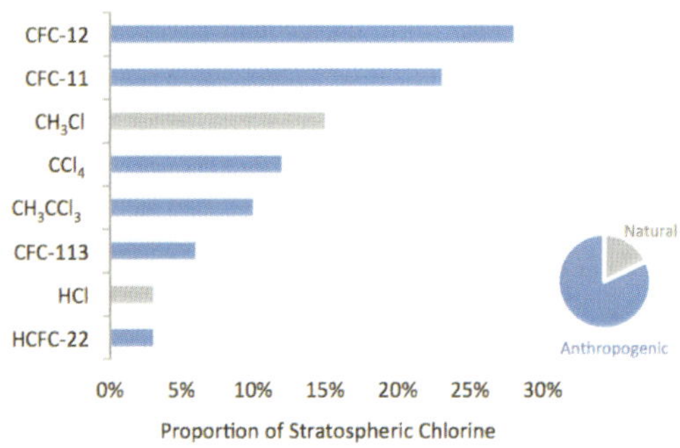

Sources of Stratospheric Chlorine

As with carbon dioxide and methane, there are some natural sources of tropospheric chlorine, such as sea spray. Chlorine from ocean spray is soluble and thus is washed by rainfall before it reaches the stratosphere. It is *stratospheric* chlorine that affects ozone depletion. Only methyl chloride, which is one of the halocarbons, has a mainly natural source, and it is responsible for about 20% of the chlorine in the stratosphere; the remaining 80% comes from man-made sources. Chlorofluorocarbons, in contrast, are insoluble and long-lived, allowing them to reach the stratosphere. In the lower atmosphere, there is much more chlorine from CFCs and related haloalkanes than there is in hydrogen chloride from salt spray, and in the stratosphere halocarbons are dominant.

The same CO_2 radiative forcing that produces global warming is expected to cool the stratosphere. This cooling, in turn, is expected to produce a relative *increase* in ozone (O_3) depletion in the polar area and in the frequency of ozone holes. Conversely, ozone depletion represents a radiative forcing of the climate system of about −0.15 ± 0.10 watts per square metre (W/m^2).

References

- Tabin, Shagoon (2008). Global Warming: The Effect Of Ozone Depletion. APH Publishing. p. 194. ISBN 9788131303962. Retrieved 12 January 2016
- "Stratospheric Ozone and Surface Ultraviolet Radiation". Scientific Assessment of Ozone Depletion: 2010 (PDF). WMO. 2011. Retrieved March 14, 2015
- Farman, J. C.; Gardiner, B. G.; Shanklin, J. D. (1985). "Large losses of total ozone in Antarctica reveal seasonal ClO_x/NO_x interaction". Nature. 315 (6016): 207–210. Bibcode:1985Natur.315..207F. doi:10.1038/315207a0
- Boyesa, Edward; Stanisstreeta, Martin (1992). "Students' perceptions of global warming". International Journal of Environmental Studies. 42 (4): 287–300. doi:10.1080/00207239208710804
- Grundmann, Reiner (2001). Transnational Environmental Policy : the ozone layer. New York: Routledge. ISBN 0-415-22423-3
- "Halocarbons and Other Gases". Emissions of Greenhouse Gases in the United States 1996. Energy Information Administration. 1997. Archived from the original on 2008-06-29. Retrieved 2008-06-24
- Grundmann, Reiner (2007). "Climate Change and Knowledge Politics" (PDF). Environmental Politics. 16 (3): 414–432. doi:10.1080/09644010701251656. Archived from the original(PDF) on August 26, 2014. Retrieved March 14, 2015
- R. P. Sinha; S. C. Singh; D. P. Häder (1999). "Photoecophysiology of cyanobacteria". Journal of Photochemistry and Photobiology. 3: 91–101
- Roan, Sharon (1989) Ozone crisis: The 15-year evolution of a sudden global emergency, New York: Wiley, p. 56 ISBN 0-471-52823-4
- Ungar, Sheldon (2000). "Knowledge, ignorance and the popular culture: climate change versus the ozone hole". Public Understanding of Science. 9 (3): 297–312. doi:10.1088/0963-6625/9/3/306. Retrieved March 14, 2015
- Solomon, Susan, et al. (June 30, 2016). "Emergence of healing in the Antarctic ozone layer". Science. 353 (6296): 269–74. Bibcode:2016Sci...353..269S. doi:10.1126/science.aae0061. PMID 27365314
- Morrisette, Peter M. (1989). "The Evolution of Policy Responses to Stratospheric Ozone Depletion". Natural Resources Journal. 29: 793–820. Retrieved April 20, 2010
- Causes and Effects of Stratospheric Ozone Reduction: An Update. National Research Council. 1982. p. Summary, 3. ISBN 0-309-03248-2

Permissions

All chapters in this book are published with permission under the Creative Commons Attribution Share Alike License or equivalent. Every chapter published in this book has been scrutinized by our experts. Their significance has been extensively debated. The topics covered herein carry significant information for a comprehensive understanding. They may even be implemented as practical applications or may be referred to as a beginning point for further studies.

We would like to thank the editorial team for lending their expertise to make the book truly unique. They have played a crucial role in the development of this book. Without their invaluable contributions this book wouldn't have been possible. They have made vital efforts to compile up to date information on the varied aspects of this subject to make this book a valuable addition to the collection of many professionals and students.

This book was conceptualized with the vision of imparting up-to-date and integrated information in this field. To ensure the same, a matchless editorial board was set up. Every individual on the board went through rigorous rounds of assessment to prove their worth. After which they invested a large part of their time researching and compiling the most relevant data for our readers.

The editorial board has been involved in producing this book since its inception. They have spent rigorous hours researching and exploring the diverse topics which have resulted in the successful publishing of this book. They have passed on their knowledge of decades through this book. To expedite this challenging task, the publisher supported the team at every step. A small team of assistant editors was also appointed to further simplify the editing procedure and attain best results for the readers.

Apart from the editorial board, the designing team has also invested a significant amount of their time in understanding the subject and creating the most relevant covers. They scrutinized every image to scout for the most suitable representation of the subject and create an appropriate cover for the book.

The publishing team has been an ardent support to the editorial, designing and production team. Their endless efforts to recruit the best for this project, has resulted in the accomplishment of this book. They are a veteran in the field of academics and their pool of knowledge is as vast as their experience in printing. Their expertise and guidance has proved useful at every step. Their uncompromising quality standards have made this book an exceptional effort. Their encouragement from time to time has been an inspiration for everyone.

The publisher and the editorial board hope that this book will prove to be a valuable piece of knowledge for students, practitioners and scholars across the globe.

Index